Insect Histology

Insect Histology

Practical Laboratory Techniques

Pedro Barbosa
Department of Entomology, University of Maryland, College Park, MD

Deborah L. Berry
Department of Oncology, Co-Director, Histopathology and Tissue Shared Resource, Lombardi Cancer Center, Georgetown University, Washington, D.C.

Christina S. Kary
Genes & Development, Cold Spring Harbor Laboratory Press, Woodbury, NY

WILEY Blackwell

Registered office: John Wiley & Sons, Ltd, The Atrium, Southern Gate, Chichester, West Sussex, PO19 8SQ, UK

Editorial offices: 9600 Garsington Road, Oxford, OX4 2DQ, UK
The Atrium, Southern Gate, Chichester, West Sussex, PO19 8SQ, UK
111 River Street, Hoboken, NJ 07030-5774, USA

For details of our global editorial offices, for customer services and for information about how to apply for permission to reuse the copyright material in this book please see our website at www.wiley.com/wiley-blackwell.

Library of Congress Cataloging-in-Publication Data

Barbosa, Pedro, 1944-
Insect histology : practical laboratory techniques / Pedro Barbosa, Deborah L. Berry, Christina S. Kary. – First edition.
pages cm
Includes bibliographical references and index.
ISBN 978-1-4443-3695-5 (cloth) – ISBN 978-1-4443-3696-2 (pbk.) 1. Entomology–Laboratory manuals. 2. Histology–Laboratory manuals. I. Berry, Deborah L., 1972- II. Kary, Christina S., 1975- III. Title.
QL464.B34 2014
595.7–dc23

2013050117

A catalogue record for this book is available from the British Library.

Wiley also publishes its books in a variety of electronic formats. Some content that appears in print may not be available in electronic books.

Cover image: Moth antenna. Autofluorescence confocal stack reconstruction. Image by Donna Beer Stolz, Ph.D. Center for Biologic Imaging. University of Pittsburgh, Pittsburgh, PA.

Typeset in 9.5/12pt BerkeleyStd by Laserwords Private Limited, Chennai, India
Printed and bound in Singapore by Markono Print Media Pte Ltd

1 2015

Contents

Preface

"The standard histological procedures of the zoologist do not as a rule give very successful results when applied to insects ... " (Van Heerden, 1945). If one adds to this statement that histological methods, entomological or otherwise, are a frustrating balance between science and art, one can understand the reluctance of entomologists to utilize histological techniques. The situation is additionally complicated by the fact that the techniques that have been formulated are scattered throughout various scientific journals and in general textbooks.

This book is designed to bring the procedures of insect histology to entomologists and others who utilize insects as experimental animals. Since no technique can be applicable to all insects, the techniques in this book are presented as guidelines. These basic methods can be easily modified to suit the characteristics of a particular insect or specific research problems.

It would be useless to present another book on the theory and use of histological methods, particularly since success in applying procedures is in part contingent upon practice and experience. In addition, there are already numerous sources of generalized information on the theoretical aspects of histological techniques. Instead, in this book the reader will fixatives, stains, procedures, and so on, which have been reported to be specifically applicable to insects. The book also presents information useful in dealing with histological problems encountered in insect tissues such as sclerotized chitin, yolk-laden eggs, chromosomes, genitalia, and so on.

* Van Heerden, H.P. 1945. Some histological methods of interest to entomologists. *Journal of the Entomological Society of South Africa* 8:157–161.

Acknowledgements

We acknowledge the support of the respective institutional affiliations of the authors of this manual, that is, the College of Computer, Mathematical, and Natural Sciences and the College of Agriculture and Natural Resources of the University of Maryland (PB), the Lombardi Comprehensive Cancer Center of Georgetown University (DB), and Cold Spring Harbor Press (CK). This work was conducted in part at the Lombardi Comprehensive Cancer Center Histopathology & Tissue Shared resource which is supported in part by NIH/NCI grant P30-CA051008. The content is solely the responsibility of the authors and does not necessarily represent the official views of the National Cancer Institute or the National Institutes of Health.

We further acknowledge and appreciate the contributions of Damien Laudier of Laudier Histology (http://www.laudierhistology.com/) for the use of key illustrations and figures. These are individually acknowledged in the legend of each submitted figure.

Introduction

The *Manual of Basic Techniques in Insect Histology* is designed as a resource for those researchers who require basic procedures and information essential for the histological display of insects, in part or in total. Specifically, it can serve as a basic laboratory reference or as an essential supplement to complement lectures in courses which deal with insect histology.

This second edition of the book extends the original histological approaches into modern applications. The manual provides a comprehensive survey of fixation techniques which are crucial to all downstream histological preparations and applications. Preparations and techniques unique to insects are provided for advanced techniques such as immunohistochemistry, in situ hybridization, TEM, SEM and whole mount preparations.

In order to permit efficient use by the reader, the information in this book is presented in a readable and consistent format. Although there are divergences where necessary or where the information is not available, most of the book follows the same format. Finally, throughout the book, the amounts of all ingredients are designated by the term, parts (pt.). In compounds which occur as solid, parts equals grams, while in those compounds occurring as liquids, parts equals milliliters. All procedural information and recommendations for the use of particular methods in this manual are taken from the literature cited. In some instances, not as much information is available as one might desire. However, these materials and procedures were included since the characteristics and limitations of a technique are a function of the insect and experimental conditions.

In histology many chemicals are used that are harsh, corrosive, potential irritants, and some (such as Dioxane or Formaldehyde) may be carcinogenic. Like most chemicals they can be absorbed through the skin or inhaled; in some cases inhaled over a period of time. Thus, one must use common sense in developing lab practices and constant vigilance and care in order to keep chemicals off the skin, or avoid inhalation. And, when in doubt, use the hood. A small amount of planning and thought can avoid a great deal of trouble and regret. Thus, safety glasses or goggles and shield, proper gloves, laboratory coat and apron, adequate ventilation, and a class B extinguisher should be used or available in the lab. Always seek expert advice when in doubt.

About the companion website

This book is accompanied by a companion website:

www.wiley.com/barbosa/insecthistology

This website includes:

- Powerpoints of all figures from the book for downloading
- PDFs of tables from the book

1 Problems of sclerotized chitin: Softening insect cuticle

1.1 Introduction

The softening and processing of heavily sclerotized specimens for subsequent histological preparations is one of the major problems in insect histology. Many approaches to the solution of this problem have been suggested. Attempts to soften and otherwise alter sections with sclerotized chitin have been incorporated at every procedural level of histological methods. Suggestions have been made for changes in fixation, clearing, mounting, and embedding. Others have also attempted prefixation, postfixation, premounting, presectioning, and so on, as additional steps geared towards improving the quality of sections.

Aside from the more detailed procedures and specific compounds that are recommended in the following pages there are other simple general methods recommended. These techniques represent basic procedures that have been used independently or in conjunction with other methods. One of the most widely used procedures is the treatment of insect specimens with sodium or potassium hydroxide. These chemicals soften sclerotized portions of specimens and dissolve the soft internal tissues. They are generally used either cold or warm at a 10% concentration. These substances are also frequently used in the preparation of insect specimens for taxonomic study.

The use of hypochlorite of soda is another alternate for softening chitin. It is suggested for the preparation of all stages, that is, larvae, pupae, and adults. The insect is usually placed in boiling hypochlorite of soda (about 25% in distilled water). It is usually left in the solution for about 24 hours or more. A third, widely used approach is the use of tenerals or newly moulted specimens. In this way, the specimens are used before the cuticle has hardened.

The elimination of certain chemical agents which tend to harden insect tissues can also be helpful. Occasionally, it is best merely to avoid long exposures to hardening compounds. For example, to avoid excess hardening, short exposures or avoidance of the higher concentrations of ethanol will aid in preventing its hardening effects. The use of n-butyl or t-butyl alcohol as a substitute dehydrating agent may avoid the hardening of tissues. Similarly, prolonged exposure to certain chemicals or fixatives containing chemicals such as acidified dichromate, mercuric chloride or chromic acid is not recommended. Prolonged heating may also cause unwanted brittleness. The choice of clearing agent may also be a key factor in brittleness of tissue preparations. Thus, the use of clearing agents other than xylene or similar compounds will result in

Insect Histology: Practical Laboratory Techniques, First Edition.
Pedro Barbosa, Deborah L. Berry and Christina S. Kary.

Companion Website: www.wiley.com/go/barbosa/insecthistology

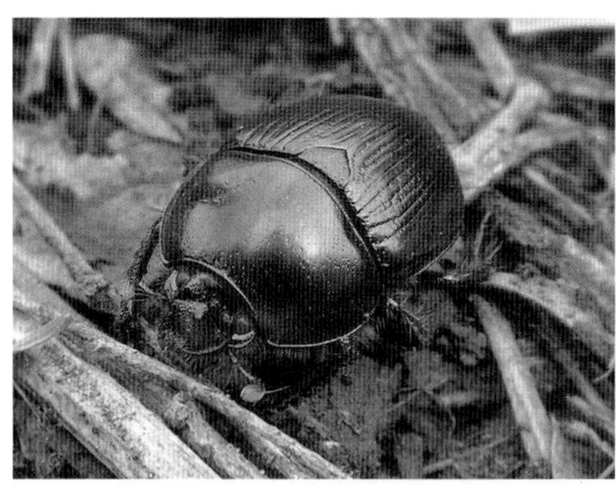

Fig. 1.1 Beetles have a hardened cuticle. (Source: © Michal Grabowski. http://commons.wikimedia.org/wiki/File:Xylena_exsoleta.jpg#filehistory/CC BY-SA 3.0.) See plate section for the color version of the figure.

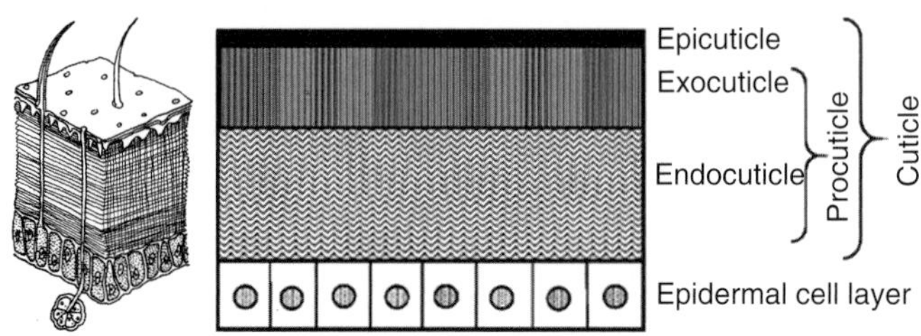

Fig. 1.2 Components of the cuticle. Procuticle – polysaccaride chitin and cross-linked proteins involved in sclerotization.

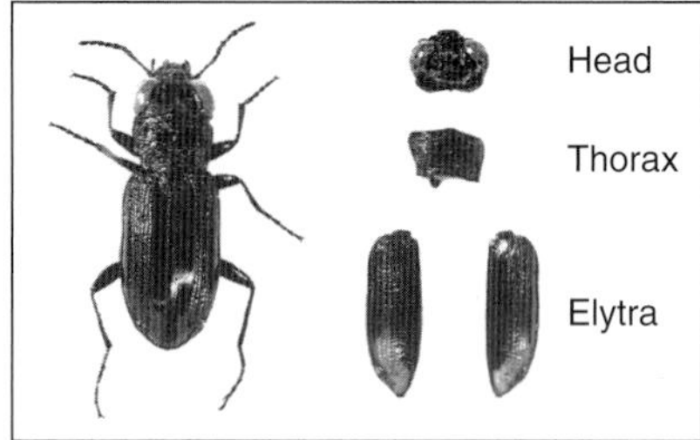

Fig. 1.3 The most sclerotized parts of beetles. (Source: Ellis 2000. Reproduced with permission of Elsevier.)

improved preparations. Finally, excessively high temperatures and prolonged periods of infiltration in wax may be another source of troublesome tissue hardening.

Another widely used procedure involves the puncturing of insect specimens before placing them in a fixative. This allows complete penetration of the fixing agent. Care must always be taken not to damage particular areas of interest on the specimen. The following procedure was suggested as an alternative to the puncturing of specimens.

1.1.1 *Gottlieb's technic [1]*

Application:
Recommended for the histological preparation of insect larvae, pupae, and adults.

Formula:
Solution A: Relaxing fluid
Drosophila Ringer's with magnesium sulphate (4%) added.

Solution B: Chrome alum fixative

Chrome Alum	3 pt
Formaldehyde (40%)	30 pt

Proprionic Acid	2 pt
Distilled Water	238 pt
Dimethyl Sulfoxide	25 pt

Procedure:

1. Rinse specimen in Ringer's solution and place in warmed solution A for 2 to 5 min.
2. Transfer to warmed solution B in a covered dish (on hot plate) for no more than 5 min.
3. Keep specimen in dish and remove from hot plate for 10 to 15 min.
4. Rinse in distilled water and dehydrate.
5. Dehydration must be slow and gentle.
6. Three methods of dehydration are recommended:
   a. Dioxane.
   b. Graded ethanol series with a benzene clearing agent.
   c. Graded tertiary butyl alcohol series.
7. In graded ethanol series a slow transfer to benzene is required before infiltration by using trichloropropane.
8. In the dioxane procedure, the following steps are necessary:
   a. Several changes of 50% dioxane for one day.
   b. Several changes of 100% dioxane for 2 to 3 days.
9. Infiltration is as follows:
   a. Transfer to solutions of increasing paraffin concentrations for 24 to 36 hr.
   b. Transfer to two baths of pure paraffin for 12 hr each.

Note:

1. Graded alcohols series consist of the following: 10–70% (in 10% steps), to 85% (in 5% steps), to 100% (in 2.5% steps).
2. A complete schedule of dehydration solutions and a timetable is available on Table 1 of Gottlieb (1966).
3. Either t-butyl alcohol or dioxane is recommended to avoid excess hardening.

1.2 General Methods

The following are general methods, fixatives, and softening agents recommended for the softening of histological preparations with extensive sclerotized chitin.

1.2.1 *Cox's technic [2]*

Application:
Recommended for small insects and insect parts, e.g., beetle elytra.

Formula:
Solution A:
Potassium Hydroxide (10%)

Solution B:
Acetic Acid (33%)

Procedure:

1. Fix insect specimen.
2. Transfer to solution A for 24 hr.
3. Wash in water for 6 hr.
4. Transfer to solution B for 24 hr.

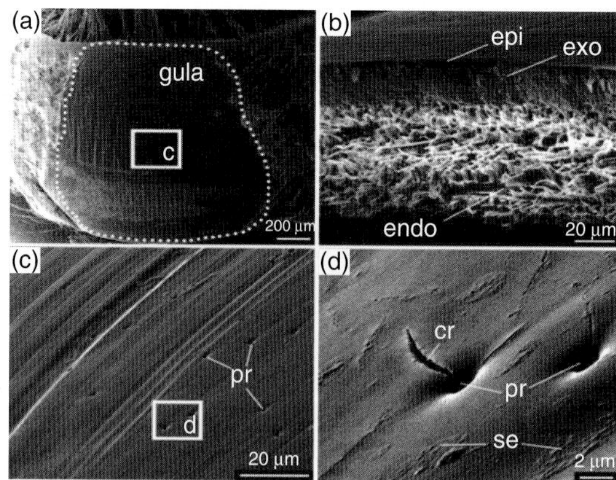

Fig. 1.4 Components of the cuticle – SEM images of the dry gula (a plate which in most insects supports the basal part of the labium). (a,c,d) Surface of the gula. (b) Cross fracture of the gula cuticle showing the epicuticle (epi), exocuticle (exo) and endocuticle (endo). Fibers of the outer part of the exocuticle are oriented perpendicular to the surface but are parallel in the deeper layers of the exocuticle and in the endocuticle. Pores (pr), dried organic substances (se) and cracks (cr) can be seen on the cuticle surface. Rectangles c and d, indicate parts of the sample magnified in c and d, respectively. (Source: Barbakadze *et al.* 2006. Reproduced with permission of the authors.)

5. Wash in water for 6 to 8 hr.
6. Dehydrate, embed, and mount.

1.2.2 *Eltringham's method II [2]*

Formula:

Solution A:

Sodium Hypochlorite (6%)

Solution B: Fixative

Water	250 pt
Picric Acid	2.6 pt
Nitric Acid	10 pt

Procedure:

1. Fix insect specimen.
2. Wash in running water for 4 hr.
3. Transfer to solution A for 60°C for 36 hr.
4. Wash in water for 4 hr.
5. Transfer to solution B at 60°C for 6 days.
6. Boil in alcohol (70%) for 1 min.
7. Let stand in alcohol (70%) for 1 min.
8. Dehydrate, clear in cedar-oil and mount.

1.2.3 *Verdcourt's nitric-ethanol [3]*

Application:

Recommended for the softening elytra of Coleoptera.

Formula:

Solution A:

Ethyl Alcohol	3 pt
Nitric Acid (concentrated)	1 pt

1.2.4 *Schultze's [3]*

Application:
Recommended for the softening elytra of Coleoptera.

Formula:

Nitric Acid (concentrated)	2 pt
Potassium Chlorate	1 pt

Note:
1. When warmed, this substance softens in a few minutes, but is harsh and must be watched.

1.2.5 *Modified Schultze's [3]*

Application:
Recommended for the softening elytra of Coleoptera.

Formula:

Nitric Acid (concentrated)	2 pt
Potassium Chlorate	1 pt

Note:
1. Softening occurs within 5 to 6 days, however, after about 12 days specimens become deformed.
2. Wash well in water after use.

1.2.6 *Verdcourt's technic [3]*

Application:
1. Recommended for the softening and histological preparation of hard chitinous insect tissues.
2. Recommended as particularly useful for softening and preparating slides of the elytra of coleopterans.

Procedure:
1. Place in softening agent (see Verdcourt's Nitric-ethanol, Schultze's or Modified Schultze's).
2. Wash well in water.
3. Dehydrate in alcohol.
4. Place in 1:1 alcohol and ether mixture for 2 days.
5. Transfer through 4, 8, and 10% solutions of celloidin for 3 days in each.
6. Prepare blocks as usual and harden in two changes of chloroform for 2 days.
7. Clear in cedarwood oil for several weeks.
8. Section.
9. Transfer to 1:1 alcohol and chloroform mixture.
10. Transfer to xylene.
11. Mount in balsam.

1.2.7 *Eltringham's method III [2]*

Formula:
Solution A: Sodium Hypochlorite

Solution B: Kleineberg's Fluid

Water	250 pt

Picric Acid	0.75 pt
Sulfuric Acid	5 pt

Procedure:
1. Wash in running water.
2. Transfer to solution A at 60°C for 24 hr.
3. Wash in water.
4. Place in solution B and boil for 1 min.
5. Let stand in solution B at 60°C for 4 days.
6. Place in alcohol (70%) and boil.
7. Dehydrate for several hours.
8. Clear in cedar-oil and mount.

1.2.8 *Eltringham's method IV [2]*

Application:
Recommended for use with live material.

Formula:
Solution A: Sodium Hypochlorite

Solution B: Kleineberg's Fluid

Water	250 pt
Picric Acid	0.75 pt
Sulfuric Acid	5 pt

Procedure:
1. Kill in solution A and boil for 10 min.
2. Let stand in solution A at 60°C for 4 days.
3. Place in alcohol (70%) and boil.
4. Dehydrate for several hours.
5. Clear in cedar-oil and mount.

1.2.9 *Henning's II [4]*

Application:
Recommended for general use as a softener of sclerotized chitin.

Formula:

Water	86 pt
Absolute Alcohol	128 pt
Mercuric Chloride	8 pt
Picric Acid	0.3 pt
Chromic Acid	0.2 pt
Nitric Acid	36 pt

1.2.10 *Kingsbury and Johannsen's fixative [5]*

Application:
Recommended for use in the prevention of excess hardening of insect tissues.

Formula:
Solution A: Perenyi's

Water	165 pt
Alcohol (90%)	75 pt
Chromic Acid	0.5 pt
Nitric Acid	10 pt

Solution B: Working Fixative

Water	185 pt
Mercuric Chloride	7.5 pt
Solution A	65 pt

1.2.11 *Murray's [4, 6]*

Application:
Recommended for softening formol-fixed insect material.

Formula:
Solution A:

Phenol	5 pt
Chloral Hydrate	5 pt

Solution B: Carnoy and Lebrun's

Absolute Alcohol	8 pt
Acetic Acid	8 pt
Chloroform	8 pt
Mercuric Chloride (to saturation)	6 pt

Procedure:

1. Primary fixation occurs in 10% formalin (in 8% sodium chloride).
2. Secondary fixation occurs in solution B.
3. Transfer to solution A for 12 to 24 hr.
4. Clear in chloroform, xylol, or carbon disulphide and mount.

1.2.12 *Sinha's fixative [7]*

Formula:

Picric Acid (saturated in 90% ethanol)	75 pt
Formalin	25 pt
Nitric Acid (concentrated)	8 pt

Procedure:

1. Fix specimens for 4 to 6 days.

1.2.13 *Modified Henning's [5]*

Formula:

Picric Acid (sat. aqueous solution)	6 pt
Sublimate (sat. solution in 60% alcohol)	12 pt
Chromic Acid (1/2% aqueous solution)	8 pt

1.2.14 *Frenzel's fluid [8]*

Formula:

Nitric Acid	1 drop
Mercuric Chloride (half-sat. solution in 80% alcohol)	1–2 pt

1.2.15 Post-fixative technic [9]

Formula:

Mercuric Chloride	4 pt
Chromic Acid	0.5 pt
Nitric Acid (concentrated)	10 pt
Ethyl Alcohol (95%)	50 pt
Distilled Water	200 pt

Procedure:
1. Fix material.
2. Transfer to above mixture until the specimen is completely dechitinized and change every 2 days.
3. Wash in running water for 3 hr.
4. Dehydrate, clear, and mount.

1.2.16 Parks and Larsen's [10]

Application:
Recommended for softening of highly sclerotized and refractory material by treatment of the tissue after embedding in blocks.

Formula:

Ethyl alcohol (70%)	Containing
Glycerine	10%
Liquid Detergent	0.1% pt

Procedure:
1. Soak block for about 1 week.

1.2.17 Slifer's and King's [11]

Application:
Recommended for softening of the highly sclerotized and refractory material by treatment of the tissue after embedding blocks.

Formula:
Phenol (4%) in alcohol (80%).

Procedure:
1. Place in phenol-alcohol mixture before embedding for 24 hr.
2. Prepare paraffin blocks.
3. Cut block to expose tissues.
4. Soak in water for 24 to 48 hr.

1.2.18 Pearlman and Cole's (I) [11]

Application:
Recommended for softening of highly sclerotized and refractory material by treatment of the tissue after embedding in blocks.

Formula:

Glycerol	1 pt
Distilled Water	2 pt

Procedure:
1. Soak entire block for 24 hr.

1.2.19 *Perlman and Cole's (II) [11]*

Application:
Recommended for softening of highly sclerotized and refractory materials by treatment of the tissue after embedding in blocks.

Formula:
One per cent solution (aqueous) of detergent (wetting agent).

Procedure:
1. Soak paraffin blocks for a few hours to overnight.

Note:
1. Procedure recommended for paraffin blocks not nitrocellulose blocks.
2. Of five detergent brands in the original publication tested, all worked equally well.

1.2.20 *Beckel's technic [12]*

Application:
Recommended as a procedure to soften cuticular areas without excessive softening of soft internal tissues.

Procedure:
1. Bisect specimen sagitally.
2. Fix, dehydrate, and infiltrate with paraffin (vacuum infiltration is suggested) for 24 hr.
3. Cast in a block and cool.
4. Crack or split blocks and wax will break away from unimpregnated cuticle.
5. Place specimen in softening agent, e.g., Mollifex™ (commercial) at 20–30°C for 24 hr.
6. Cuticle will take up agent but wax impregnated soft tissues will not.
7. Melted wax is poured over exposed cuticle of specimen.
8. Specimen is reblocked and sectioned.

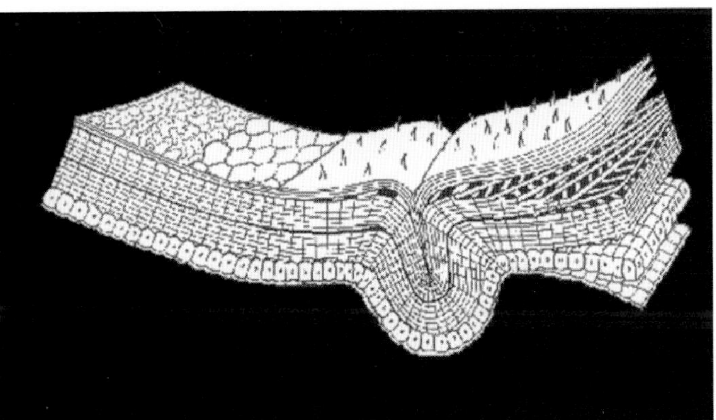

Fig. 1.5 Differentiation of exocuticle involves a chemical process (called sclerotization) that occurs shortly after each molt. During sclerotization, individual protein molecules are linked together by quinone compounds. These reactions "solidify" the protein matrix, creating rigid "plates" of exoskeleton known as sclerites. (Source: Reproduced with permission of John R. Meyer, NC State University.)

1.2.21 *Beckels' modified palmgren tape technic [13]*

Application:

1. Recommended as a means of avoiding the disintegration of sections with sclerotized chitin or heavily yolk-laden material.
2. Recommended to prevent compression and subsequent distortion of sections.

Formula:

Solution A:

Chloroform	1 pt
Glacial Acetic Acid	1 pt
Tetrahydrofuran	6 pt

Solution B:

Tetrahydrofuran	1 pt
Paraffin	1 pt

Procedure:

1. Fix with desired fixative or solution A for 1 hr.
2. Transfer to pure tetrahydrofuran for 1 hr.
3. Transfer to solution B or tetrahydrofuran and 2% celloidin, at 60°C for 1 hr.
4. Transfer to pure tissue mat. at 60°C under vacuum for 3 hr.
5. Transfer to new paraffin at lower vacuum pressure (20 in. Hg.) for 3 hr.
6. Transfer to new paraffin at 25 in. Hg. for 12 to 24 hr.
7. Embed in tissue mat. (a product of paraffin, containing rubber, with the same property as paraplast) and block.
8. Trim block and expose tissue.
9. Soften in Mollifex™ (softening agent) for 24 hr and place at 5°C for last 2 hr.
10. Section as follows:
    a. Cut foot long strip of Scotch tape No. 810 (width of block).
    b. Apply to block with even pressure.
    c. Serial sections can be achieved with practice.
11. Mount as follows:
    a. Albumized clean slide and add several drops of water.
    b. Place section(s) down on slide (in water).
    c. Blot and dry for a few hours.
    d. Place slide in tetrahydrofuran so that tape adhesive and paraffin will dissolve in about 30 min.
    e. Chloroform will also remove acetate backing (2 to 5 min) and xylol removes adhesive and paraffin (30 min).
    f. Before staining transfer from xylol or tetrahydrofuran to 1% celloidin in ether: alcohol for 30 sec.

1.2.22 *Barros-Pita technic [14]*

Application:

Recommended as a procedure which allows for total softening of cuticle without damaging the soft inner tissues.

Procedure:

1. Fix in Bouin for about 3 hr.
2. Wash in ethanol (80%) for 2 hr or until picric acid is not evident.
3. Dehydrate with ethanol up to 90%.
4. Transfer through two changes of n-butanol.

5. Place in a 1:1 mixture of n-butanol and paraffin (56–57°C) in an oven set at 57–58°C for 12 hr.
6. Transfer through two changes of paraffin for 3 hr each.
7. Wash specimen in xylene for 4 min or until paraffin is removed.
8. Hydrate the specimen in 1% aqueous detergent solution for 1 hr.
9. Rinse in tap water.
10. Transfer to concentrated sulfuric-nitric mixture for about 4 hr or until highly pigmented parts (e.g., mandible) loses black color.
11. Wash in running water for 12 hr.
12. Dehydrate in acetone.
13. Place specimen in celloidin (2% ethyl acetate) under vacuum.
14. Immerse in two changes of chloroform.
15. Re-embed in paraffin.

Note:
1. The type of fixative used is not critical.
2. An effort should be made to maintain the mouth area open for better penetration of fixative.
3. After fixation all steps are under vacuum to ensure infiltration.
4. Before placement in acid mixture, eyes are covered with paraffin to avoid damage to eyes.
5. Submergence of specimen can be performed by insertion in glass tubing.
6. When working with acid mixture, work under a hood.
7. The celloidin solution (step 13) forms an artificial cuticle to avoid the collapse of inner tissue.
8. It is suggested that preliminary runs be made with specimens using several dilutions of the acid mixture, for best results.

1.2.23 *Dioxan technic [15]*

Procedure:
1. Fix with fixative of choice, e.g., Bouin, Carnoy's, formal fixatives, etc.
2. Transfer through three changes of dioxan.
3. Transfer to a mixture of dioxan and wax.
4. Transfer to pure wax.
5. Specimens fixed in sublimate fixatives must go through alcohols as usual and iodized alcohol.
6. Then transfer to dioxan.

1.2.24 *Carlisle's technic [16]*

Application:
Recommended for chitinous material but not for heavily sclerotized integuments.

Formula:

Mushrooms	100 pt
Sodium Chloride (35% solution)	100 pt

Preparation:
1. Let mixture of shredded mushrooms and sodium chloride stand overnight.
2. Centrifuge mixture.
3. Keep stock solution in refrigerator.
4. Dilute for use with distilled water or acetate buffer (at pH 5).

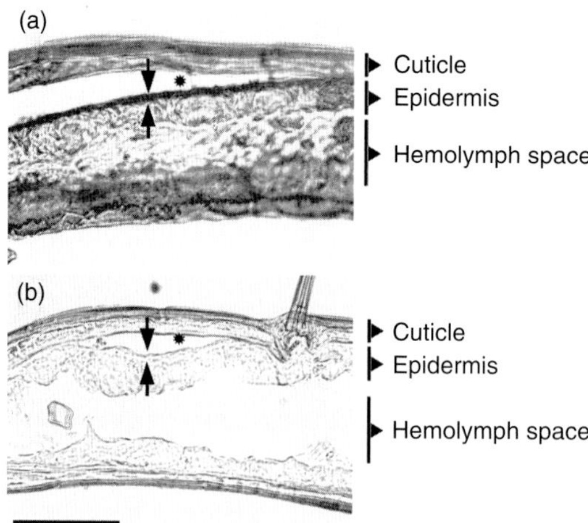

Fig. 1.6 Localization of chitin synthase in the epiproct of *Periplaneta americana*. (a) Cryosections of 10μm were stained with a polyclonal antiserum raised against a conserved region of the *Manduca sexta* chitin synthase. Visualization of primary antibodies was conducted with anti-rabbit antibodies (whole molecules) conjugated to alkaline phosphatase. Detection with 5-bromo-4-chloroindolylphosphate and nitroblue tetrazolium was carried out in the presence of 2 mmol l-1 levamisole to block endogenous alkaline phosphatase activity. (b) Control reaction performed in the absence of primary antibodies. Arrows in a and b mark the apical region of epidermal cells, which are in part detached from the endocuticle as a result of the sectioning procedure (asterisks). Scale bar, 50 μm. (Source: Merzendorfer & Zimoch 2006. Reproduced with permission of the authors.) See plate section for the color version of the figure.

Procedure:

1. Fix specimen in fixative of choice.
2. Wash in running water overnight.
3. Transfer to enzyme solution at 37°C for 12 to 24 hr.
4. Prepare for sectioning in usual way.

Note:

1. Toluene may be added to enzyme mixture to prevent bacterial activity.

1.2.25 *Weinstock and McDonald's technic] [17]*

Application:

1. Recommended for the histological preparations of sections through head and thorax (of bees, Lepidoptera, flies), through ovipositors and other body parts.
2. Recommended for use on adult and immature forms.

Formula:

Same as that recommended by Carlisle (see above).

Procedure:

1. Fix specimen in Carnoy's for 3 hr.
2. Wash in water.
3. Transfer to mushroom enzyme at 37°C for 4 weeks (change weekly).
4. Wash.
5. Embed in paraffin or paraplast.
6. Stain and mount.

1.2.26 *Rexrode and Krause's technic [18]*

Formula:

Solution A: Smith's Modified Bouin

Picric Acid (sat. aqueous solution)	45 pt
Ethyl Alcohol (95%)	45 pt
Formalin	5 pt
Glacial Acetic Acid	5 pt

Solution B: Acetate Buffer

Sodium Acetate (0.2M, 13.01 g/500 ml)	7 pt
Glacial Acetic Acid (0.2M, 6.005 g/500 ml)	3 pt

Solution C:

Carlisle's chitinase solution diluted 1:5 with solution B.

Procedure:

1. Kill and fix specimen in hot solution A at 55°C.
2. Fix in vacuum oven (55°C) with three changes, releasing vacuum slowly from 25 in. Hg. for 10 min.
3. Wash three times in ethyl alcohol (70%) at 25°C in vacuum oven (24 in. Hg.) for 30 min/wash.
4. Transfer through 95%, 100% ethyl alcohol, ethyl ether and blot, for 1 min in each.
5. Return rapidly to 100% ethyl alcohol.
6. Transfer through 100, 95, 70, 35% ethyl alcohol and distilled water for 1 min in each.
7. Transfer to a 1:1 ethyl alcohol: benzene solution at 25°C for 1 hr.
8. Transfer to benzene at 56°C and then a 1:1 benzene:paraplast solution for 1 hr in each.
9. Transfer to 100% paraplast in a vacuum oven at 56°C for 10 min.
10. Transfer through three changes in 100% paraplast in a vacuum oven at 56°C for 18 hr.
11. Embed, block, section, and stain.

1.2.27 *Dillon's technic [19]*

Formula:

	Ethanol	n-butyl
Solution A:	9 pt (45%)	1 pt
Solution B:	8 pt (60%)	2 pt
Solution C:	6 pt (80%)	4 pt
Solution D:	4 pt (95%)	6 pt
Solution E:	2 pt (95%)	8 pt

	Butanol	Terpineol
Solution F:	1 pt	1 pt
Solution G:	1 pt	2 pt
Solution H:	1 pt	3 pt

Procedure:

1. Fix in Gilson's fluid (formula in Chapter 2) (in 28 in. Hg. vacuum for larger specimens).

2. Transfer to ethanol (35%) for 30 min.
3. Transfer to solution A for 2 hr.
4. Transfer to solution B for 3 hr.
5. Transfer to solution C for 4 hr.
6. Transfer to solution D for 6 to 24 hr.
7. Transfer to solution E for 6 to 24 hr.
8. Transfer to solution F, G, and H for 1 hr each.
9. Transfer through three changes of terpineol for 1 hr each.
10. Transfer to tuluol (toluene) and then tuluol-wax.
11. Transfer to paraffin in a vacuum oven at 65°C for 24 hr to 72 hr.
12. After sectioning and before staining all fixative must be removed with iodized alcohol (70%) or lithium carbonate.
13. Stain and mount.

Note:
1. Phenol (4%) is added to each step in the dehydration process except the second.
2. If vacuum is used in fixation, pressure must be reduced slowly.

1.3 Preparations of insect eggs

There are many special problems in the preparation of insect eggs. These problems are primarily due to the presence of yolk and highly sclerotized chitin. These conditions make eggs hard, brittle, and histologically difficult to prepare. As with other highly sclerotized tissues, the penetration of fixatives can be enhanced by puncturing the eggs. However, this procedure is not always desirable since puncturing may distort or destroy many structures. Many of the materials and procedures already recommended in this chapter are useful in preparation of egg material with highly sclerotized chitin. However, the following procedures and compounds have been recommended specifically for use in the histological preparation of insect eggs. Other methods may be found in Chapters 5, 9, and 13 which have other procedures for the display of insect eggs and for the double embedding of insect tissues.

1.3.1 *Terpineol technic [15]*

Application:
Recommended for the histological preparation of yolk-laden and chitinous insect eggs.

Procedure:
1. Dehydrate up to 95% alcohol.
2. Transfer to terpineol (in a warm location for a few hours).
3. Transfer to melted paraffin.
4. Transfer through successive changes of paraffin until lilac odor is gone (or 1 to 2 hr in each).

Note:
1. Transfer specimen with forceps.
2. Do not transfer by pouring out old paraffin.

1.3.2 *Johnston's rubber paraffin technic [15]*

Application:
Recommended for the histological preparation of yolk-laden and chitinous insect eggs.

Formula:

Crude India-Rubber	1 pt
Asphalt	0.1 pt
Paraffin	99 pt

Preparation:
1. Cut rubber into small pieces.
2. Heat ingredients to 100°C for 24 to 48 hr.
3. Pour off supernatant fluid.
4. Cool immediately.

Procedure:
1. Use as ordinary paraffin.

1.3.3 *Tahmisian and Slifer technic [20]*

Application:
1. Recommended for the histological preparation of refractory materials like insect eggs and eyes.
2. Recommended as a method for obtaining thin serial paraffin sections.

Procedure:
1. Fix specimen in Bouin's fluid.
2. Transfer through two changes of dioxan over a period of 8 hr.
3. Transfer through three changes of paraffin (with 0.5% beeswax) over a period of 8 hr.
4. Embed as usual and cut block so as to expose tissue.
5. Soak block in water for at least 24 hr.
6. Section.
7. Transfer to water.
8. Cool and replace water with solution A of Mallory's triple stain (Chapter 4) for 5 min.
9. Replace first solution with solution B for 5 min.
10. Briefly replace with water.
11. Replace immediately with alcohol (95%).
12. Replace with absolute alcohol.
13. Add drops over sections of 0.5% celloidin solution (in a 1:1 ether and alcohol solution).
14. Dry for several hours or overnight.
15. Clear in xylol.
16. Mount in balsam or clarite (clarite is a thermoplastic resin commonly used in histology for mounting sections).

1.3.4 *Henking's [4]*

Application:
Recommended for softening fixed and stained insect eggs before sectioning.

Formula:

Water	30 pt
Alcohol (95%)	70 pt

Hydrochloric Acid	0.2 pt
Pepsin	0.1 pt

1.4 Double Embedding Techniques

Double embedding is another approach utilized to facilitate the preparation and sectioning of highly sclerotized chitinous materials. Other technics can be found in the chapter on embedding (Chapter 3).

1.4.1 *Double embedding technic [8]*

Formula:

Solution A:

Celloidin Flakes (air dried)	1 pt
Methyl Benzoate	100 pt

Solution B:

Paraffin	1 pt
Benzene	1 pt

Preparation:
1. Add celloidin flakes to methyl benzoate.
2. Shake and let stand for 1 hr.
3. Invert container and let stand for 1 hr.
4. Place container on side and let stand for 1 hr.
5. Repeat process until solution is completed or about ½ to 1 day.

Procedure:
1. Fix and wash specimen as desired.
2. Transfer through 50, 70, and 90% ethyl alcohol for 2 hr each.
3. Transfer to ethyl alcohol for 2 to 16 hr.
4. Transfer to solution A for 24 hr.
5. Transfer to fresh solution A for 48 hr.
6. If not clean, transfer to third change of solution A for 72 hr.
7. Transfer through three changes of benzene for 4, 8, and 12 hr.
8. Transfer to solution B in embedding oven for 1 hr.
9. Transfer through two changes of paraffin for ½ to 6 hr (3 hr for 5 mm thickness).
10. Embed (paraffin), section and mount as usual.

1.4.2 *Wilson's modified boycott technic [15]*

Application:

Recommended for small insect specimens.

Procedure:
1. Take small glass tube and stabilize upright (with clay, plasticine etc.).
2. Add small layer of celloidin on bottom.
3. Add similar layer of celloidin and clove oil (equal part) mixture.
4. Follow with layer of clove oil, then a layer of absolute alcohol.
5. Add dehydrated specimen to absolute alcohol and cork tube.
6. When specimen sinks to celloidin layer, pipette off all top layers.
7. Pour specimen on to waxed slide.
8. Expose slide to chloroform vapors.
9. Embed in wax through cedarwood oil.

1.4.3 *Sinha's modified Mukerji's [21]*

Application:
Recommended for use on specimens of families in the orders Isoptera, Anoplura, Homoptera, Lepidoptera, Coleoptera, Diptera, Hymenoptera and the hexapod, Collembola.

Formula:
Solution A: Modified Mukerji's Fixative

Picric Acid (sat. in 90% alcohol)	75 pt
Formalin	25 pt
Nitric Acid (conc.)	8 pt

Solution B:

Nitric Acid	3–6 pt
Alcohol (90%)	100 pt

Solution C:
Celloidin in aged clove oil (2 months).

Solution D:
Thin celloidin in ether-alcohol.

Solution E:

Paraffin	1 pt
Chloroform	5 pt

Procedure:
1. Fix in solution A for 4 to 6 days.
2. Or, fix in Carnoy's and Lebrun's (Chapter 2) for 24 to 48 hr followed by transfer to solution B (3 to 6 days).
3. Using both can be helpful; fix in Carnoy's and Lebrun's (2–4 hr), dip in iodine and 90% alcohol then solution A.
4. From either of above, transfer through to 90% and absolute alcohol for 16 to 24 hr.
5. Transfer to clove oil for 1 day.
6. Then, transfer to solution C for 20 to 40 days.
7. Or, to ether-alcohol for several hours.
8. Then, transfer to solution D for 7 days.
9. From either of above, transfer to chloroform for 12 to 24 hr.
10. Square off block, transfer to solution E at 78–88°F, overnight.
11. Prepare as usual (embed, block, etc.).
12. After removal of paraffin from sections, with xylene, coat with thin layer of celloidin.

Note:
1. Solution A is reported to fix soft parts of specimen while softening sclerotized areas.

1.4.4 *Agar-Ester wax double embedding technic [22]*

Application:
Recommended for cutting sections of insect cuticle.

Procedure:
1. Transfer specimen to 5% agar at 55–60°C for 1 hr or more.

2. Allow to set at room temperature (orient specimen).
3. Transfer agar block through 30, 50, and 70% ethyl alcohol for ½ hr each.
4. Transfer to a 2:1 alcohol (70%) and cellosolve solution then to another (in a 1:2 ratio) for ½ hr each.
5. Transfer through three changes of cellosolve for ½ hr each.
6. Transfer to cellosolve and ester wax (equal part) mixture for ½ hr.
7. Transfer through two changes of ester wax for ½ hr each.
8. Transfer to third change of ester wax at 55–60°C, overnight.
9. Place agar block in embedding trough and fill with ester wax.
10. Cool rapidly.
11. Mount block and trim so agar extends to the surface on all sides.
12. Section and prepare slides.

Note:
1. Material is difficult to cut.
2. Technic is reported to give unbroken series of sections with the epidermal cells to the cuticle.

1.4.5 *Tissue Mat, N-butyl methacrylate technic [23]*

Application:
Recommended for specimens of Hemiptera, Homoptera, Hymenoptera, larval Lepidoptera, Orthoptera and Coleoptera.

Procedure:
1. Fix in Carnoy's (see Chapter 3 for De Giusti and Ezman's Tissue Mat isopropyl method).
2. Wash and dehydrate up to absolute alcohol.
3. Transfer to monomer (fluid) of N-butyl methacrylate at 54°C for 1 to 1½ hr.
4. Transfer to mixture of monomer, 2% catalyst (2, 4, dimethyl benzoyl peroxide) and tissue Mat or paraffin equal amounts.
5. Polymerize in gelatin capsule at 54°C for 12 hr.
6. Remove gelatin capsule and trim block.
7. Place block in melted Tissue Mat (54–56°C), and infiltrate for 15 min.
8. Block, trim, and section.

Note:
1. Since the amount of paraffin affects the hardness of the methacrylate, the composition of mixture (step 4) depends on size of specimen.
2. Proportions used in step 4 are for large insects such as roaches, smaller hard insects would require a reduction of tissue mat to 1–2% of mixture.
3. Methacrylate sectioning causes great static electricity thus requiring neutralization.

1.4.6 *Storch and Chadwick technic [24]*

Application:
Recommended for obtaining serial sections of celloidin embedded insect specimens.

Procedure:
1. Celloidin matrix sections are placed in individual spot plate depressions filled with ethanol (50%).

2. Sections are picked up with fine forceps (dipped in turpentine) and placed on a slide with a thin coat of Mayer's albumen.
3. Slowly lower a piece of cigarette paper over the sections.
4. Using a photographic ring roller, press paper (20–25 times) with increasing pressure.
5. Place a clean slide and a 250–300 gm weight on paper and slide.
6. Heat preparation in an oven set at 55°C for 30 to 60 min.
7. Transfer to acetone after paper floats to top.
8. Change acetone hourly to remove celloidin matrix (8 to 10 times).
9. Transfer to absolute ethanol and proceed as with paraffin sections.

Note:
1. In steps 1 and 2 sections should not be allowed to dry out.
2. Wrinkled sections (step 2) can be smoothed out by brushing with turpentine.
3. It is critical that pressure be applied with roller for full 25 times. Very heavy pressure should be applied the last few times.

References

1. Gottlieb F.J. Punctureless preparation of insect specimens. *J. Roy. Micros. Soc.* **85**, 369–373 (1966).
2. Eltringham H. *Histological and Illustrative Methods for Entomologists.* Oxford University Press, London (1930).
3. Verdcourt B. The sectioning of beetle elytra. *Microsc. & Entomol. Mon.* **6**, 305–306 (1948).
4. Gray P. *The Microtomist's Formulary and Guide.* The Blakiston Company, Inc., New York (1954).
5. Kingsbury B.F. & Johannsen O.A. *Histological Technique.* John Wiley & Sons, Inc., New York (1927).
6. Murray J.A. Technique for paraffin sections of formol-fixed insect material. *J. Roy. Micros. Soc.* **57**, 15 (1937).
7. Wilson S.D. *Applied and Experimental Microscopy.* Burgess Publish Company, Minneapolis, MN (1967).
8. Humason G.L. *Animal Tissue Techniques.* W.H. Freeman & Co., San Francisco (1962).
9. Luna L.G. (ed.). *Manual of Histologic Staining Methods of the Armed Forces Institute of Pathology.* McGraw Hill Co., New York (1968).
10. Parks J.J. & Larsen J.R. A morphological study of the female reproductive system and follicular development in the mosquito *Aedes aegypti (L).* *Trans. Amer. Micros. Soc.* **84**, 88–98 (1965).
11. Pearlman R.C. & Cole B.C. Softening of hard tissue for sectioning. *Stain Technol.* **26**, 115–118 (1951).
12. Beckel W.E. Selective softening of insect cuticle after impregnation with wax. *Stain Technol.* **35**, 356–357 (1960).
13. Beckel W.E. Sectioning large heavily sclerotized whole insects. *Nat.* **184**, 1584–1585 (1959).
14. Barros-Pita J.C. Protective paraffin infiltration of soft tissues in insects to facilitate softening of hard exoskeletons. *Stain Technol.* **46**, 171–175 (1971).
15. Lee A.B. *Microtomist's Vade-Mecum.* The Blakiston Company, Philadelphia (1950).
16. Carlisle D.B. Softening chitin for histology. *Nature* **187**, 1132–1133 (1960).
17. Weinstock M. & McDonald F.J.D. The use of mushroom enzymes for softening insect cuticle. *J. Aust. Entomol. Soc.* **8**, 115–116 (1969).
18. Rexrode C.O. & Krause C.R. Serial sections of *Pseudopityopthorus spp. Ann. Entomol. Soc. Amer.* **61**, 1340–1341 (1968).

19. Dillon L.S. Some notes on preparing whole insects for sectioning. *Entomol. News* **65**, 67–70 (1954).
20. Tahmisian T.N. & Slifer E.H. Sectioning and staining refractory materials in paraffin. *Science* **95**, 284 (1942).
21. Sinha R.N. Sectioning insects with sclerotized cuticle. *Stain Technol.* **28**, 249–253 (1953).
22. Wigglesworth V.B. A simple method for cutting sections in the 0.5 to 1 μm range, and for sections of chitin. *Quart. J. Micros. Sci.* **100**, 315–320 (1959).
23. De Guisti D.L. & Ezman L. Two methods for serial sectioning of arthopods and insects. *Trans. Amer. Microsc. Soc.* **74**, 197–201 (1955).
24. Storch R.H. & Chadwick L.E. Serial sections of whole insects. *Stain Technol.* **39**, 59–60 (1964).

2 Fixation

2.1 Introduction

The choice of fixative is the critical first step in histology. Insect tissues can pose a challenge both in terms of their unique external structures and features, and their dense egg and embryologic tissues. Often, careful mechanical disruption of a cuticle or other exoskeletal tissue is required for adequate penetration of the fixative throughout the insect. Where possible, we provide specific recommendations for individual tissue types, keeping in mind that these fixatives may also have broader utility.

The type of fixative can significantly affect downstream applications and the fixative selected should be considered with the goals of the experiment in mind. For routine histology, formalin and acid based fixatives provide excellent morphology for routine staining. See Chapter 4 for morphological stains requiring specific fixatives. For immunohistochemistry, one should avoid protein denaturing fixatives such as those in the Carnoy's family. For nucleic acid studies, alcoholic fixatives are preferred, producing the best sized fragments for PCR [1]. Alcoholic and picric acid based fixatives work well for glycogen whereas detection of lipids is best in an unfixed sample. For the transmission electron microscope, ultrastructure is best preserved by an oxidizing fixative such as osmium tetroxide.

Recent technological advances have increased the need for unfixed tissues. Although rarely appropriate for intact insects or larvae, individual organs can be dissected and snap frozen in liquid nitrogen or in a dry ice/isopentane bath. These fresh frozen tissues can be cryosectioned for downstream applications including but not limited to immunohistochemistry, RNA or DNA isolation, microarray analysis, laser capture microscopy, and proteomic and metabolomic analyses.

2.2 Aldehyde based fixatives

Application:
Recommended as a fixative which causes no shrinkage to certain insect tissues.

2.2.1 *Alkaline formalin [2]*

Formula:

Formol (alkaline)	10%

Note:
1. Formalin may be made alkaline by the addition of sodium or magnesium carbonate or lithium carbonate.

Insect Histology: Practical Laboratory Techniques, First Edition.
Pedro Barbosa, Deborah L. Berry and Christina S. Kary.

Companion Website: www.wiley.com/go/barbosa/insecthistology

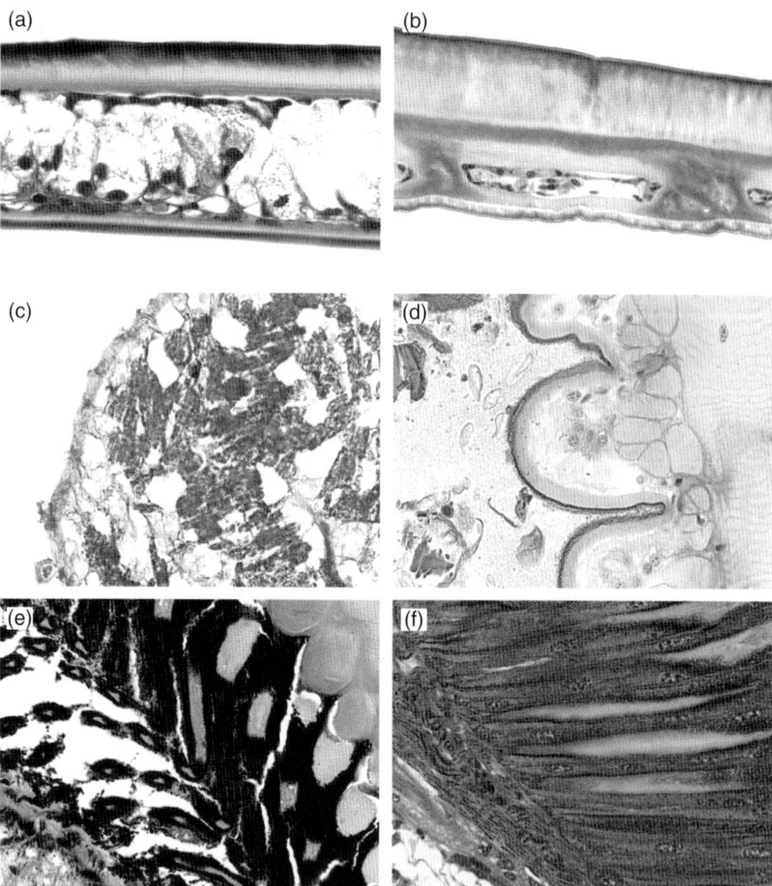

Fig. 2.1 Comparison of fixation techniques for individual tissues from the cockroach *Blatella germanica*. Routine fixation with 10% neutral buffered formalin preserves many structural features of the cuticle stained with Aniline blue (a), midgut stained with Lower's Trichrome (c), and eye stained with Toluidine Blue (e). In contrast, fixation of the cuticle with Davidson's fixative (b), the midgut with formal-acetic-alcohol fixative (d), and the eye with Davidson's AFA fixative (f) significantly improves over all morphology, providing faithful demonstration of the structures in tissue section. (Source: Damien Laudier, Laudier Histology, 2013. Reproduced with permission of Damien Laudier.) See plate section for the color version of the figure.

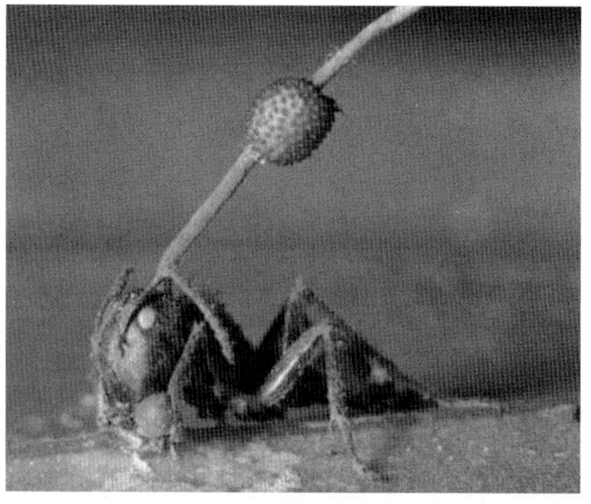

Fig. 2.2 Ant infected by the fungus *Ophiocordyceps unilateralis* bites the underside of a leaf. The fungal stroma emerges from the back of the ant's head and develops a fruiting body with a capsule full of spores. (Source: David Hughes, Pennsylvania State University. Reproduced with permission.)

2.2.2 *Ammerman's [3, 4]*

Application:
Recommended for heavily yolk-laden embryos and larvae.

Formula:

Water	238 pt
Chrome Alum (Chromium potassium sulfate)	3 pt
Formaldehyde (40%)	30 pt
Acetic Acid	2 pt

1. Fix for 1 to 4 hrs.

2.2.3 *Carl's [5]*

Formula:

Absolute Alcohol	15 pt
Formalin	6 pt
Glacial Acetic Acid	2 pt
Distilled Water	30 pt

Procedure:
1. Fix for no more than 24 hr.
2. Overfixation occurs after 24 hr and specimens must be transferred to another preservative.

Note:
1. This solution is similar to Kahle's and Bles' but is reported to be stronger and faster.
2. Washing is not necessary.

2.2.4 *Davidson's alcohol: Formalin: Acetic Acid (AFA) [6]*

Formula:

95% Ethyl Alcohol	33 pt
Formalin	22 pt
Glacial Acetic Acid	11.5 pt
Distilled Water	33.5 pt

2.2.5 *Formaldehyde/ Glutaraldehyde [7]*

Formula:

Sodium dihydrogen phosphate (anhydrous)	3.5 pt
Disodium hydrogen phosphate (anhydrous)	6.5 pt
40% Formaldehyde	100 pt
25% Glutaraldehyde	100 pt
Distilled Water	800 pt

2.2.6 *Formol-acetic-alcohol [4, 8, 9]*

Application:
Recommended for eggs, larvae, and pupae (soft bodied forms).

Formula:

	Kahle's	Bles'
Glacial Acetic Acid	5 pt	3 pt
Alcohol (70%)	90 pt	90 pt
Formaldehyde (40%)	5 pt	7 pt

Procedure:
1. Fixing time ranges from 30 min to overnight.

Note:
1. Kahle's is sometimes called Dietrich's.
2. These fluids are sometimes used for field collection and storage of specimens.

2.2.7 *Frings' and Frings' IP CFAA [10]*

Application:
Recommended as a rapid fixative which prevents excessive hardening of the cuticle.

Formula:
Solution A: Fixative

Chloroform	8 pt
Isopropyl Alcohol (100%)	45 pt
Formalin	10 pt
Water	35 pt
Glacial Acetic Acid	2 pt

Solution B: Preservative

Isopropyl Alcohol (70%)	2 pt
Water	7 pt
Glycerol (10%)	1 pt

Procedure:
1. Fix in solution A for ½ to 2 hr (large insects).
2. Transfer material within 6 to 8 hr to solution B, ethyl alcohol.

Note:
1. Fixative may be used in field for collection of specimens.
2. If separation of fixative occurs due to frequent use, more alcohol should be added until homogeneous.

2.2.8 *Hall's [11]*

Application:
Recommended for preservation of larval muscoid insects.

Formula:

Ethanol (95%)	85 pt
Formalin (40%)	10 pt
Glycerine	5 pt

2.2.9 *Heat fixation [12]*

Application:
Heat increases the speed of fixation.

Formula:
Formalin (2%) in phosphate buffered saline.

Procedure:
1. Place tissues in formaldehyde at 55°C for 6 min.

2.2.10 *Homberg's GPA [13]*

Formula:

Glutaraldehyde	6 pt
Acetic Acid	0.5 pt
Picric Acid	1 pt
Distilled Water	92.5 pt

2.2.11 *Karr and Albert's embyro technic [14]*

Application:
Recommended for embryos for removal of chorion and vitelline membranes to allow for adequate penetration of fixative, nucleic acids, and immunoglobulins.

Formula:
Solution A:

Commercial Bleach	1 pt
Distilled Water	1 pt

Solution B: PEM Buffer

1 M Pipes	1 pt
10 mM $MgCl_2$	1 pt
10 mM EGTA	1 pt
Distilled Water	7 pt
pH adjusted to 6.9 with KOH	

Solution C:

0.5M Taxol	0.01 pt
Heptane	5 pt

Solution D:

20% Formaldehyde	1 pt

Solution E:

Heptane	1 pt
90% Methanol with 0.05M Na_3EGTA	1 pt

Procedure:
1. Remove chorion in solution A for 1.5 min.
2. Rinse and remove bleach in NaCl/Triton solution.
3. Add 5 ml solution B, then add 5 ml solution C and shake vigorously for 0.5 min.
4. Add solution D and shake for 10 min.
5. Rinse embryos in phosphate buffered saline, transfer to ice cold solution E, and shake for 10 min.
6. Rinse with methanol, and rehydrate through to PBS.

2.2.12 *Kelsheimer's [15]*

Application:
Recommended originally for the display of egg masses of the European corn borer.

Formula:

Uranium Nitrate	1.5 pt
Ethyl Alcohol (95%)	90 pt
Formalin	5 pt
Glacial Acetic Acid	2.5 pt
Glycerine	2.5 pt
Copper Chloride	5 pt

Procedure:

1. Fix for 2 to 5 days.
2. Transfer to alcohol.

2.2.13 *MacGregor's [16]*

Formula:

Formalin (40%)	10 pt
Borax (5%)	10 pt
Distilled Water	80 pt
Glycerine	0.25 pt

2.2.14 *Methanol/ Formalin fixation [17]*

Application:

Recommended for improved preservation of NADPH.

Formula:

37% Formaldehyde	1 pt
Methanol	9 pt

2.2.15 *Microwave fixation [18]*

Application:

Recommended for preserving high quality ultrastructure while using only a small percentage of glutaraldehyde to facilitate antigen accessibility.

Formula:

10% Glutaraldehyde	0.5 pt
10% Formaldehyde	20 pt
1% $CaCl_2$	2.5 pt
0.2M Cacodylate Buffer pH 7.4	50 pt
Distilled Water	27 pt

Procedure:

1. Place tissue into fixative.
2. Microwave on full power for 10–12 seconds. This should bring the temperature up to 42–45°C.
3. Rinse in saline.

2.2.16 *10% neutral buffered formalin [19]*

Formula:

Sodium dihydrogen phosphate (anhydrous)	3.5 pt
Disodium hydrogen phosphate (anhydrous)	6.5 pt
40% Formaldehyde	100 pt
Distilled Water	900 pt

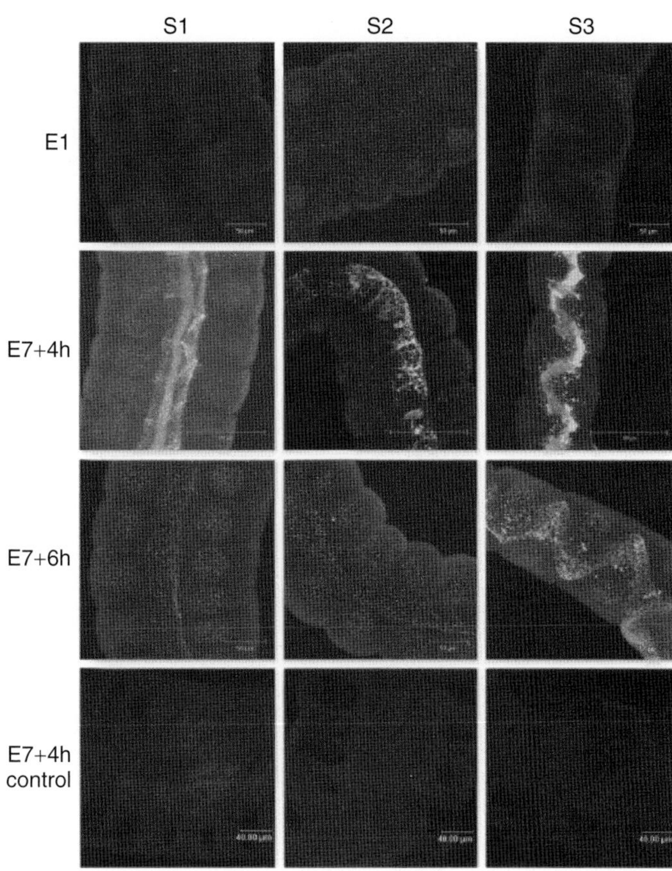

Fig. 2.3
Immunolocalization of BhC4-1 in whole mount S1, S2 and S3 salivary gland regions. Confocal images of optical longitudinal sections of salivary gland regions (S1, S2 and S3) at stages (E1), (E7+4 h) and (E7+6 h). Whole mount salivary glands were incubated with the α-BhC4-1, followed by detection with α-rabbit IgG conjugated to Oregon Green. Nuclei were stained with DAPI. (E7+4 h control) control salivary glands in which the α-BhC4-1 primary antibodies were omitted during the processing for immunolocalization. (Source: Monesi *et al.* 2004. Reproduced with permission of Elsevier.)

2.2.17 *Ott's zinc technic [20]*

Formula:

Solution A:

Paraformaldehyde	4 pt
Phosphate Buffered Saline	96 pt

Solution B:

Zinc Chloride	0.25 pt
Sodium Chloride	0.788 pt
Sucrose	1.2 pt
Formaldehyde	1 pt
Distilled Water	97 pt

Procedure:

1. Fix tissues in solution A.
2. Post fix tissues in solution B.

2.2.18 *Pampel's [16]*

Application:

Recommended for the preservation of small and large insect larvae and for the preservation of discoloration.

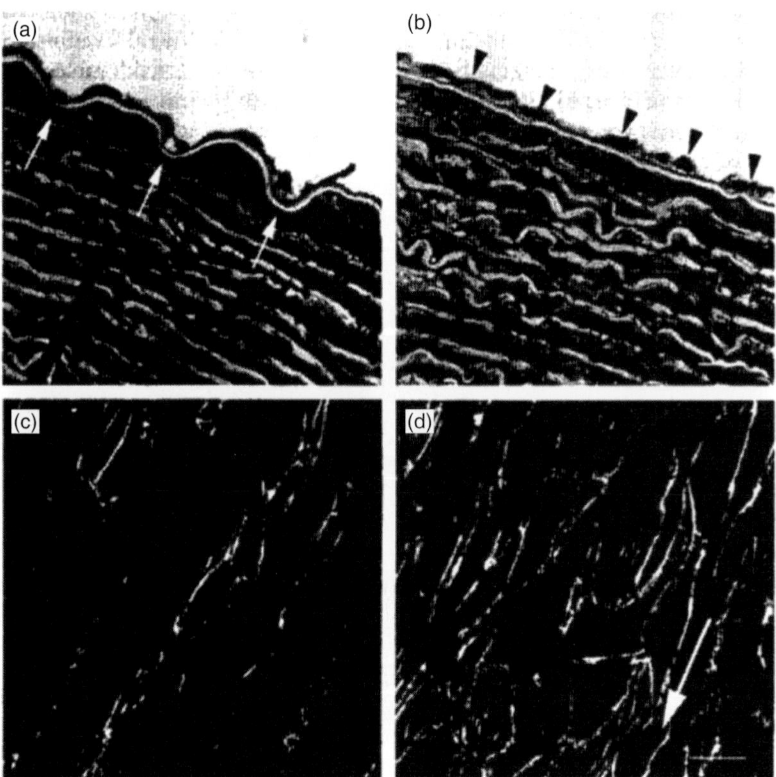

Fig. 2.4 Microwave fixation – Comparison between conventional fixation (a and c) and microwave irradiation-assisted fixation (b and d) in cryosectioned specimens (a and b) and whole mount preparations of blood vessels (c and d). Toluidine blue staining of cryosections fixed by the conventional procedure (a) and microwave irradiation-assisted fixation (b) are shown. Microwave irradiation-assisted fixation showed well preserved morphology of blood vessels including endothelial cells and vascular smooth muscle (b). Conventionally fixed specimens showed marked shrinkage, especially in the vascular smooth muscle cell layer (a). Whole mount preparations using microwave irradiation-assisted fixation showed a well preserved endothelial cell layer without shrinkage of vessels (d).Conventionally fixed specimens showed undulating morphology with vessel shrinkage (c). Microwave irradiation during fixation preserved the morphology of blood vessel structures as seen by both conventional light microscopy and confocal laser scanning microscopy. Bar = 20 μm. (Source: Katoh *et al.* 2009. Reproduced with permission of Biotechnic & Histochemistry.)

Formula:

Glacial Acetic Acid	4 pt
Distilled Water	30 pt
Formalin (40%)	6 pt
Alcohol (95%)	15 pt

2.2.19 *Preparation for cryosectioning [21]*

Formula:

Gelatin/Albumin blocks:

Gelatin	2.4 pt
Albumin	15 pt
dH_2O	Up to 50 pt

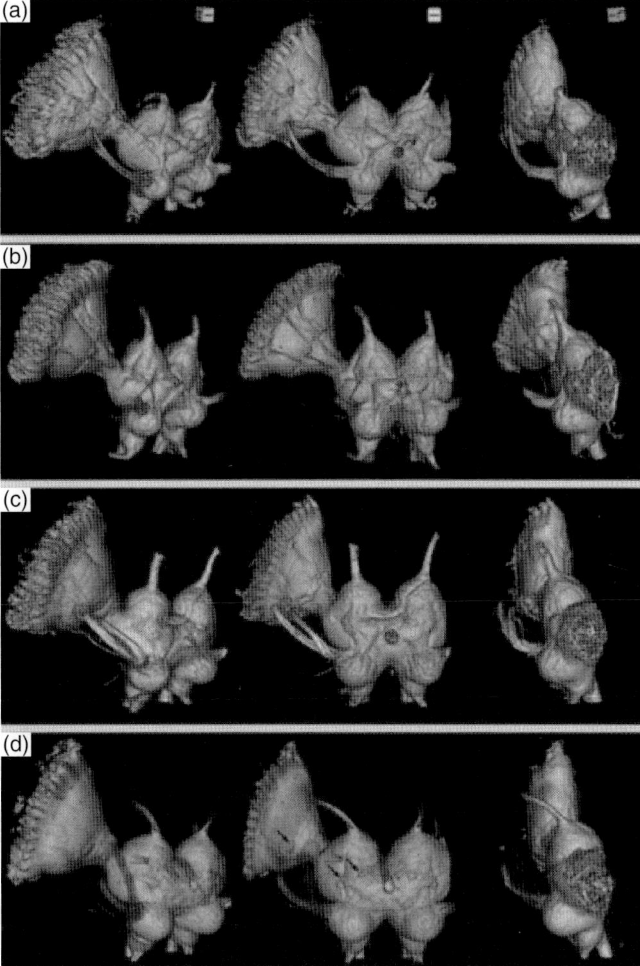

Fig. 2.5 The effects of fixation and dehydration on the preservation of morphology in locust brains that have been processed for synapsin immunohistochemistry. 3D perspective views were obtained by volume rendering of confocal stacks. The opacity was set to a value that corresponded to, and therefore made visible, the surface of the brain rather than the synapsin staining. Each brain with only the right optic lobe is shown in frontal view (center), tilted 30° to the left (views on the left), and 45° to the right (views on the right). In the latter views, the left hemisphere of the brain has been digitally trimmed away. (a) Fixation in Paraformaldehyde Fixative (PFA) for 20 hr, dehydration in ethanol grades. (b) Fixation in PFA for 2 hr, dehydration in ethanol grades. In both (a) and (b), considerable shrinkage is evident from the heavy creasing of the brain surface. The shrinkage has also lead to a distortion of shape. For example, in life, the optic lobe is tilted backwards out of the frontal plane of the brain; in both (a) and (b) the optic lobe is warped forwards into the frontal plane. (c) Fixation in zinc-formaldehyde (ZnFA) for 20 hr, dehydration in ethanol grades. The brain surface shows less creasing than in (a) and (b) and the distortions of morphology are less severe. (d) Fixation in ZnFA (20 hr), dehydration in glycerol grades. The brain surface shows very little creasing and the shape is well preserved, with the optic lobe retaining its natural proportions and orientation. Particularly brightly stained, superficial neuropile structures (including the accessory medulla, anterior optic tubercle, and tip of the α-lobe) bled through the surface optically, creating the impression of bumps on the surface. These artifacts in the volume rendering arise from imperfect confocality and are indicated by black arrows in the frontal view. (Source: Ott 2008. Reproduced with permission of Elsevier.)

Procedure:
1. Fix dissected tissues in 4% paraformaldehyde in phosphate buffered saline (PBS) overnight at 4°C.
2. Rinse the tissues three times in PBS.
3. Incubate tissues in 30% sucrose in PBS overnight at 4°C.
4. Embed the tissues in the gelatin/albumin blocks.
5. Fix blocks in 10% formaldehyde in PBS overnight at 4°C.
6. Rinse the tissues three times in PBS.
7. Incubate tissues in 30% sucrose in PBS overnight at 4°C.
8. Mount blocks in OCT, freeze, and section.

2.2.20 *Serra's fixative [22]*

Formula:

Glacial Acetic Acid	1 pt
Formalin (40%)	3 pt
Alcohol (95%)	6 pt

2.2.21 *Weaver's formol [23]*

Formula:

Formaldehyde(40%)	5 pt
Glacial Acetic Acid	2.5 pt
Chloral Hydrate	20 pt

Preparation:
1. Mix the above and dilute with distilled water up to 100 ml.

Note:
1. A mid-dorsal incision in the abdomen before fixation is suggested.
2. Placement of material under vacuum with several releases is suggested.
3. It is suggested that specimens not be placed in ethyl alcohol before or after fixation.

2.3 **Protein denaturing**

Application:
Recommended as a general insect fixative.

2.3.1 *Carnoy's A [4, 24]*

Formula:

Alcohol (absolute)	6 pt
Chloroform	3 pt
Glacial Acetic Acid	6 pt

Procedure:
1. Fix for 1 to 2 hr.

Note:
1. Carnoy's is a nuclear fixative which preserves (cell elements) but is not suggested for cytoplasm.

2.3.2 *Carnoy's acetic-alcohol [15]*

Formula:

Alcohol (absolute)	3 pt
Glacial Acetic Acid	1 pt

Table 2.1 Aldehyde Fixatives: Many of the types of fixatives are variations on a central list of primary ingredients. In this and the following tables, each category of fixatives is summarized to provide the reader an overview of the components and how they can be varied to optimize fixation of the tissue to be studied.

	Formalin/ Formaldehyde	Ethanol	Acetic Acid	Glycerin	Other
Carl's	4%	28%	0.2%		
Davidson's AFA	8%	32%	11.5%		
Formol-Acetic-Alcohol	2%	61%	5.0%		
Frings' and Frings' IP CFAA	4%	45%	2.0%	1%	Chloroform
Kelsheimer's	2%	81%	2.5%	2.5%	Uranium Nitrate
Pampel's	5%	30%	2.0%		
Serra's Fixatve	12%	57%	10.0%		
Hall's	4%	80%		5%	
Karr and Albert's Embryo Technic	2%				Bleach
Methanol/Formalin	4%	90%*			
Microwave Fixation	2%				
Ammerman's	4%		0.1%		Chrome Alum
Weaver's Formol	2%		2.5%		Chloral Hydrate
Formaldehyde/ Glutaraldehyde	4%				Glutaraldehyde
Heat Fixation	2%				
Ott's Zinc Technic	4%				Zinc Chloride
MacGregor's	4%			0.25%	Borax
Neutral Buffered Formalin	10%				Phosphate Buffer
Alkaline Formalin	10%				
Homburg's GPA	1.5%**		0.5%		Picric Acid

*Methanol
**Glutaraldehyde

2.3.3 *Carnoy's B*

Formula:

Alcohol (absolute)	6 pt
Chloroform	3 pt
Glacial Acetic Acid	1 pt

Procedure:

1. Fix for 4 to 6 hr.
2. Rinse several times in absolute ethanol.
3. Place in absolute ethanol for 2 hr.
4. Transfer through at least four changes of butanol for 2 hr each.

2.3.4 *Carnoy-Lebrun [2–4, 8]*

Application:
Recommended for arthropods with highly sclerotized cuticle.

Formula:

Chloroform	33 pt
Alcohol (absolute)	33 pt
Glacial Acetic Acid	25 pt

Procedure:

1. Fix for 10 to 15 min or until specimen sinks.
2. Wash well in several changes of alcohol (95%).
3. Transfer to iodine alcohol (85%).

Note:

1. This fixative is considered by some researchers as one of the quickest penetrating fluids.

2.3.5 *Farmer's [25]*

Application:

Recommended for use in the histological preparation of aphid material.

Formula:

Ethyl Alcohol (absolute)	2 pt
Glacial Acetic Acid	1 pt

2.3.6 *Peterson's (commonly referred to as K.A.A.) [26]*

Application:

Recommended for killing and fixing insect larvae and other immature stages.

Formula:

Kerosene (commercial)	1 pt
Ethyl (or isopropyl) alcohol (95%)	10 pt
Glacial Acetic Acid	2 pt

Note:

1. Excessive swelling may be eliminated by reducing the amount of kerosene.

2.3.7 *K.A.A.D. [27]*

Application:

Commonly used for preservation of insect larvae for curative purposes.

Formula:

Kerosene (commercial)	1 pt
Dioxane	1 pt
Ethyl (or isopropyl) alcohol (95%)	10 pt
Glacial Acetic Acid	2 pt

Note:

1. Place specimens into K.A.A.D. while still alive to allow for full distension.

2.3.8 *K.A.S.A. [28]*

Formula:

Kerosene (commercial)	3 pt
Ethyl (or isopropyl) alcohol (95%)	9 pt
Sec-butyl Alcohol	5 pt
Glacial Acetic Acid	2 pt

Note:

1. The solution should be water clear, if not, add additional sec-butyl alcohol until the solution clears.

2.3.9 *Tetrahydrofuran Carnoy's [29]*

Formula:

Tetrahydrofuran	6 pt
Chloroform	3 pt
Glacial Acetic Acid	1 pt

Table 2.2 Protein denaturing fixatives.

	Chloroform	Ethanol	Acetic Acid	Other
Farmer's		66%	33%	
Carnoy's A	20%	40%	40%	
Carnoy's B	30%	60%	10%	
Carnoy-Lebrun	36%	36%	27%	
Tetrahydrofuran – Carnoy's	30%		10%	60% Tretrahydrofuran
Carnoy's Acetic-Alcohol		75%	25%	
KAAD	7%*	68%	14%	Dixoane
KASA	16%*	45%	10.5%	Sec-butyl Alcohol
Peterson's	8%*	73%	15%	

*Kerosene

Procedure:

1. After fixation transfer to pure tetrahydrofuran.

Note:

1. Use of tetrahydrofuran avoids the hardening effects of higher alcohols.

2.4 Picric acid based

2.4.1 *Allen's PFA [30]*

Formula:

Picric Acid (saturated aqueous)	75 pt
Formalin	15 pt
Glacial Acetic Acid	10 pt
Urea (immediately before use)	1 pt

2.4.2 *B-15 [2]*

Application:

Recommended as a fixative which causes only moderate shrinkage to histological preparations of insect tissue.

Formula:

Solution A:

Picric Acid (saturated aqueous)	47.5 pt
Glacial Acetic Acid	5 pt
Chromic Acid Crystals	1 pt

Solution B:

Picric Acid (saturated aqueous)	27.5 pt
Formalin (acid)	25 pt
Urea Crystals	1 pt

Note:

1. Both types are modifications of the original Bouin's formula.

2.4.3 *Bouin's [8, 15]*

Formula:

Picric Acid (saturated aqueous solution)	75 pt
Formalin (acid)	25 pt
Glacial Acetic Acid	5 pt

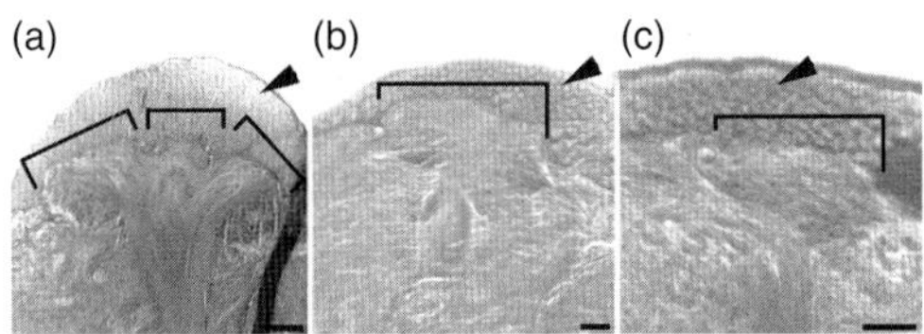

Fig. 2.6 Differences in calyx structure in generalist and specialist predators. (a) Tripartite gyrencephalic calyx (brackets) of the generalist predator Scarites subterraneus (Carabidae). (b) Lissencephalic calyx of the carabid Scaphinotus elevatus, a specialist predator of snails and slugs. (c) Minute lissencephalic calyx of the ladybird beetle Harmonia axyridis (Coccinellidae), a specialist predator of aphids and scale insects. Arrowheads, Kenyon cell body regions. (Scale bars: a, 50 μm; b and c, 20 μm. Note: Fixed in Carnoy's at 20°C. Fixation was allowed to proceed for a duration ranging from 1 h 15 min to 1 h 45 min to keep tissue shrinkage between brains relatively constant. After fixation, brains were stored in 70% ethanol at 4°C overnight. The next day, dehydration through a graded ethanol series (for 10 min each) was followed by clearing in xylene (twice for 10 min) and paraffin embedding in a frontal orientation. Embedded brains were sectioned at 10 μm on a rotary microtome and stained by using Cason's stain according to the protocol of Kiernan (26) to reveal brain architecture. (Source: Farris & Roberts 2005. Reproduced with permission of Elsevier.) See plate section for the color version of the figure.

Procedure:

1. Fix 12 to 24 hr.
2. Wash with alcohol (70%) until yellow color is no longer visible.

Note:

1. Fluid can be used hot.

2.4.4 *Brazil's [26]*

Formula:

Picric Acid	1 pt
Glacial Acetic Acid	15 pt
Formalin (40%)	60 pt
Ethyl Alcohol (80%)	150 pt

Procedure:

1. Fix for 24 hr.
2. Wash several times in alcohol (70%) or until yellow color disappears.

Note:

1. This fixative can be used hot or cold.

2.4.5 *Duboscq-Brasil fluid (alcoholic Bouin's) [31]*

Formula:

Picric Acid	1 pt
Formalin (40%)	60 pt
Glacial Acetic Acid	15 pt
80% ethyl alcohol	150 pt

Procedure:

1. Heat insect in the above solution on a hot plate until vapor appears (do not boil).
2. Slit insect and cool in solution for 3–24 hr in a fume hood.

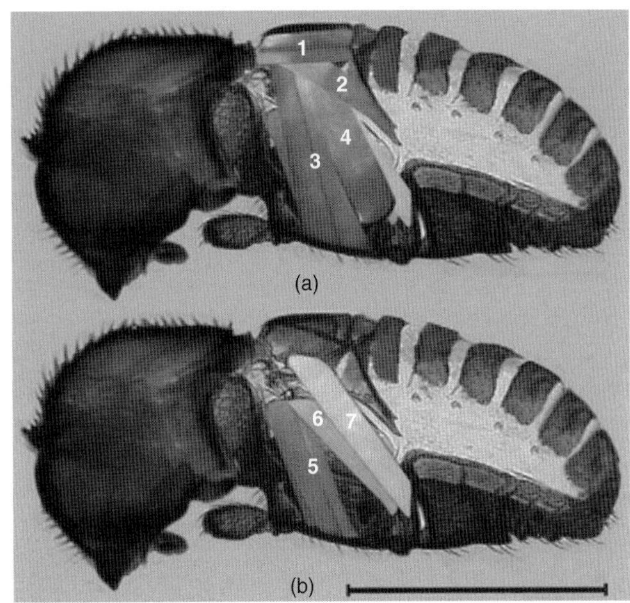

Fig. 2.7 Specimens prepared in Duboscq-Brasil fluid histological fixative. Longitudinal section of the flight muscle of adult female of coffee berry borer. a, dorsal muscles; and b, tergosternal muscles. 1 = median dorsolongitudinal muscles, 2 = oblique lateral dorsal muscles, 3 = tergosternal muscles, 4 = tergo-trochantinal muscles, 5 = sternobasalar muscles, 6 = coxobasalar muscles, and 7 = coxosubalar muscles. Scale bar = 1 mm. (Source: López-Guillén *et al.* 2011. Reproduced with permission of Environment Entomology.)

2.4.6 *Hollande Bouin's fixative [32]*

Formula:

Copper Acetate	2.5 pt
Picric Acid Crystals	4 pt
Formalin	10 pt
Distilled Water	100 pt
Glacial Acetic Acid	1.5 pt

Procedure:

1. Dissolve copper acetate in water with heat.
2. Add picric acid slowly with stirring.
3. Once dissolved, filter.
4. Add formalin and acetic acid.

2.4.7 *Modified Hollande-Bouin solution [33, 34]*

Formula:

Copper Acetate	2.5 pt
Picric Acid Crystals	4 pt
Formalin	10 pt
Distilled Water	100 pt
Mercuric Chloride	0.7 pt

Procedure:

1. Dissolve copper acetate in water with heat.
2. Add picric acid slowly with stirring.
3. Once dissolved, filter.
4. Add formalin and mercuric acid.

2.4.8 *Improved synthetic alcoholic Bouin [31]*

Formula:

Paraformaldehyde (40%)	15 pt
Ethanol	25 pt
Acetic Acid	5 pt
Ethyl Acetate	5 pt
Diethoxymethane	15 pt
Picric Acid	0.5 pt
Distilled Water	34.5 pt

2.4.9 *Kramer's Bouin [3]*

Application:
Recommended for use in the preparation of insect whole mounts.

Formula:

Water	190 pt
Picric Acid	2.4 pt
Formaldehyde	60 pt
Glacial Acetic Acid	12.5 pt

2.4.10 *Mayer's picric hydrochloride [9, 35]*

Formula:

Water	100 pt
Hydrocholoric Acid (25% of N_2O_5)	8 pt
Picric Acid	To saturation

2.4.11 *Mayer's picro-nitric acid technic [8, 9, 35]*

Application:
Recommended for yolk-laden insect eggs.

Formula:

Water	100 pt
Nitric Acid (25% of N_2O_5)	5 pt
Picric Acid	To saturation

Procedure:
1. Use cold for 1 to 5 hr.
2. Wash in several changes of alcohol (70%) over several days until yellow color is gone.

2.4.12 *Picro-chloro acetic [15]*

Application:
Recommended as a general fixative for most insect tissues.

Formula:

Picric Acid (1% in 96% alcohol)	60 pt
Chloroform	10 pt
Acetic Acid	5 pt

Procedure:
1. Fix for 12 hr.
2. Wash in several changes of alcohol (70%) until yellow color is gone.

2.4.13 *Picro-sulfuric acid technic – better known as Kleinenberg's fluid/solution [15, 36]*

Application:
Recommended for use in insect embryology for the effective penetration of eggs.

Formula:

Distilled Water	100 pt
Sulfuric Acid	2 pt
Picric Acid	To saturation

Procedure:
1. Fix for 12 to 24 hr.
2. Transfer through two to three changes of alcohol (70%) for 12 hr.
3. Transfer through one to two changes of alcohol (85%).

2.4.14 *Van Leeuwen's technic [8]*

Application:
Recommended for fixing fluid for insect embryos.

Formula:

Picric Acid (in 1% in absolute alcohol)	12 pt
Chloroform	2 pt
Formalin (strong)	2 pt
Glacial Acetic Acid	1 pt

Note:
1. After fixation, wash in alcohol (80%).

2.4.15 *Zamboni's fixative [37]*

Formula:

Paraformaldehyde	20 pt
Distilled Water	150 pt
Picric Acid	To saturation
Sodium Hydroxide	Add slowly until solution clears
Phosphate Buffered Saline	830 pt

2.5 **Mercuric chloride based**

Application:
Recommended for use on small insects and invertebrate eggs.

2.5.1 *Corrosive acetic (Sherlock's) [15, 38]*

Formula:

Mercuric Chloride (saturated solution)	95 pt
Acetic Acid	5 pt

Procedure:
1. Fix material (warm or cold) for 2 to 3 hr.
2. Remove excess mercuric chloride with iodine alcohol.

2.5.2 *Corrosive sublimate (mercuric chloride) [2]*

Application:
Recommended as a fixative which causes only moderate shrinkage to histological preparations of insect tissue.

Table 2.3 Picric acid based fixatives.

	Picric Acid	Acetic Acid	Formalin	Other (1)	Other (2)
Brazil's	Picric Acid	Acetic Acid	Formalin	EtOH	
Improved Synthetic Alcoholic Bouin's	Picric Acid	Acetic Acid	Formalin	EtOH	Ethyl acetate, diethoxymethane
Allen's PFA	Picric Acid	Acetic Acid	Formalin	Urea	
Bouin's	Picric Acid	Acetic Acid	Formalin		
Hollande Bouin's Fixative	Picric Acid	Acetic Acid	Formalin		Copper Acetate
Modified Hollande Bouin's Fixative	Picric Acid	Acetic Acid	Formalin	Mercuric Chloride	Copper Acetate
Kramer's Bouin	Picric Acid	Acetic Acid	Formalin		
Van Leeuwen's Technic	Picric Acid	Acetic Acid	Formalin	Chloroform	
Picro-Chloro Acetic	Picric Acid	Acetic Acid		Chloroform	
B-15	Picric Acid	Acetic Acid		Urea	
Mayer's Picro-nitric acid technic	Picric Acid	Nitric Acid			
Zamboni's Fixative	Picric Acid	Sodium Hydroxide	Formaldehyde		Phosphate Buffer
Picro-Sulfuric Acid technic	Picric Acid	Sulfuric Acid			
Mayer's Picric Hydrochloride	Picric Acid	Hydrochloric Acid			

Formula:

Bichloride of Mercury	Saturated aqueous solution

2.5.3 *Eltringham's [3, 15]*

Application:
Recommend for use on chitin free soft insect parts.

Formula:

Water	225 pt
Mercuric Chloride	15 pt
Formaldehyde (40%)	25 pt

2.5.4 *Gilson's M-N [38]*

Formula:

Mercuric Chloride	5 pt
Nitric Acid (80%)	4 pt
Glacial Acetic Acid	1 pt
Alcohol (95%)	15 pt
Distilled Water	230 pt

Procedure:
1. Let stand for 3 to 4 days and filter.
2. Fix material for 15 min to 6 hr (Function of size).
3. Transfer to iodine alcohol to remove excess mercuric chloride.

2.5.5 *Movat's FSA [39]*

Application:
Recommended as a fixative that will give consistently reproducible staining reactions in all insect tissue. Recommended as useful for staining with hematoxylin-eosin and Wismar's quadrachrome.

Formula:

Formalin	20 pt
Mercuric Chloride	4 pt
Glacial Acetic Acid	5 pt
Distilled Water	80 pt

Procedure:
1. Fix 15 mm tissue blocks; should be fixed for a maximum of 24 hr.
2. Rinse quickly in a 15% iodine alcohol solution (70%).
3. Store in alcohol (70%).

Note:
1. Store in non-metallic containers (indefinitely if necessary).

2.5.6 *Petrunkevitch's [40]*

Formula:

Mercuric Chloride	5 pt
Nitric Acid	5 pt
Glacial Acetic Acid	45 pt
Absolute Alcohol	100 pt
Distilled Water	150 pt

Procedure:
1. Fix for 30 min (for small specimen) or up to 6 hr (for larger specimens).
2. Wash for a few hours in running water.
3. Transfer to several changes of iodine alcohol (70%).

2.5.7 *Schaudin's [30]*

Formula:

Mercuric Chloride (saturated aqueous)	66 pt
Alcohol (95%)	33 pt
Glacial Acetic Acid (immediately before use)	5 pt

Note:
1. This fluid is generally used hot.

2.5.8 *Tetrahydrofuran-Eltringham [29]*

Application:
Recommended for precise nuclear staining with minimum hardening of tissues.

Formula:

Distilled Water	145 pt
Tetrahydrofuran	100 pt
Glacial Acetic Acid	10 pt
Mercuric Chloride	10 pt

Procedure:
1. After fixation transfer to a 1:1 mixture of water and tetrahydrofuran for about 1 hr.
2. Transfer to pure tetrahydrofuran.

2.5.9 *Tetrahydrofuran-sublimate [29]*

Application:
Recommended as a generalized fixative for insect tissue.

Formula:

Tetrahydrofuran	95 pt
Mercuric Chloride	8 pt
Glacial Acetic Acid	5 pt

Procedure:
1. After fixation transfer to pure tetrahydrofuran.
2. Use of tetrahydrofuran can eliminate both dehydration and clearing.
3. Tetrahydrofuran has the advantages of being relatively non-toxic and miscible in water and wax.

2.5.10 *Zenker's [8, 26]*

Formula:

Bichromate of Potassium (A)	2.5 pt
Mercuric Chloride (B)	5 pt
Sodium Sulfate	1 pt
Distilled Water	100 pt
Glacial Acetic Acid (C)	5 pt

Preparation:
1. A and B are dissolved in heated water.
2. C is added when fluid is to be used.

Procedure:
1. Fix tissues for 6 to 24 hr.
2. Wash in running water for 12 to 24 hr.
3. Transfer to alcohol (35%) for 20 min.
4. Transfer to alcohol (50%) for 20 min.
5. Transfer to iodine alcohol (70%).

2.6 **SEM/TEM**

2.6.1 *Carbon tetrachloride [41, 42]*

Procedure:
1. Immerse tissues in Carbon Tetrachloride (CCl_4) overnight at room temperature.
2. Boil tissues in CCl_4, three times for 1 min each.
3. Rinse in 100% ethanol.

2.6.2 *Glutaraldehyde/osmium tetraoxide*

Formula:
Solution A:

0.2M Cacodylate buffer pH 7.4	50 pt
25% aqueous glutaraldehyde	8 pt
Distilled/deionized water	42 pt

Solution B:
1% Osmium tetroxide

Table 2.4 Mercuric chloride based fixatives.

	Mercuric Chloride	Acetic Acid	EtOH	Other
Corrosive Sublimate (Mercuric Chloride)	Saturated			
Eltringham's	5.7%			4% Formalin
Corrosive-Acetic (Sherlock's)	95%	5%		
Movat's FSA	4%	5%		8% Formalin
Gilson's M-N	2%	0.4%	6%	1.25% Nitric Acid
Petrunkevitch's	1.6%	15%	33%	1.6% Nitric Acid
Schaudin's	63%	5%	30%	
Tetrahydrofuran-Eltringham	4%	4%		38% Tertrahrdrofuran
Tetrahydrofuran-sublimate	7.4%	4.6%		88% Tertrahrdrofuran
Zenker's	4.4%	4.4%		

Procedure:

1. Fix sample in solution A, 4 hr to overnight.
2. Post-fix in solution B.

2.6.3 *Glutaraldehyde/uranyl acetate*

Formula:

Solution A:

0.1M Phosphate buffer pH 7.4	50 pt
25% aqueous glutaraldehyde	8 pt
Distilled/deionized water	42 pt

Solution B:

1% Uranyl Acetate

Procedure:

1. Fix sample in solution A, 4 hr to overnight.
2. Post-fix in solution B.

2.6.4 *Kalt and Tandler technic [43, 44]*

Formula:

0.2 M Cacodylate pH 7.4	25 pt
Paraformaldehyde	2 pt
Aqueous glutaraldehyde (25%)	12 pt
Acrolein	1 pt
Dimethylsulfoxide	2.5 pt
Distilled/deionized water	34.5 pt

2.6.5 *Karnovsky's solution [45, 46]*

Formula:

0.2 M Cacodylate pH 7.4	50 pt
Paraformaldehyde	2 pt
Aqueous glutaraldehyde (25%)	10 pt
Distilled/deionized water	38 pt

2.6.6 *McDowell-trump solution [47]*

Formula:

Distilled Water	86 pt
Paraformaldehyde (40%)	10 pt

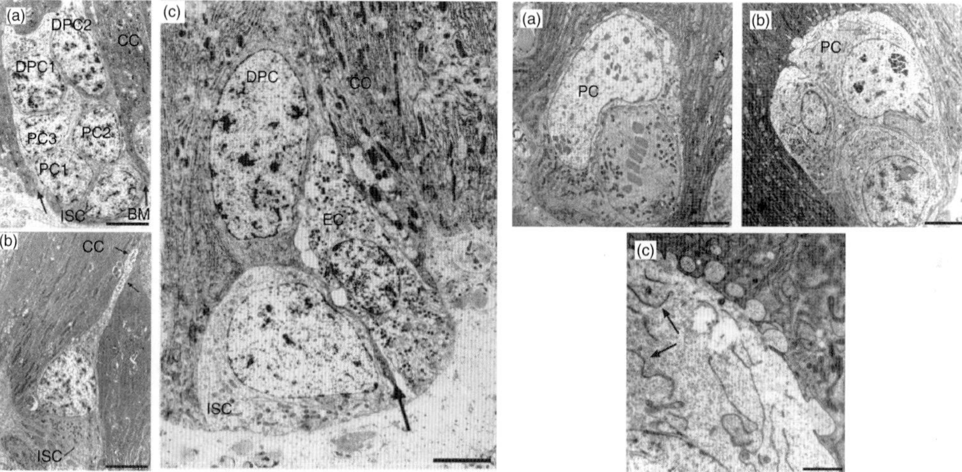

Fig. 2.8 The fixative used for these tissues consisted of glutaraldehyde (3%), formaldehyde (2%), acroleine (1%), dimethylsulfoxide (2.5%), and 0.2 mol l-1 sucrose in 0.05 mol l-1 cacodylate buffer (pH 7.4). Midgut nidi (a) The nidus shows a most basal cell that has a much more extended contact area with the basement membrane than the other nidus cells: it conceivably represents the intestinal stem cell (ISC). The three cells above it may represent a pool of columnar cells (CC) progenitor cells (PC1–PC3). DPC1 and DPC2 represent CC progenitor cells that have commenced differentiation, which is indicated by the patchy distribution of the chromatin in their nuclei. CC, differentiated CC; BM, basement membrane. Scale bar 5 6.2 lm. (b) Nidus with ISC and differentiating CC progenitor cell. Arrows point to its slender growth cone pushing its way to the luminal surface. Scale bar 5 5.7 lm. (c) Nidus that encloses an ISC, an endocrine cell (EC), and a differentiating CC progenitor cell (DPC). Note the patchy chromatin distribution in its nucleus. The arrow points to the process of the DPC that extends to the basement membrane. Scale bar 5 3.4 lm. (Source: Hoffmann *et al.* 2009. Reproduced with permission of John Wiley & Sons.)

Glutaraldehyde (25%)	4 pt
NaH_2PO_4 H_2O	1.16 pt
NaOH	0.27 pt

2.6.7 *Mosca and Schwarz [48]*

Formula:

0.2 M Cacodylate pH 7.4	50 pt
Paraformaldehyde	2.5 pt
Aqueous glutaraldehyde (25%)	20 pt
Picric Acid	0.06 pt
Distilled Water	27.5 pt

Procedure:

1. Immerse in fixative overnight at 4°C.
2. Post-fix in 2% cacodylate buffer for 2 hr on ice.

2.6.8 *Preparation of* Drosophila *eye specimens for scanning electron microscopy – critical point dried [49]*

Application:
Recommended to prevent collapse of *Drosophila* eyes during critical point drying.

Formula:
SEM Fixative:

Glutaraldehyde	1%
Formaldehyde	1%
1 M Sodium Cacodylate (pH 7.2)	To volume

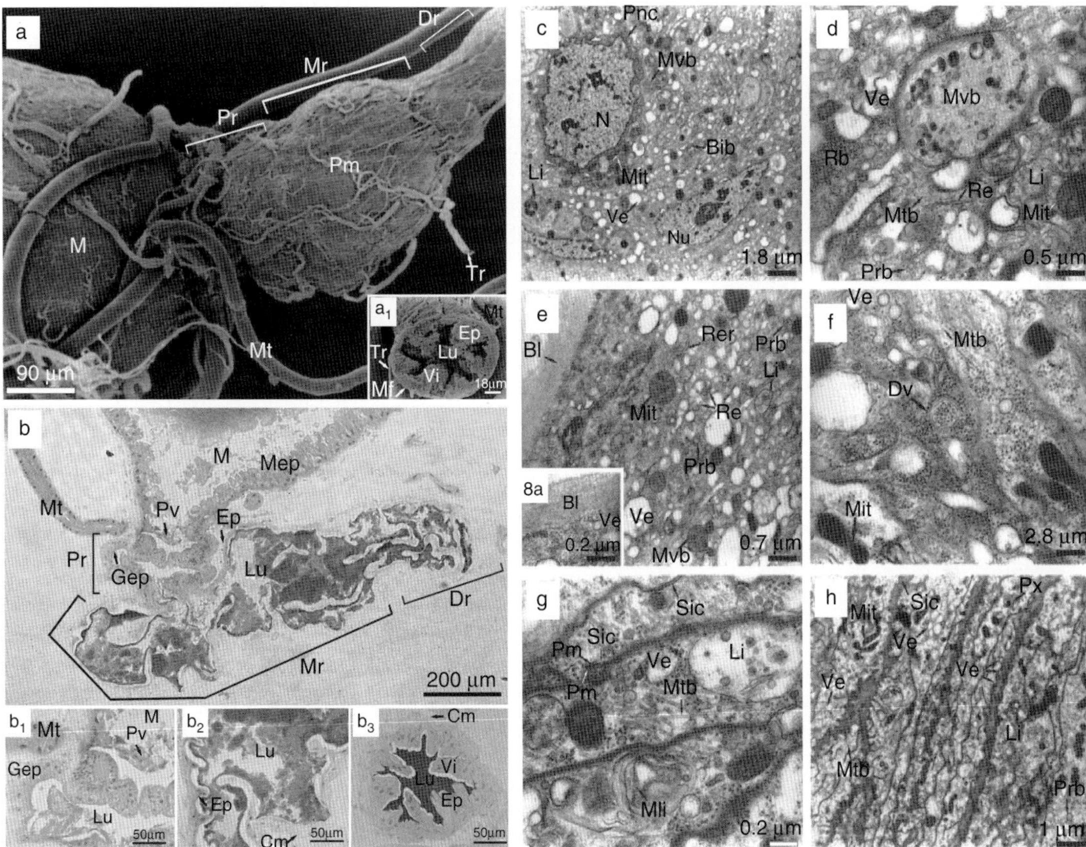

Fig. 2.9 The ileum was fixed in 4% paraformaldehyde for 2 h and then buffered in sodium phosphate pH 7.4 for 24h. Right - Ultrastructure of the basal region of the glandular epithelium (pylorus).The perinuclear cytoplasm (Pnc) is clearly more electron dense than the rest of the cytoplasm. Several vesicles (Ve), large quantities of lysosomes (li), and a complex basal labyrinth (Bib) are typical of the epithelium of this region of the ileum of *C. clypeatus*. (d) (f) Micrographs showing the cytoplasm of epithelial cells of the proximal region of the ileum of *C. clypeatus* with digestive vesicles (Dv), lysosomes (Li), and multivesicular bodies (Mvb). Polyribosomes (Prb), smooth (Ser) and rough endoplasmic reticulum (Rer), microtubules (Mtb), mitochondria (Mit), several vesicles (Ve), and ribosomes (Rb) are observed. (g)-(h) Details of the ultrastructure of epithelial cells of the proximal region of the ileum of *C. clypeatus*. Infoldings in the epithelium and intercellular spaces (lcs) are observed, as well as peroxisomes (Px). On Left: (a) Scanning electron micrograph (SEM) showing the ultramorphology of the ileum of *C. pusillus* queens. SEM figure (Fig. a,a_1) was fixed in Karnovsky fluid for 24 hr. Note the distal region of the midgut (M), the attachment site of the Malphigian tubules (Mt) and the three regions of the ileum: proximal region (Pr) (or pylorus), medial region (Mr), and distal region (Dr). (a_1) shows details of the distal region of the ileum, with infoldings of the epithelium (Ep) or villi (Vi) extending toward the lumen (Lu) as well as tracheoles (Tr) and adjacent muscle fibers (Mf). (b) Longitudinal section of the ileum of *C. clypcatus* (H.E.). General aspect of the ileum with part of the midgut (M) followed by the pyloric valve (Pv) that precedes the proximal region of the ileum (Pr) (or pylorus). The glandular epithelium (Gep) of the proximal region is continuous with the opening of the Malpighian tubules (Mt). (b_1) shows details of this region. The medial region (Mr) includes the pouch of microorganisms. Details of the medial region are shown in (b_2).The posterior region of the ileum (Pr) is illustrated in details in (b_3). (Source: Bution & Caetano 2010. Reproduced with permission of Elsevier.) See plate section for color version of the figure.

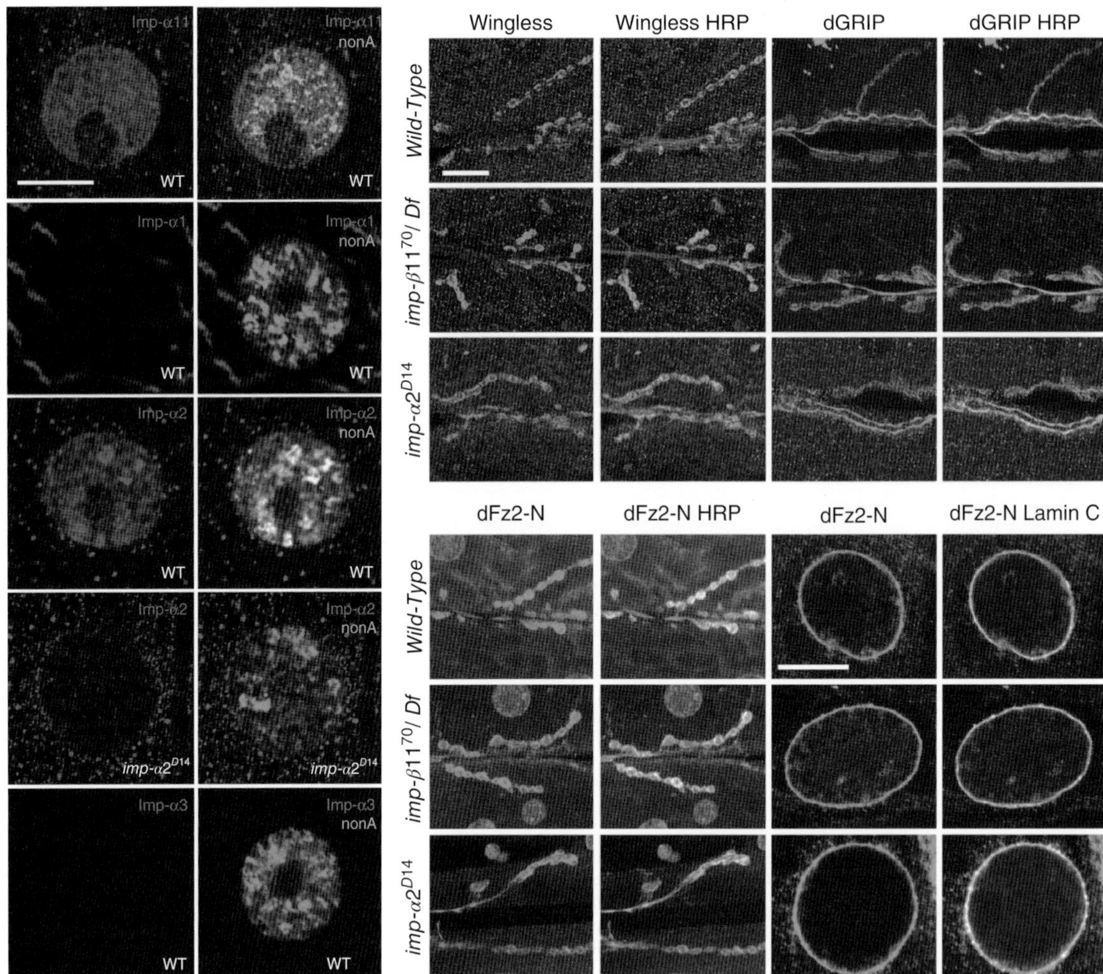

Fig. 2.10 Importin-β11 and Importin-α2 are expressed in *Drosophila* muscle cell nuclei. (**a–j**) Representative confocal images of wild-type (*y,w; FRT42D*; +; +) nuclei (**a–f**, **i**, **j**) or *imp*-α2 null mutant (*y,w; imp*-$\alpha 2^{D14}$; +; +) nuclei (**g**, **h**) stained with antibodies to an importin (magenta) and to the nuclear marker nonA (green). Importin-β11 staining was observed at the nuclear envelope and in the nucleus (**a**, **b**). Antibodies to Importin-α1 nonspecifically labeled Z-bands but staining was undetectable in the nucleus (**c**, **d**). Importin-α2 staining was observed in the nucleus (**e**, **f**). In *imp*-α2 null mutant nuclei, antibodies to Importin-α2 did not label muscle nuclei, indicating that the nuclear staining in **e** and **f** was specific for Importin-α2 (**g**, **h**). Nonspecific puncta in the cytoplasm persisted, however, in the mutant. (**i**, **j**) Antibodies to Importin-α3 did not stain wild-type muscles, although the antibody did recognize Importin-α3 in neuronal tissues that are known to express Importin-α3 (data not shown). Scale bar represents 10 μm. Figure 3 Wnt pathway components are properly localized in importin mutants. (**a–r**) Representative confocal images of the NMJ at muscles 6 and 7 in the indicated genotypes, immunostained as indicated with antibodies to HRP (magenta) to mark the neuronal endings. Neither importin mutant prevented Wingless ligand localization to synaptic boutons, dGRIP localization to synapses or Frizzled2 receptor localization to the synapse. Antibodies to the N terminus (Fz2-N) and C terminus (Fz2-C; data not shown) gave equivalent results. (**s–x**) Single confocal slices through the equator of individual third-instar muscle nuclei in the indicated genotypes and stained with the antibody to Fz2-N (green) and an antibody to Lamin C (magenta) to mark the nuclear envelope. Perinuclear staining in each genotype indicates endocytosis and trafficking of the Fz2 receptor to the nuclear envelope. Scale bar represents 10 μm. Genotypes: wild type (*y,w; FRT42D*; +; +), *imp*-$\beta 11^{70}$ / *Df* (*y,w; imp*-$\beta 11^{70}$ / *Df*; +; +), *imp*-$\alpha 2^{D14}$ (*y,w; imp*-$\alpha 2^{D14}$; +; +). (Source: Mosca & Schwarz 2010. Reproduced with permission of Nature.) See plate section for the color version of the figure.

Procedure:

1. Immerse whole *Drosophila* in SEM fixative for 2 hr.
2. Rinse in distilled water.
3. Dehydrate through a graded ethanol series (25%, 50%, 75%, and 100% (twice)), each step for 12 hr at room temperature.
4. Critical point dry and sputter-coat.
5. Stick carbon conductive tabs onto aluminum specimen mounts.
6. Mount flies onto conductive tab.

2.6.9 *Preparation of* Drosophila *eye specimens for scanning electron microscopy – trichlorofluoroethane [49]*

Application:

Recommended for preparation in the absence of critical point drier.

Procedure:

1. Dehydrate *Drosophila* samples through a graded ethanol series (25%, 50%, 75%, and 100% [twice]), each step for 12 hr at room temperature.
2. Pass samples through a series of trichlorofluoroethane diluted in 100% ethanol (25%, 50%, 75%, and 100% trichlorofluoroethane, [twice]) each step for 12 hr at room temperature.
3. Remove the trichlorofluoroethane and dry the samples under a vacuum.
4. Stick carbon conductive tabs onto aluminum specimen mounts.
5. Mount flies onto conductive tab.

2.6.10 *Preparation of* Drosophila *eye specimens for scanning electron microscopy – HMDS [49]*

Application:

Recommended for preparation in the absence of critical point drier.

Procedure:

1. Place *Drosophila* in 25% ethanol at room temperature overnight.
2. Dehydrate in graded alcohol series (50%, 75%, and 100%, three times) for 2 hr to overnight at room temperature.
3. Place in Hexamethyldisilazane (HMDS) for 3 min at room temperature, twice.
4. Dry flies overnight under vacuum.
5. Stick carbon conductive tabs onto aluminum specimen mounts.
6. Mount flies onto conductive tab.

2.6.11 *Preparation of* Drosophila *eye specimen for transmission electron microscopy [50]*

Formula:

Fixative for TEM

1M Sodium cacodylate (pH 7.4)	1 pt
20% Formaldehyde	2 pt
50% Glutaraldehyde	0.7 pt
dH_2O	6.3 pt

Procedure:

1. Inject TEM fixative into *Drosophila* larval, pupal or adult head, incubate animal for 10 min at room temperature.
   a. For larvae and pupae, remove the retinal tissue from the head and place in fresh TEM fixative for 4 more hr at room temperature.
   b. For adult eyes, add a drop of TEM fixative to the top of the head and carefully cut the head in half. Place each head half into fresh TEM fixative for 4 more hr at room temperature.

2. Fix in 1% tannic acid in TEM fixative overnight.
3. Rinse in 01.M sodium cacodylate (pH 7.4) for 30 min.
4. Transfer to 2% OsO_4/0.1M sodium cacodylate for 2 hr.
5. Rinse in dH_2O.
6. Transfer to 2% uranyl acetate overnight at room temperature.
7. Dehydrate through graded ethanol series for 5 min each (30%, 50%, 70%, 90%, 100%).
8. Transfer to propylene oxide for 5 min.
9. Infiltrate with increasing graded changes of resin in propylene oxide (2 hr each) to 100% resin overnight.
10. Fill molds with resin and push eye to the floor of the mold with the eye facing out toward the cutting surface. Orient the eye as required.
11. Bake at 70°C for 25 min or until resin is tacky. Re-orient eye as necessary.
12. Bake at 70°C for 24 hr or until fully hardened.

2.6.12 *Preparation of* Rhodnius prolixus *for phase contrast X-ray synchrotron imaging [51]*

Formula:

Fixative:

Glutaraldehyde	1%
Sucrose	5%
0.1M cacodylate buffer pH 7.2	To volume

Procedure:

1. Feed fifth instar nymphs on human blood and collect the insects 3 days after the blood meal.
2. Immobilize the insects at 4° C for 10 min and transversally cut at the junction between the prothoracic and mesothoracic segments of the body.
3. Fix the anterior fragments in fixative at room temperature.
4. Samples should be imaged within a month.

2.7 **Other**

2.7.1 *Champy's [15]*

Application:

Recommended for chitin free cytological preparations.

Formula:

Potassium Bichromate (3%)	7 pt
Chromic Acid (1%)	7 pt
Osmic Acid (2%)	4 pt

Procedure:

1. Fix for an average of 12 hr.
2. Wash in water.
3. Transfer to alcohol (50%).

2.7.2 *Chromic acid [2]*

Application:

Recommended as a fixative which causes no shrinkage to certain insect tissues.

Formula:

Chromic Acid	1 pt

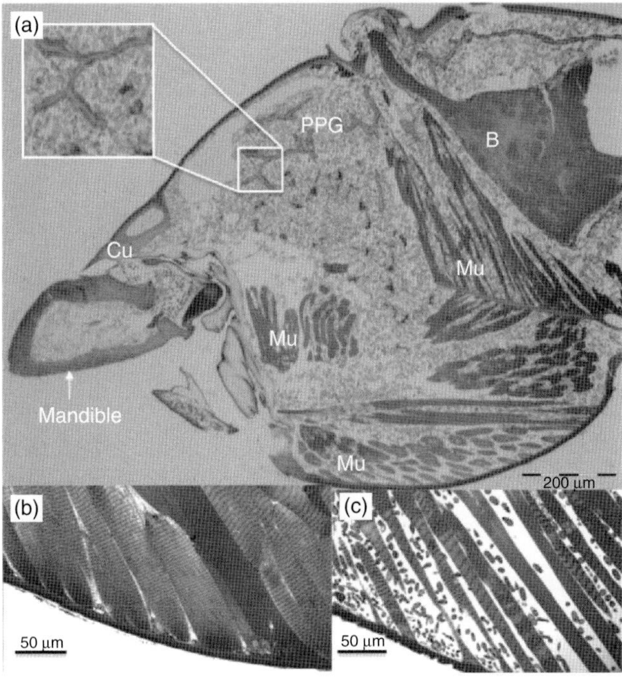

Fig. 2.11 Simultaneous Fixation of Tissue Types - Heads of manipulated ants colonized by the fungus *Ophiocordyceps unilateralis* s.l,. a (top panel) is a light micrograph (LM) saggital section through the head of an ant infected by the fungus, that was biting a leaf at the moment of fixation (i.e., alive). The small grey blobs are fungal hyphal bodies that fill the head and mandibles. Note the spacing between the muscle fibers. The insect shows a close up of hyphal bodies around the post-pharyngeal gland (PPG). b is the brain, Mu, muscles and Cu is cuticle. b) is a LM of healthy muscle and c) is a LM of muscle from a behaviorally manipulated ant that was biting a leaf and alive when removed for fixation. The small blobs between the fibers are fungal cells. Fixed in 2% glutaraldehyde (buffered at pH 7.3 with 50 mM sodium cacodylate and 150 mM saccharose) and post-fixed in 2% osmium tetroxide in the same buffer. (Source: Hughes *et al.* 2011. Reproduced with permission of the authors.) See plate section for the color version of the figure.

2.7.3 *Dioxane [24]*

Application:
Recommended for the fixation of embryonic tissue in the preparation of insect eggs and young larvae.

Formula:
Solution A:

Dioxane	5 pt
Soft Paraffin (46–48 mp)	5 pt
Xylol	1 pt

Procedure:
1. Place specimens into dioxane with a bottom layer of calcium chloride in the vessel.
2. Transfer to solution A, heated to 37°C, for 3 to 4 hr.
3. Embed in fresh, hard paraffin.

Note:
1. It is suggested that the specimen not be left in paraffin over long periods.
2. Care should be exercised when using dioxane because the vapors of dioxane are toxic.

2.7.4 *EDC [52]*

Application:
Recommended as a fixative for antigens commonly rendered undetectable by other fixation methods.

Procedure:
1. Pre-fix tissues in 4% 1-ethyl-3-(3-deimethylaminopropyl) carbodiimide in deionized water for 4 hr at 4°C.
2. Fix tissues in 4% paraformaldehyde overnight at 4°C.
3. Embed tissues in 5% low-melting-point agarose for sectioning on vibratome.

2.7.5 *Flemming's strong solution [2]*

Application:
Recommended as a fixative which causes only moderate shrinkage to histological preparations of insect tissue.

Formula:
Solution A: Stock

Chromic Acid (1% aqueous)	11 pt
Glacial Acetic Acid	1 pt
Distilled Water	4 pt

Solution B:

Osmic Acid (aqueous)	2%
Chromic Acid	1%

Solution C: Working Fluid

Solution A	4 pt
Solution B	1 pt

Note:
1. Solution C is used as a fixative.

2.7.6 *Helly's Fixative (Zenker-formol)*

Formula:

Zinc Chloride	15 pt
Potassium dichromate	7.5 pt
Sodium Sulfate	3 pt
Distilled water	300 pt
Formalin (A)	5 pt

Procedure:
1. Mix first four components together.
2. Add 5 pt formalin to 95 pt mixture.

2.7.7 *Henning's [40]*

Application:
Recommended for fixation as well as the softening of sclerotized chitin.

Formula:

Corrosive Sublimate (saturated in 60% alcohol)	24 pt
Nitric Acid	16 pt
Absolute Alcohol	42 pt
Water (saturated with picric acid)	12 pt
Chromic Acid (0.5%)	16 pt

Procedure:
1. Add sublimate and aqueous picric to chromic acid.
2. Cool.
3. Add nitric acid.
4. Fix specimens for 12 to 24 hr.
5. Wash with iodine alcohol.

Note:
1. This fluid may fix in shorter period if used hot.

2.7.8 *Modified Zenker's fixative*

Formula:

Zinc Chloride	15 pt
Potassium Dichromate	7.5 pt
Sodium Sulfate	3 pt
Distilled water	300 pt
Glacial Acetic Acid (A)	5 pt

Procedure:
1. Mix first four components together.
2. Add 5 pt glacial acetic acid to 95 pt mixture.

2.7.9 *Perenyi's chromo-nitric [8]*

Formula:

Nitric Acid (10%)	4 pt
Alcohol (95%)	3 pt
Chromic Acid (0.5%)	3 pt

Procedure:
1. Fix for 1 to 5 hr.
2. Wash in alcohol (70%) for 24 hr.
3. Embed soon after washing.

2.7.10 *Petrunkevitch's cupric-paranitrophenol [53]*

Formula:

Alcohol	100 pt
Nitric Acid	3 pt
Ether	5 pt
Cupric Nitrate Crystals	2 pt
Paranitrophenol Crystals	5 pt

Procedure:
1. Wash several times in alcohol (70%).

Note:
1. There is no time limitation on fixation or extensive storage (e.g. several months).
2. The rate of penetration is approximately one half millimeter per hr.

2.7.11 *Petrunkevitch's cupric-phenol [53]*

Formula:

Solution A: Stock

Distilled Water	100 pt
Nitric Acid	12 pt
Cupric Nitrate Crystals	8 pt

Solution B: Stock

Alcohol (80%)	100 pt
Phenol Crystals	4 pt
Ether	6 pt

Solution C: Fixative

Solution A	1 pt
Solution B	3 pt

Procedure:
1. Fix for 12 to 24 hr in solution C.
2. Wash in several changes of alcohol (70%).

Note:
1. Mixture must be used within a few hours and fixation time must not exceed 48 hr.
2. Specimens must remain soft, don't harden in alcohol.

2.7.12 *Steck and Wharton's technic [54]*

Procedure:
1. Boil insects in water.
2. Fix in 70% ethanol.
3. Treat overnight in 10% postassium hydroxide.

2.7.13 *Nässel/Elekes fixative [55]*

Procedure (A):

1. Fix either in 6% or 2.5% glutaraldehyde and 1% sodium metabisulfite in 0.1 M sodium cacodylate buffer (pH 7.2) for 4 hr at 4°C.

2. Thereafter, thoroughly rinse in cacodylate buffer containing 1% sodium metabisulfite.

Procedure (B):
1. Fix in 4% paraformaldehyde in 0.1 M phosphate buffer for 4 hr at 4°C.
2. Or, fix in 4% w/v EDCDI (Sigma®) for 12 hr at 4°C.

Notes:

1. (A) is for dopamine (DA) fixation of immunoreactive neurons in insects.

2. (B) is for tyrosine hydroxylase (TH) fixation of immunoreactive neurons in insects.

3. Originally described for the fixation in preparation of immunohistochemical preparation of the brains of blowflies and *Drosophila*.

References

1. Quicke D.L.J., Lopez-Vaamonde C. & Belshaw R. Preservation of hymenopteran specimens for subsequent molecular and morphological study. *Zoologica Scripta* **28**, 261–267 (1999).
2. Scott A.C. Variables of tissue fixation in the grasshopper *Melanoplus femur rubrum*. Trans. *Amer. Microsc. Society* (1931).
3. Gray P. *The Microtomist's Formulary and Guide*. The Blakiston Company, Inc., New York (1954).
4. Wilson S.D. *Applied and Experimental Microscopy*. Burgess Publish Company, Minneapolis, MN (1967).
5. Dodd F.O. *Histological and Illustrative Methods for Entomologists*. Oxford University Press, London (1930).
6. Lightner, D.V. *A Handbook of Pathology and Diagnostic Procedures for Diseases of Penaid Shrimp*. World Aquaculture Society (1996).
7. Berry R.P., Warrant E.J. & Stange G. Form vision in the insect dorsal ocelli: An anatomical and optical analysis of the locust ocelli. *Vision Res.* **47**, 1382–1393 (2007).
8. Kennedy C.H. *Methods for the Study of the Internal Anatomy of Insects*. Department of Entomology, Ohio State University (Mineographed) (1932).
9. Peacock H.A. *Elementary Technique*. Edward Arnold Ltd., London (1966).
10. Frings H. & Frings N. A fixing agent for insect tissues. *Entomol. News* **82**, 157–159 (1971).
11. Hall D.G., Hull J.B. & Dive W.E. New method in sand fly studies (Diptera:Chironmidae). *Entomol. News.* **44**, 29–32 (1933).
12. Mito T. *et al.* Involvement of hedgehog, wingless, and dpp in the initiation of proximodistal axis formation during the regeneration of insect legs, a verification of the modified boundary model. *Mech. Dev.* **114**, 27–35 (2002).
13. Homberg U., Kingan T.G. & Hildebrand J.G. Immunocytochemistry of Gaba in the Brain and Subesophageal Ganglion of *Manduca sexta*. *Cell Tissue Res.* **248**, 1–24 (1987).
14. Karr T.L. & Alberts B.M. Organization of the cytoskeleton in early *Drosophila* embryos. *J. Cell Biol.* **102**, 1494–1509 (1986).
15. Eltringham H. *Histological and Illustrative Methods for Entomologists*. Oxford University Press, London (1930).
16. Lane J. *The Preservation and Mounting of Insects of Medical Importance*. World Health Organization Vector Control (1965).
17. Ott S.R. & Elphick M.R. Nitric oxide synthase histochemistry in insect nervous systems: Methanol/formalin fixation reveals the neuroarchitecture of formaldehyde-sensitive NADPH diaphorase in the cockroach *Periplaneta americana*. *J. Comp. Neurol.* **448**, 165–185 (2002).
18. Monesi N. *et al.* Immunocharacterization of the DNA puff BhC4-1 protein of *Bradysia hygida* (Diptera: Sciaridae). *Ins. Biochem. and Mol. Biol.* **34**, 531–542 (2004).
19. Silva-Zacarin E.C.M. *et al.* Protocol for optimization of histological, histochemical and immunohistochemical analyses of larval tissues: application to histopathology of honey bee. *Curr. Micro. Contri. to Adv. in Sci and Tech.* 696–703 (2012).
20. Ott S.R. Confocal microscopy in large insect brains: Zinc-formaldehyde fixation improves synapsin immunostaining and preservation of morphology in whole-mounts. *J. Neurosci. Meth.* **172**, 220–230 (2008).
21. Clark J., Meisner S. & Torkkeli P.H. Immunocytochemical localization of choline acetyltransferase and muscarinic ACh receptors in the antenna during development of the sphinx moth *Manduca sexta*. *Cell Tissue Res* **320**, 163–173 (2005).
22. Tolbert L.P., Matsumoto S.G. & Hildebrand J.G. Development of synapses in the antennal lobes of the moth *Manduca sexta* during metamorphosis. *J. Neurosci.* **3**, 1158–1175 (1983).
23. Weaver N. & Thomas R.C. Jr. A fixative for use in dissecting insects. *Stain Technol.* **31**, 25–26 (1956).

24. Van Heerden H.P. Some histological methods of interest to entomologists. *J. Entomol. Soc. S. Africa* **8**, 157–161 (1945).
25. Davidson J. Biological studies of *Aphis rumicis L.* The penetration of plant tissue and the course of the food supply of aphids. *Ann. Appl. Biol.* **10**, 34–54 (1923).
26. Peterson A. *Entomological Techniques. How to work with Insects*. Edwards Brothers Inc., Ann Arbor, Michigan (1964).
27. Sutrisno H. & Suwito A. *Collecting and Preserving of Insects (Lepidoptera, Coleoptera, and Hymenoptera)*. Research Center for Biology – LIPI, Cibinong, Indonesia (2005).
28. McFarland N. Additional notes of rearing and preserving larvae of macrolepidoptera. *Journal of the Lepidoterists's Society* **19**, 233 (1965).
29. Salthouse T.N. Tetrahydrofuran and its use in insect histology. *Canad. Entomol.* **90**, 555–557 (1958).
30. Galigher A.E. & Kozloff E.N. *Essentials of Practical Microtechnique*. Lea & Febiger, Philadelphia (1964).
31. Gregory G.E., Greenway A.R. & Lord K.A. Alcoholic bouin fixation of insect nervous systems for Bodian silver staining .1. Composition of aged fixative. *Stain Technol.* **55**, 143–149 (1980).
32. Zavodska R., Sehadova H., Sauman I. & Sehnal F. Light-dependent PER-like proteins in the cephalic ganglia of an apterygote and a pterygote insect species. *Histochem. Cell Biol.* **123**, 407–418 (2005).
33. Watson C.A., Sauman I. & Berry S.J. Actin is a major structural and functional element of the egg cortex of giant silkmoths during oogenesis. *Dev. Biol.* **155**, 315–323 (1993).
34. Zavodska R., Sauman I. & Sehnal F. The cycling and distribution of PER-like antigen in relation to neurons recognized by the antisera to PTTH and EH in *Thermobia domestica. Ins. Biochem. and Mol. Biol.* **33**, 1227–1238 (2003).
35. Lee A.B. *Microtomist's Vade-Mecum*. The Blakiston Company, Philadelphia (1950).
36. Comstock J.H. & Kellogg V.L. *The Elements of Insect Anatomy*. Comstock Pub. Co., Ithaca, NY (1899).
37. Stefanini M., DeMartino C. & Zamboni L. Fixation of ejaculated spermatozoa for electron microscopy. *Nature* **216**, 173–174 (1967).
38. Becker E.R. & Roudabush R.L. *Brief Directions in Histological Technique*. Collegiate Press, Inc., Ames, Iowa (1935).
39. Wismar B.L. Quad-type stain for the simultaneous demonstration of intracellular and extracellular and tissue components. *Stain Technol.* **41**, 309–313 (1966).
40. Levenbook L. Outline for the study of insect histology. *Microsc. & Entomol. Mon.* **4**, 212–220 (1940).
41. Bleeker M.A.K., Smid H.M., Van Aelst A.C., Van Loon J.J.A. & Vet L.E.M. Antennal sensilla of two parasitoid wasps: A comparative scanning electron microscopy study. *Micros. Res. and Tech.* **63**, 266–273 (2004).
42. Cuperus P.L. Inventory of pores in antennal sensilla of *Yponomeuta spp.* (Lepidoptera: Yponomeutidae) and *Adoxophyes orana F.v.R.* (Lepidoptera: Tortricidae). *Intl. J. Ins. Morph. and Embry.* **14**, 347–359 (1985).
43. Meola S.M. *et al.* Ultrastructural analysis of neurosecretory cells in the antennae of the mosquito, *Culex salinarius* (Diptera: Culicidae). *J. Mol. Neurosci.* **14**, 17–25 (2000).
44. Kalt M.R. & Tandler B. A study of fixation of early amphibian embryos for electron microscopy. *J. Ultrastruct. Res.* **36**, 633–645 (1971).
45. Oliveira D.C., Magalhaes T.A., Carneiro R.G.S., Alvim M.N. & Isaias R.M.S. Do *Cecidomyiidae* galls of *Aspidosperma spruceanum* (Apocynaceae) fit the pre-established cytological and histochemical patterns? *Protoplasma* **242**, 81–93 (2010).
46. Bution M.L. & Caetano F.H. Symbiotic bacteria and the structural specializations in the ileum of *Cephalotes* ants. *Micron* **41**, 373–381 (2010).

47. Khoo C.C.H. & Tan K.H. Rectal gland of *Bactrocera papayae*: Ultrastructure, anatomy, and sequestration of autofluorescent compounds upon methyl eugenol consumption by the male fruit fly. *Microsc. Res Tech.* 67, 219–226 (2005).
48. Mosca T.J. & Schwarz T.L. The nuclear import of Frizzled2-C by Importins-beta 11 and alpha 2 promotes postsynaptic development. *Nature Neurosci.* **13**, 935–950 (2010).
49. Wolff T. Preparation of *Drosophila* eye specimens for scanning electron microscopy. *Cold Spring Harbor Protoc.* **11**, 1383–1385 (2011).
50. Wolff T. Preparation of *Drosophila* eye specimens for transmission electron microscopy. *Cold Spring Harbor Protoc.* **11**, 1386–1388 (2011).
51. de Almeida A.P. *et al.* Phase contrast X-ray synchrotron imaging for assessing external and internal morphology of *Rhodnius prolixus*. *Appl. Radiat. Isot.* 70, 1340–1343 (2011).
52. Dacks A.M., Reisenman C.E., Paulk A.C. & Nighorn A.J. Histamine-immunoreactive local neurons in the antennal lobes of the hymenoptera. *J. Comp. Neurol.* **518**, 2917–2933 (2010).
53. Petrunkevitch A. New fixing fluids for general purposes. *Science* 77, 117–118 (1933).
54. Steck G.J. & Wharton R.A. Descriptions of immature stages of eutreta (Diptera, Tephritidae). *J. Kansas Entomol. Soc.* 59, 296–302 (1986).
55. Nassel D.R. & Elekes K. Aminergic neurons in the brain of blowflies and *Drosophila*: dopamine- and tyrosine hydroxylase-immunoreactive neurons and their relationship with putative histaminergic neurons. *Cell and Tissue Res.* 267, 147–167 (1992).

3 Dehydrating, clearing, and embedding

3.1 Dehydration

N-butyl alcohol has been commonly used by plant histologists. It apparently has also been somewhat successfully used with animal tissues. This alcohol is reported to prevent hardening of tissues usually caused by other substances such as high percentage ethyl alcohol, xylol, and benzol. Thus, it has been particularly useful in histological preparations of insects and ticks.

3.1.1 *Acetone [1, 2]*

Application:
Particularly effective dehydration reagent for SEM and TEM.

Procedure:
1. Place fixed tissue in 50% acetone for 15 min.
2. Transfer to 70% acetone for 15 min.
3. Transfer to 90% acetone for 15 min.
4. Immerse in 100% acetone three times for 10 min.

3.1.2 *Oil-methyl salicylate technic*

Application:
Recommended as a procedure which avoids the hardening effects of xylol by dehydrating in anilin oil and clearing in oil of wintergreen.

Formula:

Solution A:

Alcohol (70%)	2 pt
Anilin Oil	1 pt

Solution B:

Alcohol (95%)	1 pt
Anilin Oil	2 pt

Solution C:

Anilin Oil	1 pt
Oil of Wintergreen	1 pt

Insect Histology: Practical Laboratory Techniques, First Edition.
Pedro Barbosa, Deborah L. Berry and Christina S. Kary.

Companion Website: www.wiley.com/go/barbosa/insecthistology

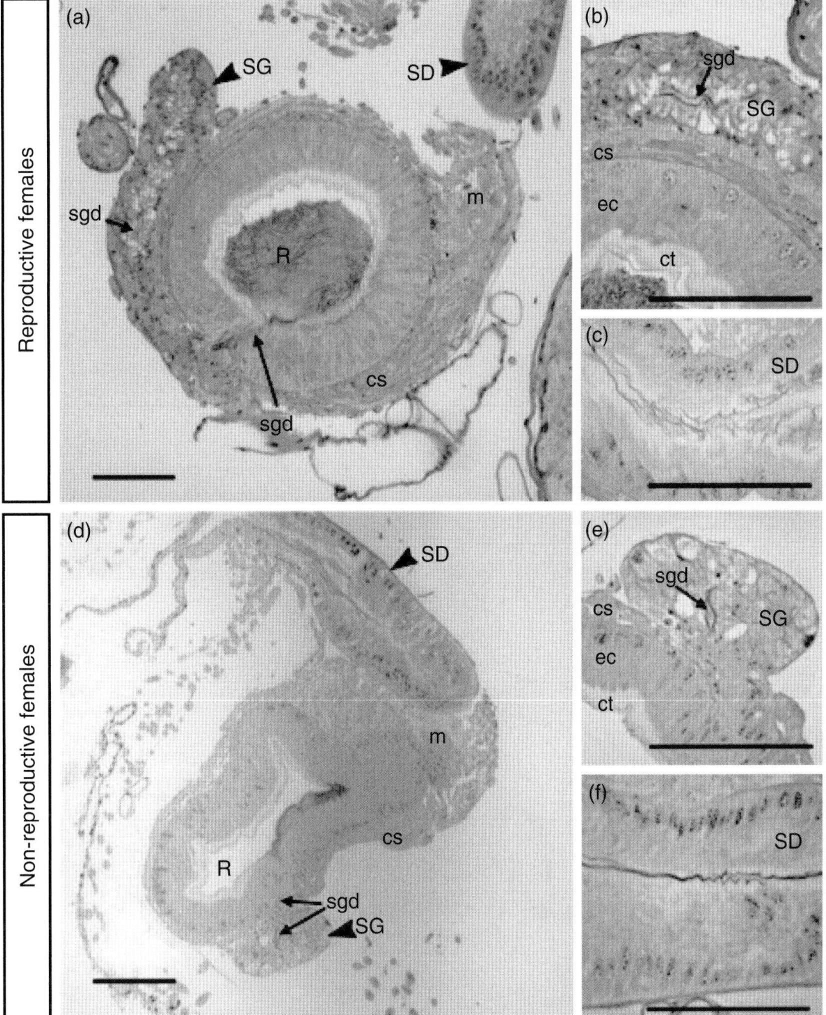

Fig. 3.1 Tissues dehydrated in a graded acetone series and embedded in araldite. Cross-section through the spermatheca of reproductive (a, c) and non-reproductive females (d, f) of *Vespa mandarinia*. Survey of spermatheca. (b, e) shows details of reservoir wall and spermathecal gland. (c, f) Spermathecal duct. Scale bar=100 µm, ct=cuticle, ec=epithelial cells, m=muscle, R=reservoir, SD=spermathecal duct, SG=spermathecal gland, sgd=spermathecal gland duct, cs=surrounding connective tissue sheath. (Source: Gotoh *et al.* 2008. Reproduced with permission of Elsevier.)

Procedure:

1. Fix material.
2. Transfer to ethyl alcohol (70%) for 30 min.
3. Transfer to solution A for 2 hr.
4. Transfer to solution B for 2 hr.
5. Transfer to anilin for 2 hr.
6. Transfer to solution C for 2 hr.
7. Transfer through two changes of oil of wintergreen for 2 hr each.
8. Transfer through three changes of paraffin for 1 hr each.
9. Embed and section.

3.1.3 *Dimethyl formamide [3]*

Procedure:

1. Dehydrated through graded series of dimethyl formamide.
2. Miscible with resin.

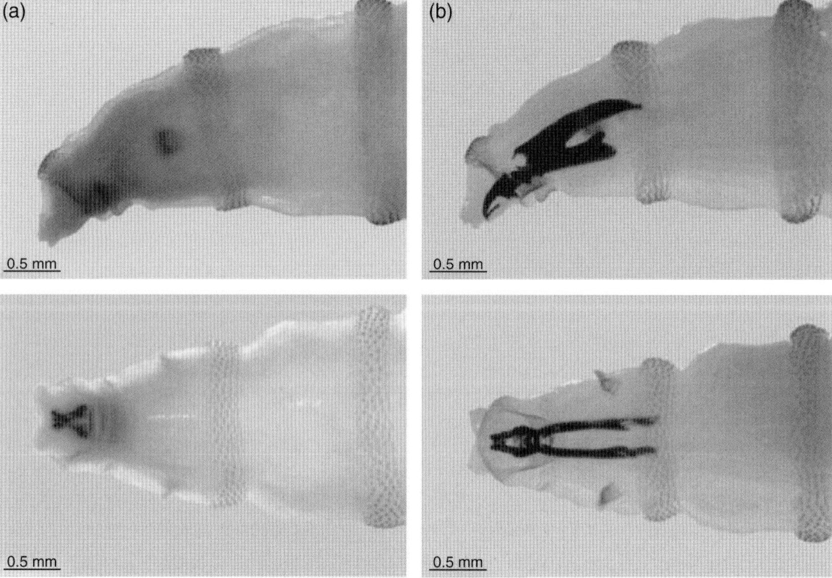

Fig. 3.2 Before (a) and after (b) clearing in methyl salicylate. Lateral (top) and ventral (bottom) views of a third instar *Calliphora vomitoria* maggot. (Source: Niederegger *et al.* 2011. Reproduced with permission of Elsevier.) See plate section for the color version of the figure.

3.1.4 *Fassig's iso-amyl alcohol technic [4]*

Application:
Recommended for fourth instar mosquito larvae.

Formula:

Solution A:

Ethyl Alcohol (95%)	2 pt
Iso-amyl Alcohol	1 pt

Solution B:

Ethyl Alcohol (95%)	1 pt
Iso-amyl Alcohol	1 pt

Solution C:

Ethyl Alcohol (95%)	1 pt
Iso-amyl Alcohol	2 pt

Solution D:

Iso-amyl Alcohol	1 pt
Xylol	1 pt

Procedure:
1. Kill and fix in formalin.
2. Transfer to ethyl alcohol (50%) for 20 min.
3. Transfer to ethyl alcohol (70%) for 10 min.
4. Transfer to ethyl alcohol (95%) for 10 min.
5. Transfer to solution A for 7 min.
6. Transfer to solution B for 7 min.
7. Transfer to solution C for 7 min.

8. Transfer through two changes of iso-amyl alcohol for 10 min each.
9. Transfer to solution D for 7 min.
10. Transfer through two changes of xylol for 15 min each.
11. Mount.

3.1.5 *Romeis' dioxane technic [5]*

Application:
Recommended as a dehydration procedure which also clears specimens.

Formula:
Solution A:

Paraffin (56–58°C m.p.)	2 pt
Dioxane	1 pt

Procedure:
1. Proceed with usual post-fixing treatment.
2. Transfer through three to four changes of dioxane for a total period of 12 to 24 hr.
3. Transfer to solution A in an oven for 1 to 2 hr.
4. Embed in final bath of paraffin.
5. Complete as usual.

Note:
1. Specimens fixed in Carnoy or other alcoholic fixatives may be transferred to dioxane.
2. The dioxane method is less successful than the use of N-butyl alcohol.

Tetrahydrofuran (tetramethylene oxide) has been very strongly recommended as an all purpose clearing and dehydration agent. It is reported to be both a great deal less toxic than dioxane and also miscible with a great many chemical substances used in histology techniques. It is, for example, miscible with water, molten paraffin, ethyl alcohol, chloroform, xylene, toluene, clove oil, and cedar oil. It is, in addition, soluble in mercuric chloride, picric acid, and iodine. It has been incorporated in the following procedure and suggested for dehydration and embedding.

3.1.6 *Tetrahydrofuran [6]*

Application:
Miscible with wax, water, and resinous mounting materials.

Procedure:
1. Dehydrate through increasing graded concentrations of tetrahydrofuran until in 100% tetrahydrofuran.

3.1.7 *Salthouse's dehydration and embedding technic [7]*

Procedure:
1. Place material in tetrahydrofuran for 2 hr.
2. Transfer to new tetrahydrofuran for 1 hr.
3. Transfer to a mixture of equal parts of paraffin and tetrahydrofuran (b.p. = 65°C) in a closed jar (in oven) for ½ hr.
4. Transfer to pure molten paraffin (in open jar) for 1 hr.
5. Transfer to new paraffin for ½ hr.
6. Place in a block and section.

3.1.8 *Stiles' N-butyl alcohol technic [8]*

Application:
Recommended as particularly useful for small insects.

Formula:

Solution A:

Ethyl Alcohol (45%)	90 pt
N-butyl Alcohol	10 pt

Solution B:

Ethyl Alcohol (62%)	80 pt
N-butyl Alcohol	20 pt

Solution C:

Ethyl Alcohol (77%)	65 pt
N-butyl Alcohol	35 pt

Solution D:

Ethyl Alcohol (90%)	45 pt
N-butyl Alcohol	55 pt

Solution E:

Ethyl Alcohol (absolute)	25 pt
N-butyl Alcohol	75 pt

Solution F:

Paraffin (m.p. 56–58°C)	2 pt
N-butyl Alcohol	1 pt

Procedure:
1. After fixation (e.g., in Gilson's) place in ethyl alcohol (35%) for 30 min to 1 hr.
2. Transfer to solution A for 2 hr.
3. Transfer to solution B for 2 hr.
4. Transfer to solution C for 4 hr.
5. Transfer to solution D for 6 hr to several days.
6. Transfer to solution E for 6 hr to overnight (with one change).
7. Transfer through two changes of n-butyl alcohol (at several hours interval).
8. Transfer to solution F in covered dish (in oven) for 12 to 24 hr.
9. Uncover dish until n-butyl odor dissipates.
10. Transfer to paraffin (optional).
11. A long infiltration period (4–5 days) is recommended for best results (it causes less hardening).
12. Embed in paraffin wax (56–58°C m.p.) with slight addition of bayberry wax (< 3%).

Note:
1. The choice of fixative is important so that no hardening factors are added to the procedure.
2. Specimens may be left in n-butyl (step 7) for a very long period of time.

3.1.9 *Becker's modified stiles technic*

Procedure:

1. Slit insect, fix (Gilson's) and remove excess sublimate in 70% iodine alcohol.
2. Transfer to ethyl alcohol (70%) for ½ hr.
3. Transfer to ethyl alcohol (50%) for ½ hr.
4. Transfer to ethyl alcohol (35%) for ½ hr.
5. Follow steps 2–12 (Stile's Technic).

Note:

1. This technic utilizes the same solutions indicated in the previous method (Stiles).
2. If material is highly sclerotized it is recommended that paraffin block be soaked in water for 2 to 3 weeks.

3.1.10 *Bartlett's N-Butyl technic [9]*

Application:

Recommended for yolk-laden insect eggs or brittle insect tissues.

Formula:

Solution A:

N-butyl Alcohol	1 pt
Ethyl Alcohol (50%)	2 pt

Solution B:

N-butyl Alcohol	2 pt
Ethyl Alcohol (50%)	1 pt
Molten Phenol	4.5 pipettes full/150 pt

Solution C:

N-butyl Alcohol	As needed
Molten Phenol	4.5 pipettes full/150 pt

Solution D:

N-butyl Alcohol	1 pt
Melted paraffin (m.p. 50–52°C)	1 pt

Procedure:

1. Drop insect in boiling water for 15 to 30 sec.
2. Cut slits in thorax and abdomen and inject a small amount of fixative with a capillary pipette.
3. Place insect in fixative.
4. Transfer to ethyl alcohol (35%) after post-fixative treatment, for 2 hrs to overnight (or if post-fixative treatment involves 70% or higher percent alcohols go directly to step 5).
5. Transfer to solution A for 2 to 4 hr.
6. Transfer to solution B for 24 to 48 hr.
7. Transfer to n-butyl alcohol for 4 to 8 hr.
8. Transfer to solution C for 12 to 24 hr.
9. Transfer to solution D for 1 to 2 hr.
10. Transfer to paraffin wax (50–52°C) for 1 ½ to 3 hr.
11. Embed in paraffin wax (58–60°C).

Note:
1. It is suggested that for best results fixatives containing nitric acid be used.

3.2 Clearing

Clearing consists of removal of the dehydrant (e.g., alcohol) with a substance that will be miscible with the embedding medium (such as paraffin or paraplast). There are, of course, many clearing agents available for use in histological preparations. The type of clearing agent selected is primarily dependent upon the choice of dehydrating agent and mountant. Thus, the most preferred clearing agent would be one that is miscible in both the dehydrating agent and the mounting medium. Many of these agents can be found listed in any basic text on histological techniques. The third consideration in choosing a clearing agent is the type of tissue to be prepared. Due to the many special problems caused by certain insect tissues, various clearing agents have been recommended as useful for specific insect mounts.

3.2.1 *Berlese's fluid [10]*

Application:
Recommended for both clearing and mounting small insects.

Formula:
Solution A:

Glucose Syrup	
Glucose	1 pt
Water	1 pt

Solution B:

Glacial Acetic Acid	5 pt
Chloral Hydrate	40–160 pt
Solution A	10 pt
Gum Arabic	5 pt
Distilled Water	20 pt

Preparation:
1. Dissolve gum Arabic in distilled water.
2. Add solution A.
3. Add chloral hydrate (to saturation).
4. Heat slowly, agitate, and filter.

Procedure:
Incubate specimen in prepared solution.

3.2.2 *Carbol-xylol [11, 12]*

Application:
Recommended for whole mounts.

Formula:

Carbolic Crystals	1 pt
Xylene	1 pt

Note:
1. Hardens tissues quickly.
2. Rapid clearing from 95% alcohol or lower percent alcohol.
3. All the carbolic acid must be removed before mounting in balsam.

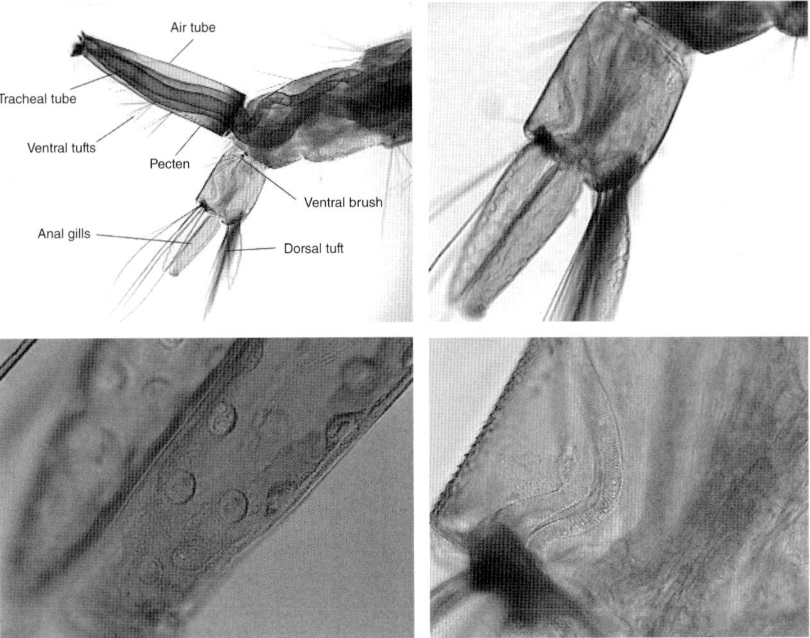

Fig. 3.3 The clearing power of Lactoglycerol. The mosquito larva was fixed in 70% alcohol, and cleared using the solvent density gradient of Pantin. The solution layers used were alcohol 70% – glycerin – lactoglycerol. The sunken larva was mounted in pure lactoglycerol. The third picture shows the nucleus of the cells inside the caudal paddles. The fourth one shows two muscular cells at the end of the body. The left one shows the nucleus. The right one shows that these are striated muscle cells. (Source: Walter Dioni.)

3.2.3 *Cedar oil [11–13]*

Notes:
1. Generally works slowly.
2. Clears from 95% alcohol but is best when transferred from 100% alcohol.
3. Prolong exposure of tissue to oil produces hardening, but less than xylol.

3.2.4 *Methanol/cedar oil clearing [14]*

Formula:
Solution A:

Absolute Methanol	3 pt
Glacial Acetic Acid	1 pt

Procedure:
1. Transfer fixed tissues into solution A for 15–30 min.
2. Immerse in absolute methanol three times for 15 to 30 min each.
3. Immerse in cedar oil for 30–60 min.

3.2.5 *Clove oil [11–13]*

Notes:
1. Clears from 95% alcohol.
2. High index of refraction.
3. Recommended for genitalia and other chitinous preparations.

3.2.6 *Creosote [15, 16]*

Application:
Does not harden tissues as much as ethanols can.

Procedure:
1. Immerse tissues in creosote overnight.

3.2.7 *Dioxane [17]*

Application:
Recommended for aphids.

Procedure:

1. Fix in Bouin's, 1 hr to days, as needed.
2. Immerse in dioxane for 1 hr.
3. Transfer to fresh dioxane and incubate overnight.
4. Transfer to Tissue Mat for 2 hr to overnight.
5. Embed in 56–58°C Tissue Mat.

3.2.8 *Glycerol [14]*

Procedure:

1. Clear tissues in 9 pt glycerol and 1 pt 0.1M phosphate buffer.

3.2.9 *Hetherington's [13]*

Application:

Recommended for clearing insect larvae. Recommended for larval muscoid Diptera and other soft bodied forms.

Formula:

Absolute Alcohol	20 pt
Chloroform	15 pt
Glacial Acetic Acid	5 pt
Phenol Crystals	10 pt

Procedure:

1. Kill specimen in hot water or alcohol.
2. Transfer to fresh Hetherington's solution for 2 days.
3. Clear further in oil of wintergreen for a few hours.
4. Mount in Canada balsam (with a few drops of oil of wintergreen added).

Note:

1. This fluid is not good for external features of soft larvae.

3.2.10 *Methylbenzoate [18]*

Procedure:

1. Immerse dehydrated specimens in methylbenzoate until cleared.
2. Transfer to a mixture of 1 pt methylbenzoate and 1 pt paraffin.
3. Infiltrate fully and embed in paraffin.

3.2.11 *Methylsalicylate [19, 20]*

Procedure:

1. Fixed tissues are dehydrated to 100% ethanol.
2. Transfer tissues to a mixture of 1 pt 100% ethanol and 1 pt methylsalicylate for 10 min.
3. Transfer to pure methylsalicylate for 10 min.

3.2.12 *Propylene oxide [21]*

Procedure:

1. Dehydrate fixed tissues through graded ethanol series to 100% ethanol.
2. Immerse in propylene oxide.
3. Infiltrate with graded series of propylene oxide and resin.
4. Mount in pure resin.

3.2.13 *Stapp and Cumley's technic [22]*

Application:

Recommended for highly pigmented tissues.

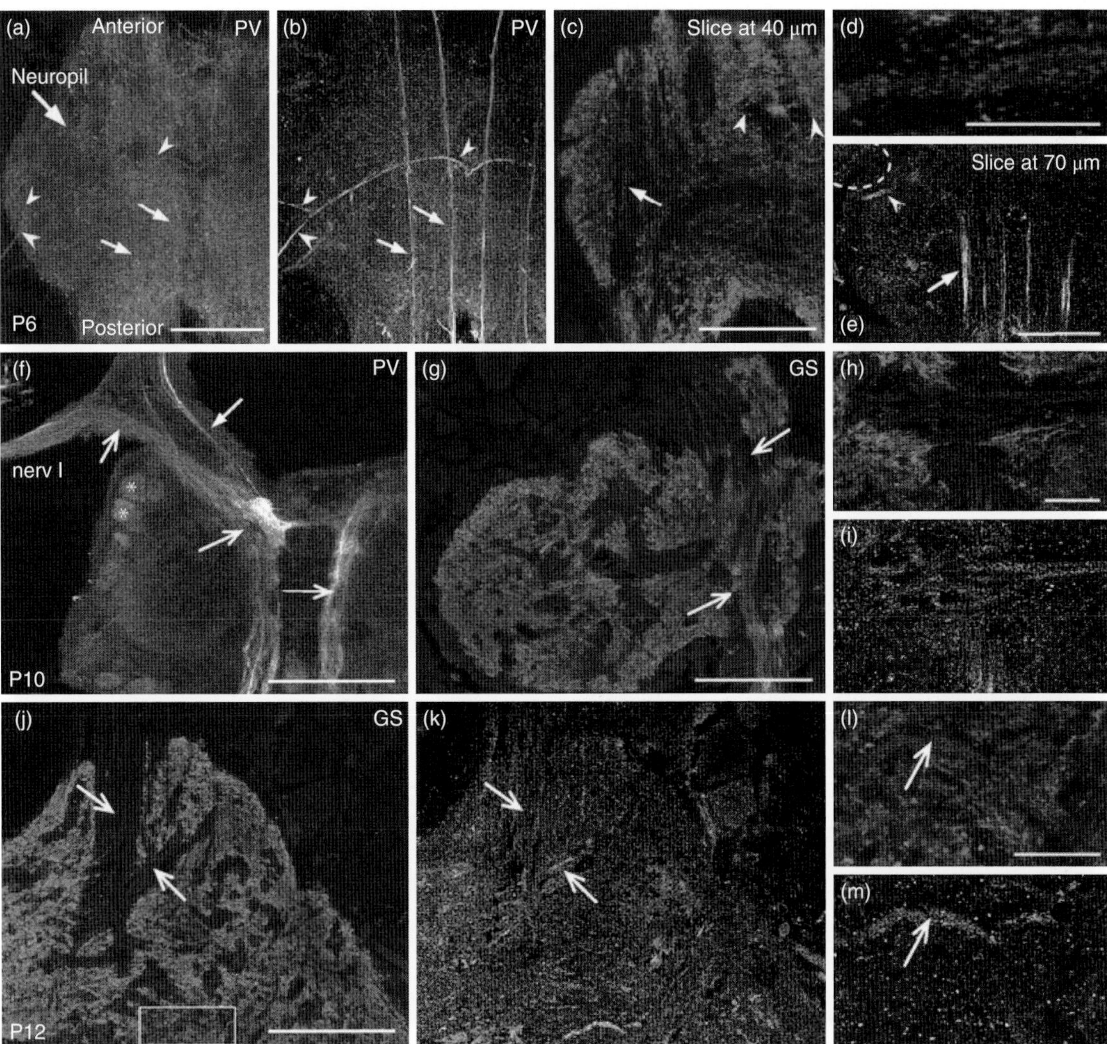

Fig. 3.4 TM-MFasII expression in the mesothoracic ganglion in pupal stages P6, P10, and P12. At all stages older than P10, antibody penetration through whole-mounts was incomplete, and therefore, at these stages additional gelatine sections (50 μm thick) were evaluated. Confocal images shown are either projection views (PV) from a whole-mount preparation or single optical sections from a gelatine section (GS). TM-MFasII immunochemistry (red) was carried out with anti-MSYT staining (green) to label all synaptic terminals. (a–e) In stage P6, TM-MFasII expression is restricted to a few intersegmental axons (filled arrows) and the efferent processes of an UM neuron (arrowheads; a, c, d double-label merge; b, e TM-MFasII). Few axons projecting through the neuropil are immunopositive, and the TM-MFasII signal appears granular (c merge, d selective enlargement). (f) At stage P10, sensory axons projecting through nerve 1 (open arrows), a few intersegmental axons (filled arrow), and some somata (asterisks) are TM-MFasII-positive. (g) Single optical section depicting axons labeled for TM-MFasII (arrows). (h) Selective enlargement of a neuropil region from (g). Commissures and tracts contain some TM-MFasII immunoreactivity, but areas with dense MSYT staining are devoid of TM-MFasII. No overlap is detectable between MSYT and TM-FasII immunostaining (g, h double-label merge, f, i TM-MFasII). (j–m) At stage P12, the intensity of the TM-MFasII immunoreactivity is comparable to stage P10 (see f–i), but sensory axons are out of focus. (j) Axons in the commissures and tracts are weakly labeled (open arrows). (k) Same as (j) but TM-MFasII channel only. TM-MFasII staining in the CNS is restricted to areas devoid of MSYT staining. (l, m) Selective enlargement of the area indicated in (j) by white box depicts an axon with granular TM-MFasII-positive staining (open arrow). No co-localization between the MSYT and TM-MFasII staining is detectable. (l double-lable merge, m TM-MFasII). Bars 100 μm (a–c, f, g, j, k), 20 μm (d, e), 10 μm (h, i, l, m). (Source: Kuehn & Duch 2008. Reproduced with permission from Springer.) See plate section for the color version of the figure.

Procedure:

1. Place in absolute alcohol for 5 to 15 days.
2. Transfer through 95, 85, 70, and 50% alcohol for about 15 min in each.
3. Transfer to 35% alcohol for 30 min.
4. Place in a solution of H_2O and H_2O_2 (50:50) with a trace of NH_4OH for 12 to 24 hr.
5. Transfer through two to three changes of absolute alcohol for a minimum of 3 days.
6. Transfer to tolulene for 10 days to 3 weeks.

3.2.14 *Terpineol [12, 13]*

Notes:

1. This compound is slower than xylol or cedar oil.
2. Clears minute insect parts overnight.
3. Clears from 95% alcohol.

3.2.15 *Terpineol-xylene [11, 12]*

Formula:

Terpineol	3 pt
Xylene	1 pt

Notes:

1. Clears from 95% alcohol.
2. It is faster than Terpineol alone.

3.2.16 *Tetrahydro-furan [6]*

Application:

Miscible with wax, water, and resinous mounting materials.

Procedure:

1. Dehydrate through increasing graded concentrations of tetrahydrofuran until in 100% tetrahydrofuran.

3.2.17 *Trimethylsilane [23]*

Application:

Recommended for clearing tissues for SEM.

Procedure:

1. Dehydrate fixed tissues through graded ethanol series to 100% ethanol.
2. Immerse tissue in a mixture of 1 pt 100% ethanol to 1 pt TMS for 10 min.
3. Transfer to pure TMS and allow it to evaporate.

3.2.18 *Turpentine [24]*

Application:

Recommended for lice and other small insects which are normally semi-transparent when alive.

Note:

When left for 2 to 3 months in this substance the tracheal system is well illustrated.

3.2.19 *Xylene [11–13]*

Notes:

1. Clears quickly from 100% alcohol.
2. Tissues must be absolutely dehydrated or cloudiness forms.
3. Causes rapid tissue hardening.
4. Suggested as a second clearing fluid.

3.3 Embedding General

3.3.1 *Apathy's double-embedding technic [13]*

Formula:

Apathy's Oil Mixture	
Chloroform	4 pt
Origanum Oil	2 pt
Cedar Oil	4 pt
Absolute Alcohol	1 pt
Carbolic Acid Crystals	1 pt

Preparation:
1. All ingredients are measured out by weight.

Procedure:
1. Fix, wash, and dehydrate as usual.
2. Transfer specimen through three changes of absolute alcohol.
3. Transfer to a tube with a 1:1 ether:alcohol mixture for at least 5 hr.
4. Keep specimen high in tube with cotton wool plug.
5. Transfer to a tube with 2% celloidin solution (in 1:1 ether:alcohol) for 24 hr.
6. Transfer to a tube with a 4% celloidin solution (in 1:1 ether:alcohol) for 24 hr.
7. Embed in 4% celloidin in mold.
8. Harden in chloroform vapors for 2 hr.
9. Trim the block without exposing tissue.
10. Immerse in a tube of chloroform for 12 hr.
11. Transfer to Apathy's Oil mixture for 3 to 7 days or until the block is clear and sinks.
12. Wash in three or more changes of benzol (which is a crude form of benzene, containing toluene, xylene, and other hydrocarbons).
13. Infiltrate in paraffin and embed in paraffin.
14. Trim block so as to leave layer of paraffin.
15. Section and mount.

Note:
Some dried sodium sulfate should be placed on the bottom of the tube with solution A to take up excess water of celloidin block.

3.3.2 *Bayberry-paraffin medium [13]*

Application:
Recommended as a general embedding medium.

Formula:

Bayberry Wax	1 pt
Hard paraffin (52–56°C)	50 pt

Notes:
1. Wax is denser and has a finer grain than just paraffin.
2. The bayberry wax addition does not alter the paraffin melting point.
3. Mixture holds better than paraffin.

3.3.3 *Bayberry, paraffin, rubber medium [13]*

Application:
Recommended as an embedding medium that is particularly useful in holding sclerotized larvae such as those of *Rhagoletis* sp.

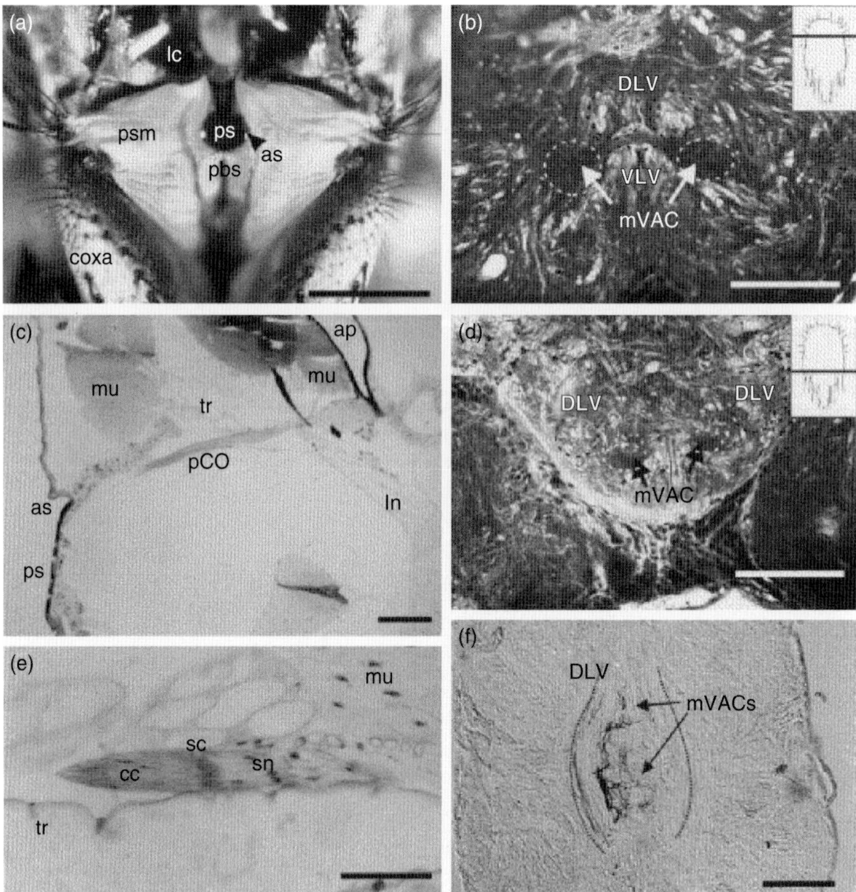

Fig. 3.5 Photomicrographs of the sensory system and the central nervous system. (a) Frontal view of the prothorax of *S. bullata*. The arrowhead points towards the attachment site of one prosternal chordotonal organ (as). lc: lateral cervicale, pbs: probasisternum, ps: presternum, psm: prosternal membrane. Scale: 500 μm. (b) Horizontal section of the prothorax showing the position of the prosternal chordotonal organ (pCO). ap: apodeme, ln: legnerve, mu: muscle, tr: trachea. Scale: 100 μm. (c) Longitudinal section of the prosternal chordotonal organ. cc: cap cells, sc: scolopidia, sn: sensory neuron. Scale: 50 μm. (d, e) Transverse sections of the thoracico-abdominal ganglion complex at two different positions indicated on the right. DLV: dorsolateral tract of the ventral fasciculus, VTV: ventromedian tract of the ventral fasciculus, mVAC: median ventral association centre. (f) Horizontal section of the thoracico-abdominal ganglion with labelled sensory fibers. The DLV is outlined and the mVAC of the pro- and mesothoracic neuromere is indicated. Scales D–F: 100 μm. Dehydration in a rising ethanol series, and specimens were embedded in agar. (Source: Stolting *et al.* 2007. Reproduced with permission of Elsevier.)

Formula:

Solution A: Rubber-Paraffin

Preparation:

1. Cut 1/8″ strips of crude rubber, place in a beaker of hard paraffin (52–56°C) and fill 1/3 full with strips.
2. Keep melted in a warm oven for one week with occasional stirring.
3. Pour off the clearer top fluid for use in solution B.

Solution B: Bayberry, Paraffin, Rubber

Solution A	40 pt
Hard Praffin (52–56°C)	450 pt
Bayberry Wax	50 pt

Procedure:

1. Infiltrate and embed as with paraffin.

3.3.4 *Boycott's technic [25]*

Application:

Recommended as a double embedding procedure for general use in insect histology.

Formula:

Solution A:

Celloidin Shreds (dessicated)	1.5 pt
Clove Oil	50 pt

Preparation:

1. Dissolve celloidin in clove oil for several days in a warm oven.
2. Or, dissolve celloidin in 1:1 ether and alcohol mixture.
3. Then add clove oil and place in a warm oven overnight.

Procedure:

1. Fix specimens in absolute alcohol or if fixed in another material, bring up to absolute alcohol.
2. Transfer to clove oil.
3. Transfer to solution A for about 24 hr (for specimen of size of a flea).
4. On a coverslip (which has been dipped in paraffin) place the specimen in a drop of celloidin.
5. Invert slip onto chloroform for ½ hr or more.
6. When drop falls away transfer to paraffin for about 20 min (e.g., for fleas and lice).
7. Embed and section as usual.

Note:

1. In the preparation of solution A (step 3), the ether and alcohol will evaporate.

3.3.5 *Dobell's embedding technic [26]*

Formula:

Solution A:

Picric Acid (sat. solution in 90% alcohol)	75 pt
Formalin	25 pt
Glacial Acetic Acid	5 pt
Chloroform (added before using)	1–2 drops

Solution B:

Absolute Alcohol	1 pt
Chloroform	1 pt

Procedure:

1. Fix in solution A for 24 hr (1 hr at 38°C and 23 hr at room temperature).
2. Transfer through several changes of ethyl alcohol (90%) for 7 days.

3. Transfer to absolute alcohol for 24 hr.
4. Transfer to solution B for 24 hr.
5. Transfer to chloroform for 24 hr.
6. Transfer to a saturated mixture of paraffin in chloroform for 24 hr.
7. Transfer to paraffin (m.p. 56°C) for 5 to 6 hr.

3.3.6 *Eltringham's double embedding technic [25]*

Application:
Recommended as an embedding procedure for cutting sclerotized specimens.

Formula:
Solution A: Celloidin Stock

Ether	3 pt
Absolute Alcohol	1 pt
Celloidin	Up to molasses consistency

Solution B:

Ether	3 pt
Absolute Alcohol	1 pt

Procedure:
1. Dry celloidin by shaking with absolute alcohol and warming on filter paper.
2. Thoroughly dehydrate the specimens to be embedded.
3. Place specimen in small test tube filled (3/4 inch) with solution A.
4. Dilute by adding solution B until it has watery consistency.
5. Let open tube stand for several days until it regains original consistency.
6. Transfer specimen to mold, ensuring a margin of celloidin is left around specimen.
7. Expose celloidin to chloroform vapors in a closed vessel for a few hours.
8. In another tube, add chloroform and saturate with paraffin shreds.
9. Warm slightly to dissolve paraffin.
10. Transfer celloidin blocks to this tube, cork and put aside.
11. When blocks sink the usual infiltration procedures are followed.
12. Remove all chloroform by placing blocks through three changes of paraffin *in vacuo* for 5 min each.
13. When ribbon is mounted place in xylol, absolute alcohol and in solution B.
14. Return to absolute alcohol and hydrate if aqueous stain is to be used.

Note:
1. The amount of solution in step 4 is suggested for two specimens of the size of a house fly.
2. In step 5 the solution should be shaken several times a day.
3. A mold may be made of aluminum foil or stiff paper.
4. In step 8, the paraffin added should have the same melting point as the paraffin to be used for infiltration.
5. An adequate margin of paraffin should be left around the block when it is trimmed.

3.3.7 *MC Indoo's technic [25]*

Application:
Recommended as a double embedding procedure for general use in insect histology.

Formula:

Solution A: Carbol-Xylol

	Phenol Crystals	1 pt
	Xylol	3 pt

Solution B:

	Ether	1 pt
	Absolute Alcohol	1 pt

Procedure:

1. Fix specimen in Bouin's, Carnoy's or Lebrun's fluid.
2. Wash in several changes of alcohol (95%).
3. Materials fixed in sublimate should be washed in iodine alcohol (85%).
4. If specimen is small, place in alcohol with a little eosin added.
5. Transfer to solution B with a little eosin for 5 to 10 min.
6. Transfer to celloidin (2%) solution in solution B overnight.
7. The celloidin in its container is held in melted paraffin for 5 min (sic).
8. Transfer specimen from celloidin into chloroform for ½ hr or until firm.
9. Trim and place in paraffin (58–60°C) on a hot plate for a few min.
10. When bubbles disappear embed as usual.
11. Mount sections with thick layer of Mayer's egg albumen.
12. Set aside and dry for 48 hr.
13. Stain in Delafield's or Ehrlich's haematoxylin for 4 to 5 min.
14. Set in water for about an hour.
15. Dehydrate through alcohol series up through 50, 70, 85, and 95% with eosin.
16. Wash out stain in alcohol (95%).
17. Transfer slide to solution A for a few seconds.
18. Drop xylol on slide until paraffin dissolves.
19. Drain off xylol and mount in balsam.

Note:

1. Fixatives with corrosive sublimate should be stored in a glass stoppered bottle.

3.3.8 *Salthouse's double embedding technic [7]*

Application:

Recommended for insects with hard cuticle.

Procedure:

1. Fix material.
2. Place in tetrahydrofuran for 2 hr.
3. Transfer to new tetrahydrofuran for 1 hr.
4. Transfer to parloidin (parloidin, also known as cellulose nitrate) (1% in tetrahydrofuran) for 24 to 48 hr.
5. Transfer to parloidin (2% in tetrahydrofuran) for 24 to 48 hr.
6. Rinse (rapidly) in tetrahydrofuran for several minutes.
7. Transfer through two changes of molten paraffin (in oven) for 1 hr each.
8. Block and section.

Note:

1. Addition of 10% beeswax to embedding wax is suggested.

3.3.9 *Smith's embedding technic [27]*

Application:
Recommended for insect cytology and embryology, e.g., the study of meiosis in insect eggs.

Formula:

Solution A:

Water	95 pt
Ethyl Alcohol	5 pt

Solution B:

Water	90 pt
Ethyl Alcohol	10 pt

Solution C:

Water	80 pt
Ethyl Alcohol	20 pt

Solution D:

Water	65 pt
Ethyl Alcohol	35 pt

Solution E:

Water	50 pt
Ethyl Alcohol	40 pt
N-butyl alcohol	10 pt

Solution F:

Water	30 pt
Ethyl Alcohol	50 pt
N-butyl alcohol	20 pt

Solution G:

Water	15 pt
Ethyl Alcohol	50 pt
N-butyl alcohol	35 pt
Phenol	4 pt

Solution H:

Water	5 pt
Ethyl Alcohol	40 pt
N-butyl alcohol	55 pt

Solution I:

Ethyl Alcohol	25 pt
N-butyl alcohol	75 pt

Solution J:

N-butyl alcohol	100 pt
Phenol	4 pt

Procedure:
1. Transfer through solutions A to C for ½ hr each.
2. Transfer through solutions D to F for 1 hr each.
3. Transfer to solution G for 24 hr.
4. Transfer through solutions H to J for 1 hr each.
5. Transfer to a mixture of equal parts solution J and paraffin (oven) for 16 hr.
6. Transfer to paraffin for 4 hr.
7. Embed and block.
8. Trim block and soak in 4% phenol solution in 80% ethyl alcohol for 24 hr.
9. Section.

Note:
1. Percentages for ethanol or N-butyl alcohol were not given by Smith (1940).
2. Dehydration does not have to begin with solution A. This depends on the fixative used.

3.3.10 *Woodworth's embedding technic [28]*

Application:
Recommended as a technique that avoids layering and in which small crystals form as paraffin cools.

Procedure:
1. Fix, dehydrate, and infiltrate in paraffin as usual.
2. The embedding paraffin and a pair of forceps are heated to at least 80°F.
3. Place the paraffin filled embedding boat on an ice cube to cool.
4. Remove the specimen with forceps out of the infiltration chamber.
5. Place specimen in hot paraffin for a moment to remove hardened paraffin.
6. Transfer to an embedding boat.
7. Orient specimen and allow boat to cool.

3.3.11 *De Guisti and Ezman's Tissue Mat-isopropyl technic [29]*

Application:
1. Recommended for use on mites and insects such as Collembola, Thysanura, *Drosophila*, Simuliidae, Mallophaga, and roaches.
2. Recommended for small specimens.

Formula:
Modified Carnoy's:

Isopropyl Alcohol (absolute)	6 pt
Chloroform	3 pt
Formic Acid	1 pt

Procedure:
1. Fix with warmed fixative under slight vacuum for 1 to 6 hr.
2. Wash in several changes of isopropyl alcohol (70%).
3. Dehydrate through a graded series of isopropyl alcohol.
4. Warm specimen in absolute isopropyl, then transfer to isopropyl and Tissue Mat liquid mixture (50:50) for 20 to 25 min.
5. Transfer to Tissue Mat for 20 to 25 min.

6. Transfer to second change of Tissue Mat for 15 min.
7. Block and cool rapidly.
8. Section and mount.

3.3.12 *Agarose [30]*

Application:
For immunocytochemical preparations of tissues or embryos whose embryonic membranes have been removed.

Procedure:
1. Fixed tissues are embedded in 5% agarose in phosphate buffered saline at 55–60°C.
2. The agarose block is cooled and sectioned on a Vibratome at 40–50 mm.

3.3.13 *Ripper cyro-technic [31]*

Formula:
Solution A:

Distilled Water	2–4 pt
OsO_4	0.1 pt
Uranyl Acetate	0.1–0.2 pt
Methanol	0.5–2.5 pt
Acetone	To 100 pt

Solution B:

Distilled Water	2–4 pt
Glutaraldehyde	0.5 pt
Acetone	To 100 pt

Procedure:
1. Place tissue in 1-hexadecene and high pressure freeze.
2. Remove tissue from 1-hexadecene and transfer to solution A at -90°C for 50–60 hr.
3. Bring tissues to −60°C for 8 hr.
4. Add 0.5 pt glutaraldehyde.
5. Bring tissues to −35°C for 8–10 hr.
6. Wash samples in solution B, four times for 30–45 min each.
7. Bring tissues to −20°C.
8. In a graded series, bring samples from solution B to 0.25% glutaraldehye in water. As the acetone is removed in graded fashion, similarly increase the temperature in graded fashion so that the final solution is added at 0°C.
9. Incubate in 0.25% glutaraldehyde for 90 min.
10. Wash tissues in water.
11. Wash tissues two times in 50 mM glycine in water.
12. Snap freeze specimen onto stubs in liquid nitrogen.

3.3.14 *Sucrose cryo-protection technic [3, 32]*

Application:
For improved histology of frozen tissues.

Procedure:
1. After fixation, immerse tissue in 10% sucrose in phosphate buffer for 15 min.
2. Transfer to fresh 10% sucrose three times for 15 min each.
3. Immerse tissue in 30% sucrose overnight.
4. Place tissue in cryo-embedding media and freeze in melting iso-pentane bath.

3.3.15 *Tobie, Burgdorfer, and Larson's freezing technic [33]*

Application:
Recommended for histological preparations and localization of disease agents in arthropods.

Procedure:
1. Fill small Dewar flask with a mixture of powdered dry ice and alcohol.
2. Place microtome object holder (shank end) in the dry ice.
3. Form heavy duty aluminum foil into cylinder (size depends on specimen).
4. Freeze upright to object holder with drops of water at base.
5. Place gelatin solution (10%) in cylinder (1/3 full).
6. Hold specimen (oriented) in gelatin solution until frozen.
7. Pipette gelatin solution around specimen until fully covered.

3.4 Embedding – Ester Wax

Another embedding medium which has been recommended for work in insect histology is ester wax. It was originally reported by Steedman[34] and modified by Chesterman and Leach.[35]

3.4.1 *Ester wax original [34]*

Formula:

Diethylene Glycol Distearate	73 pt
Ethyl Cellulose	4 pt
Stearic Acid	5 pt
Castor Oil	8 pt
Diethylene Glycol Monostearate	10 pt

3.4.2 *Modified ester wax technic [35]*

Formula:

Diethylene Glycol Monostearate (purified)	100 pt
Diethylene Glycol Distearate (purified)	100 pt
Caster Oil	5 pt

Procedure:
1. Fix with any fixative and wash.
2. Transfer to alcohol (70%) for 12 hr.
3. Transfer to alcohol (90%) for 12 hr.
4. Transfer to alcohol (96%) for 12 hr.
5. Transfer through two changes of absolute alcohol for 12 hr each.
6. Transfer through two changes of an equal part solution of absolute alcohol and xylene, overnight.
7. Transfer to xylene overnight or until cleared.
8. Transfer to one or two changes of modified ester wax.
9. Place in fresh wax for 1 to 2 hr to block.

Note:
1. Modified ester wax should be melted in an oven at 53°C the day before use.
2. Cellosolve may be substituted as a dehydrating and clearing agent and thus one can go directly from fixative (or post-fixative treatment) to cellosolve.
3. From four changes in cellosolve for 24 hr specimens may be transferred to modified ester wax.

3.4.3 *Akesson's polyester wax technic [36]*

Application:

1. Recommended for the embedding of small specimens (e.g., eggs, embryos and small larvae).
2. Recommended as particularly useful where shrinkage is a major problem.

Formula:

Solution A: Wax Mixture

Polyester Wax	8 pt
Ester Wax	2 pt

Preparation:

1. Melt waxes separately, polyester wax at 37°C and ester wax at 60°C.
2. Mix together, then place in oven at 37°C until wax temperature is 37°C.
3. Filter through coarse filter paper.
4. If yellow or yellow-brown oil substance is found at bottom of filtrate, remove it.

Procedure:

1. Specimens preserved in Bouin's are transferred to alcohol (50%).
2. Add an equal volume of cellosolve, dropwise, and stir periodically.
3. Decant most of the fluid and add more cellosolve.
4. Two hours should be the minimum time to attain pure cellosolve.
5. Transfer to saturated acid fuschin (in cellosolve) for 10 min.
6. Infiltrate by making a hold in solid wax placing the specimen in it and add enough cellosolve to cover specimen.
7. Cover wax block with glass slip and place under 60 W lamp bulb (at about 15 cm distance).
8. After wax is melted (1/2 to 2 hr) stir gently with warm pipette.
9. Place in oven at 37°C for ½ hr.
10. Transfer to second wax bath for 2 to 4 hr.
11. Transfer to third wax bath for 4 to 6 hr.
12. Place thick metal washer on glycerine coated glass slip.
13. Fill washer with embedding wax and keep liquid with adjustable microscope lamp.
14. Orient specimen.
15. Let solidify for 10 min and let stand for 1 day.

Note:

1. If large quantities of wax mixture are filtered, stratification during hardening may occur.
2. When cutting pieces, slices must thus go through entire block.
3. Specimens are stained with acid fuchsin because of opaque nature of wax mixture.
4. Before infiltration specimens may be stored in cellosolve for 1 to 2 days (at room temperature), a week (at 4°C) or for months in sealed vessels (at 20°C).
5. The embedding wax should be in liquid form for a minimum time for better sections.
6. Sectioning properties increase during 3 to 4 days after embedding.

3.5 **Embedding – Methacrylate**

The following deals with the use of thermoplastics as embedding media. The thermoplastics specifically suggested for use in the following techniques are the butyl and ethyl methacrylates, either alone or mixed. The technic usually limits tissue size to about one millimeter but produces very thin sections (0.05 to 0.05 u) suitable for electron microscopy as well as light microscopy.

3.5.1 *Beckel and Habowsky's [37]*

Formula:
Solution A:

Ethyl Methacrylate	75% (of mixture)
Butyl Methacrylate	25% (of mixture)

Procedure:
1. Dehydrate specimen in tetrahydrofuran.
2. Transfer to solution A.
3. Place under pressure (50 mm Hg) at room temperature for 24 hr (two changes).
4. Place in gelatin capsule for polymerization (curing) at 60°C for 4 hr.
5. Soften gelatin capsule with water and seal off.
6. Section.

Note:
1. Mixture is freed of inhibitor and catalyzed by 2% benzoyl peroxide.
2. Increasing percentage of ethyl methacrylate will harden block but makes it shatter prone.

3.5.2 *Modified Woodring and Cook technic [38]*

Application:
1. Recommended as an embedding procedure for dipteran eggs and larval head capsules.
2. Recommended as a generalized technique for the preparation of insect specimens in which strongly sclerotized areas alternate with membranous structures.
3. Recommended as particularly useful for larval crane flies.

Formula:
Solution A: N-butyl Methacrylate Mixture (N-b.m.)

Preparation:
1. Make a mixture of N-butyl monomer, 2 to 3% benzoyl peroxide accelerator (vol. to wt.) and 2% Tissue Mat (vol. to wt.).

Procedure:
1. Dehydrate alcohol preserved specimens by transfers through ethanol in 15% steps up to distilled water for 10 min each.
2. Dehydrate in isopropyl alcohol in 15% steps up to absolute for 3 min each.
3. Transfer through two changes of N-butyl alcohol for 30 min each.
4. Infiltrate in monomer on N-b.m. (see note).
5. Place gelatin capsule with a small amount of solution A into an oven at 60°C for 2 hr (Polymerization I).
6. Place specimen in capsule, add more solution A and return to oven (Polymerization II: see note).
7. Remove gelatin capsule immediately by softening in water.

8. Trim plastic block and place in molten Paraplast.
9. Let harden at room temperature.
10. Trim and mount for sectioning.

Note:
1. Infiltration and polymerization times vary with size of specimen (in this case, head capsules) and were reported by the authors as follows:

	Infiltration		Polymerization II	
Diameter	Time (hr)	Temp (°C)	Time (hr)	Temp (°C)
50	4	40	11–12	63
100	8	40	11–12	63
150	12	40	11–12	63
200	16	41	12	63
250	20	41	13	62
300	24	41	13	62
350	28	42	14	60
400	32	42	14	60
450	36	42	14	60
500	40	43	15	60
550	44	43	16	60
600	48	43	16	60

2. Larger head capsules were found to require longer polymerization times at lower temperatures.
3. Sections of 4–10 μm and thinner were attainable using a slide microtome.

3.5.3 *Rohde's [39]*

Application:
Recommended for thin sections of insects and mites but generally good for all refractory or brittle material, e.g., yolky eggs.

Formula:
Solution A:

Ethyl Methacrylate	1 pt
Butyl Methacrylate	9 pt

Solution B:

Tetrahydrofuran	1 pt
Solution A	1 pt

Solution C:

Solution A	as needed
Benzoyl Peroxide (2%)	as needed

Procedure:
1. Transfer through two changes of solution B for 1 to 2 hr.
2. Transfer to solution A for 1 to 2 hr.
3. Transfer to solution C for 2 to 4 hr.
4. Transfer through two changes of solution C for 1 hr each.

5. Transfer specimen to gelatin capsule with polymer and catalyst at 60°C (in oven) for about 12 hr.
6. Let stand (outside oven) for 48 hr.
7. Remove gelatin capsule.
8. Block may be cemented to Lucite stick and sectioned.

Note:
1. Benzoyl peroxide is the catalyst.

References

1. Yang H., Yan S.C. & Liu D. Ultrastructural observations on antennal sensilla of *Coleophora obducta* (Meyrick) (Lepidoptera: Coleophoridae). *Micron* **40**, 231–238 (2009).
2. Bution M.L., Caetano F.H. & Fowler H.G. Proventriculus of *Cephalotes* ants: a structural and comparative analysis. *Micron* **41**, 79–83 (2010).
3. Klein U., Loffelmann G. & Wieczorek H. The midgut as a model system for insect K+-transporting epithelia – Immunocytochemical localization of a vacuolar-type H+ pump. *J. Exp. Biol.* **161**, 61–75 (1991).
4. Fassig W.W. Iso-amyl alcohol as a dehydrating agent in the preparation of mosquito larvae and larval exuviae slides. *J. Military Med. Prac.* **1**, 51–52 (1945).
5. Van Heerden H.P. Some histological methods of interest to entomologists. *J. Entomol. Soc. S. Africa* **8**, 157–161 (1945).
6. Marshall A.T. Spittle-production and tube-building by cercopoid nymphs (Homoptera). I. The cytology of the Malpighian tubules of spittle-bug nymphs. *Quart J. Microsc. Sci.* **105**, 257–262 (1964).
7. Salthouse T.N. Tetrahydrofuran and its use in insect histology. *Canad. Entomol.* **90**, 555–557 (1958).
8. Stiles K.A. Normal butyl alcohol technic for animal tissues with special references to insects. *Stain Technol.* **9**, 97–100 (1934).
9. Bartlett L.M. A dehydration and embedding schedule for insects. *Entomol. News* **53**, 109–110 (1942).
10. Peacock H.A. *Elementary Technique*. Edward Arnold Ltd., London (1966).
11. Eltringham H. *Histological and Illustrative Methods for Entomologists*. Oxford University Press, London (1930).
12. Peterson A. *Entomological Techniques. How to work with Insects*. Edwards Brothers Inc., Ann Arbor, Michigan (1964).
13. Kennedy C.H. *Methods for the Study of the Internal Anatomy of Insects*. Department of Entomology, Ohio State University (Mineographed) (1932).
14. Ott S.R. & Elphick M.R. New techniques for whole-mount NADPH-diaphorase histochemistry demonstrated in insect ganglia. *J. Histochem. & Cytochem.* **51**, 523–532 (2003).
15. Huang F., Yu R.X. & Chen X.X. Ultrastructure of the male reproductive system of *Cotesia vestalis* (Hymenoptera : Braconidae) with preliminary characterization of the secretions. *Microscope . Res Tech.* **70**, 563–571 (2007).
16. Stirling W. *Outlines of Practical Histology*. Charles Griffon and Company, London (1890).
17. Kitzmiller J.B. The use of dioxane in insect microtechnique. *Trans. Amer. Microsc.Soc.* **67**, 227 (1948).
18. Zachepilo.T.G. *et al*. Comparative analysis of the locations of the NR1 and NR2 NMDA receptor subunits in honeybee (*Apis mellifera*) and fruit fly (*Drosophila melanogaster*, Canton-S wild-type) cerebral ganglia. *Neurosci. Behav. Physiol.* **38**, 369–372 (2008).
19. Kuehn C. & Duch C. Expression of two different isoforms of fasciclin II during postembryonic central nervous system remodeling in *Manduca sexta*. *Cell Tissue Res.* **334**, 477–498 (2008).

20. Nishino H., Yoritsune A. & Mizunami M. Different growth patterns of two adjacent glomeruli responsible for sex-pheromone processing during postembryonic development of the cockroach *Periplaneta americana*. *Neurosci. Letters* **462**, 219–224 (2009).
21. Nardi J.B., Bee C.M. & Miller L.A. Stem cells of the beetle midgut epithelium. *J. Insect Phys.* **56**, 296–303 (2010).
22. Stapp R. & Cumley R.W. A technic for clearing large insects. *Stain Technol.* **11**, 105–106 (1936).
23. Martin A.C., Gelbart M., Fernandez-Gonzalez R., Kaschube M. & Wieschaus E.F. Integration of contractile forces during tissue invagination. *J. Cell Biol.* **188**, 735–749 (2010).
24. Walsh G.B. On the preparation of insects as microscopic mounts. *Vasculum* **2**, 75–81 (1916).
25. Levenbook L. Outline for the study of insect histology. *Microsc. & Entomol. Mon.* **4**, 212–220 (1940).
26. Becker E.R. & Roudabush R.L. *Brief Directions in Histological Technique*. Collegiate Press, Inc., Ames, Iowa (1935).
27. Smith S.G. A new embedding schedule for insect cytology. *Stain Technol.* **15**, 175–176 (1940).
28. Woodworth C.E. A useful embedding technic. *Stain Technol.* **16**, 124 (1941).
29. De Guisti D.L. & Ezman L. Two methods for serial sectioning of arthopods and insects. *Trans. Amer. Microsc. Soc.* **74**, 197–201 (1955).
30. Herbert Z. *et al.* Developmental expression of neuromodulators in the central complex of the grasshopper *Schistocerca gregaria*. *J. Morph.* **271**, 1509–1526 (2010).
31. Ripper D., Schwarz H. & Stierhof Y.D. Cryo-section immunolabelling of difficult to preserve specimens: advantages of cryofixation, freeze-substitution and rehydration. *Biol. of the Cell* **100**, 109–123 (2008).
32. Zimoch L. & Merzendorfer H. Immunolocalization of chitin synthase in the tobacco hornworm. *Cell and Tissue Res.* **308**, 287–297 (2002).
33. Tobie J.E., Burgdorfer W. & Larson C.L. Frozen sections of arthropods for histological studies and fluorescent antibody investigations. *Exp'tl Parasitol.* **11**, 50–55 (1961).
34. Steedman H.F. Ester wax: A new embedding medium. *Quart. J. Microsc. Sci.* **88**, 123–133 (1957).
35. Chesterman W. & Leach E.H. A modified ester wax for embedding tissues. *Quart. J. Microsc. Sci.* **97**, 593–597 (1956).
36. Akesson B. Orientation and embedding of small objects in Steedman's polyester wax for sectioning in definite planes. *Arkiv For Zoologi* **19**, 247–249 (1967).
37. Beckel W.E. Selective softening of insect cuticle after impregnation with wax. *Stain Technol.* **35**, 356–357 (1960).
38. Smith H.G. & Hynes C.D. Modifications of the N-butyl methacrylate embedding method as applied to insect larval head capsules. *Ann. Entomol. Soc. Amer.* **59**, 230 (1966).
39. Rohde C.J. & Oemick D.A. Anatomy and the digestive and reproductive systems in an acarid mite (Sarcoptiformes). *Acarologia* **9**, 608–616 (1967).

4 Staining

This chapter includes staining procedures which have been recommended for the histological preparation of insect tissues. Other staining methods representing part of more complete histological procedures have also been included here. Staining techniques for specific tissues or whole specimens may also be found in other appropriate sections.

Insect Histology: Practical Laboratory Techniques, First Edition.
Pedro Barbosa, Deborah L. Berry and Christina S. Kary.

Companion Website: www.wiley.com/go/barbosa/insecthistology

4.1 Single-contrast staining – Carmines

4.1.1 *Mayer's acidulated carmine [1, 2]*

Formula:

Carminic Acid	4 pt
Distilled Water	15 pt
Hydrocholoric Acid	30 drops
Alcohol (85–90%)	95 pt

Preparation:

1. Boil until carmine is dissolved.
2. Cool and add alcohol.
3. Filter.
4. Add drops of ammonia to neutralize acid solution resulting in a deep red stain.

Procedure:

1. Transfer slide to stain for 5 to 20 min.
2. Wash in alcohol (70%).
3. Transfer to alcohol (96%).
4. Differentiation can be achieved by transfer to alcohol (96%) with a little HCl (1 to 1000).
5. Transfer to alcohol (96%).

Fig. 4.1 Coplin jars for starting a slide through the reagent series in procedures such as staining. (Source: Gray 1964.)

Note:
1. Picric acid may be used if counterstain is desired.

4.1.2 *Mayer's carmalum [2]*

Formula:

Carminic Acid	1 pt
Alum	10 pt
Distilled Water	200 pt

Preparation:
1. Heat to dissolve and filter.
2. Add 1 pt formalin to store fluid.

Procedure:
1. After staining, wash with water or an alum solution.

4.1.3 *HCL carmine [2]*

Formula:

Solution A: Stain

Carmine	2 pt
HCL (concentrated)	3 pt
Alcohol (70%)	100 pt

Preparation:
1. To dissolve carmine, boil for 15 to 20 min.
2. Filter when cold.

Solution B:

Alcohol (95%)	100 pt
HCL (concentrated)	0.1 pt

Procedure:
1. Stain (solution A) sections for 5 to 15 min.
2. Rinse in alcohol (70%).
3. Differentiate in solution B for 10 to 60 sec.
4. Transfer to alcohol (95%) for further differentiation.

4.1.4 *Mayer's paracarmine [3]*

Formula:

Carminic Acid	1 pt
Aluminum Chloride	0.5 pt
Ethyl Alcohol (70%)	100 pt
Calcium Chloride	4 pt

Preparation:
1. Dissolve ingredients in cold or warm solution.
2. Filter.

Procedure:
1. Wash in 70% alcohol (with 2.5% acetic acid) after staining.

Note:
1. Tissue to be processed should not be alkaline.

4.1.5 *Grenacher's [3]*

Formula:

Carmine	3 pt
Borax	4 pt
Distilled Water	93 pt

Preparation:

1. Boil carmine, borax, and water for 1 hr.
2. Let stand for 72 hr (occasional stirring).
3. Add equal volume of alcohol (70%) and filter.

Procedure:

1. Stain tissue.
2. Transfer to 70% acidulated alcohol (4 to 6 drops HCL/100 pt alcohol) for several hours.
3. Wash in neutral alcohol.

4.2 **Single contrast staining – Nuclear Stains**

4.2.1 *Delafield's hematoxylin [1]*

Formula:

Ammonia Alum (sat. solution)	400 pt
Hematoxylin Crystals (dissolved in 25 pt of strong alcohol)	4 pt

4.2.2 *Acid delafield's hematoxylin [2]*

Application:

1. Recommended for use in embryological studies.
2. Recommended as a stain for sharper differentiation of nuclei than possible with Delafield's.

Formula:

Delafield's Hematoxylin	1 pt
Distilled Water	10–20 pt
Acetic Acid	add until obtain red tint

4.2.3 *Ehrlich's acid hematoxylin [2]*

Application:

Recommended for good nuclear differentiation.

Formula:

Distilled Water	100 pt
Absolute Alcohol	100 pt
Glycerine	100 pt
Glacial Acetic Acid	10 pt
Hematoxylin Crystals	2 pt
Alum	in excess

Preparation:

1. Ripen in light until dark red and open periodically to admit air.

Procedure:

1. Stain sections for 5 to 10 min.
2. Wash in water.

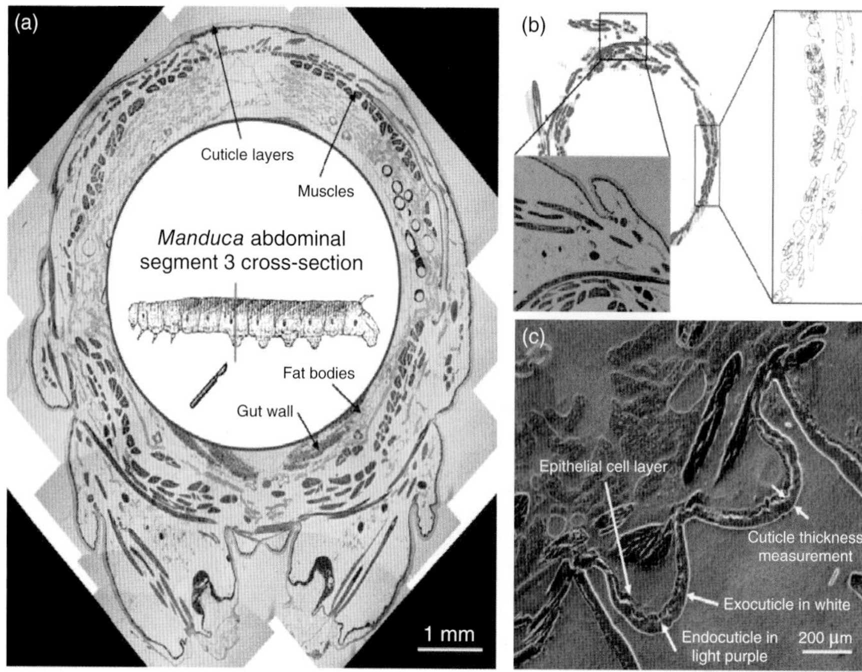

Fig. 4.2 Histology of a *Manduca sexta* caterpillar. (a) Hematoxylin and eosin (H&E) stained most of the soft tissues inside the body. Adjacent histology images of abdominal segment 3 were obtained by merging multiple microscopic images. The example cross-section came from a fifth instar animal. (b) Blue light excitation emphasized muscle fibers so that they could be identified by image analysis. (c) The endocuticle and exocuticle layers are easily distinguished from the soft flaky epithelial cell layer. The gut wall is orders of magnitude softer than the muscle fibers in the longitudinal direction and cuticle in the circumferential direction. (Source: Lin *et al.* 2011. Reproduced with permission of the authors.) See plate section for the color version of this figure.

4.2.4 *Mayer's hemalum [2]*

Application:
Recommended as a general nuclear stain useful to the beginner.

Formula:

Hematoxylin Crystals	1 pt
Distilled Water	1000 pt
Sodium Iodate ($NaI0_3$)	0.2 pt
Potash Alum	50 pt
Chloral Hydrate	50 pt
Citric Acid	1 pt

Preparation:
1. Mix the first two ingredients.
2. Dissolved the next two ingredients.
3. Filter.
4. Add chloral hydrate and citric acid to preserved fluid.

Note:
1. Dilute one to several times with water.
2. No counterstaining is necessary.

4.2.5 *Mayer's alcoholic cochineal [2]*

Application:
Recommended as an arthropod nuclear stain.

Formula:

Cochineal Powder	10 pt
Alcohol (70%)	100 pt

Preparation:
1. Shake fluid at frequent intervals over several days.
2. Filter.

Procedure:
1. Wash in alcohol (70%).
2. Stain for a few minutes.
3. Mount in balsam.

Note:
1. Specimens must be acid free and in alcohol (70%).

Result:
1. Depending on the salts in the tissues, staining may result in a red, blue, green or purple color.

4.2.6 *Heidenhain's iron hematoxylin [1]*

Formula:

Hematoxylin (16% alcoholic solution)	3 pt
Chloral Hydrate	2 pt
Distilled Water	97 pt

Procedure:
1. Transfer sections to 2 to 4% ferric alum solution (aqueous) for 1 to 2 hr.
2. Wash in water for 5 to 10 min.
3. Stain in a ½% solution of Hematoxylin (aqueous) for ½ to 2 hr.
4. Wash in water.
5. Differentiate with ferric alum solution for a few seconds.

4.2.7 *Weigert's iron hematoxylin [4]*

Formula:
Solution A: Stock

Ferric Chloride (10% aqueous)	5 pt
Distilled Water	94 pt
HCL	1 pt

Solution B: Stock
Hematoxylin (1% in 95% alcohol)

Solution C: Working Fluid

Solution A	1 pt
Solution B	1 pt

Note:
1. Solution B should be well ripened.
2. The working solution should be used when fresh but stays in good condition for 24 hr.

4.2.8 *Modified weigert's [5]*

Formula:

Solution A: Stock

Ferric Ammonium sulfate crystals	15 pt
Ferrous sulfate crystals	15 pt
Distilled Water	100 pt

Solution B: Stock

Hematoxylin	1 pt
Alcohol (95%)	50 pt
Glycerol	50 pt

Solution C: Fluid

Solution A	1 pt
Solution B	1 pt

Procedure:
1. Stain for about 5 min.

4.3 **Single contrast staining – General Stains**

4.3.1 *Bromophenol blue [6]*

Procedure:
1. Stain slides in bromophenol for 1 hr at room temperature.
2. Immerse in aqueous solution of 0.5% acetic acid for 5 min.
3. Immerse in tertiary butyl alcohol for 5 min.
4. Clear in xylol and mount in Canada balsam.

4.3.2 *DMB technic [7]*

Application:
Recommended for the detection of sulfated compounds. Polysaccharide glycosaminoglycan chains appear purple.

Formula:

1,9-dimethylmethylene blue (DMB)	1 pt
1N HCl	1 pt
0.01 M Glycine	4 pt
0.01 M NaCl	4 pt

4.3.3 *Nile blue [6]*

Procedure:
1. Stain in Nile blue for 5 min at 37°C.
2. Rinse in tap water.
3. Immerse in 1% acetic acid for 1 min.

4.3.4 *Oil red O [8]*

Application:
For the detection of lipids.

Formula:

Solution A:

Oil Red O Solution	0.5 pt
Isopropyl Alcohol	99.5 pt

Solution B:

Dextrin	1 pt
Distilled Water	100 pt

Solution C: Staining medium

Solution A	60 pt
Solution B	40 pt

Procedure:
1. Freeze dissected tissue in dry ice cold isopentane.
2. Section tissue.
3. Prepare solution C fresh.
4. Stain in solution C for 20 min at room temperature.
5. Rinse briefly in running distilled water.
6. Counterstain in hematoxylin for 30 sec.

4.3.5 *Crystal violet [9]*

Application:
Staining and observation of hemocytes.

Formula:

Crystal Violet	0.5 pt
NaCl	1.09 pt
KCl	0.157 pt
$CaCl_2$	0.085 pt
$MgCl_2$	0.017 pt
Distilled Water	to 100 pt
Glacial Acetic Acid	to pH 2.9

Procedure:
1. Immerse in stain/fixative until color is differentiated.

4.3.6 *Safranin [2]*

Formula:

Solution A: Aniline water

Anilin Oil	4 pt
Distilled Water	90 pt

Preparation:
1. Mix and shake well.
2. Filter through wet filter.

Solution B: Stain

Safranin	1 pt
Anilin Water	90 pt
Alcohol (95%)	10 pt

4.3.7 *Gentian violet [2]*

Formula:

Solution A: Stain

Gentian violet	1 pt
Anilin water (see above)	80 pt
Alcohol (95%)	20 pt

Solution B: Gram's

Iodine	1 pt
Iodine of Potassium	2 pt
Distilled Water	300 pt

Procedure:

1. Stain sections for 5 to 10 min.
2. Transfer to solution B for 2 to 3 min or until black.
3. Decolorize in absolute alcohol until gray.

4.3.8 *Safranin-gentian violet technic [2]*

Application:

Recommended for specimens fixed in Flemming's.

Procedure:

1. Stain sections in safranin for 36 to 48 hr.
2. Differentiate in alcohol.
3. Stain in gentian violet for 5 to 10 min.
4. Transfer to gram's solution (see above) for 1 to 3 hr.
5. Differentiate in absolute alcohol until clouds cease coming off from sections.
6. Place a few drops of clove oil on sections.
7. Transfer to xylol and mount.

4.3.9 *Endogenous alkaline phosphatase [10]*

Application:

Recommended for the detection of endogenous alkaline phosphatase.

Formula:

BCIP/NBT labeling mix:

BCIP	0.17 mg/ml
NBT	0.33 mg/ml
Tris-HCl pH 9.5	100 mM
NaCl	100 mM
$MgCl_2$	5 mM

Protocol:

1. Collect whiteflies (adults and nymphs) and aphids in 70% ethanol.
2. Fix overnight in ethanolic formaldehyde (3:1 ratio ethanol:formaldehyde).
3. Dehydrate through a graded series of ethanols and clear.
4. Infiltrate and embed in paraffin.
5. Cut 10 µm sections and mount onto glass slides.
6. De-paraffinize and hydrate through xylenes and a graded ethanol series into water.
7. Stain with BCIP/NBT labeling mix until color develops.
8. Dehydrate, clear and coverslip slides.

4.4 **Single contrast staining – Golgi**

4.4.1 *Golgi technic [3]*

Formula:

Solution A:

Potash Solution (3%)	3 pt
Osmic acid Solution (1%)	1 pt

Solution B: Silver Nitrate Solution (0.5–1%).

Procedure:

1. Fix in solution A.
2. Wash in solution B.
3. Embed in celloidin.

Result:

1. Nerve cells and tracheoles are black, other tissues transparent.

4.4.2 *Modified Colonnier-rapid golgi method [11, 12]*

Application:

Demonstrates arbors of Kenyon cells and other neuronal cells.

Formula:

Solution A:

Potassium dichromate	2.5 pt
Sucrose	1.3 pt
Distilled Water	Up to 100 pt

Solution B:

Solution A	5 pt
25% Glutaraldehyde	1 pt

Solution C:

2.5% Potassium dichromate	100 pt
1% Osmium tetroxide	1 pt

Procedure:

1. Dissect tissue in solution A.
2. Fix tissue in solution B for 6 days at 4°C.
3. Rinse tissue in 2.5% potassium dichromate.
4. Stain tissues in solution C for 5 days at 4°C.
5. Rinse in distilled water.
6. Stain in 0.75% silver nitrate for 3 days.
7. Dehydrate, clear, section, and mount.

4.5 **Single contrast staining – Eggs**

Application:

For the identification of forensically important fly eggs.

4.5.1 *Potassium permanganate technic [13]*

Formula:

Potassium Permanganate	1 pt
Distilled Water	99 pt

Procedure:

1. Immerse eggs in potassium permanganate solution for 1 min.
2. Remove stain.
3. Dehydrate eggs through alcohols.
4. Clear in xylenes.
5. Mount onto glass slide.

4.5.2 *Thionine staining [14]*

Application:
For the morphological observation of embryonic stages.

Formula:

Thionine	0.3 pt
70% Ethanol	100 pt

Procedure:

1. Remove egg chorion with fine forceps in Carnoy's solution.
2. Stain with 0.3% thionine.
3. Dehydrate through graded ethanols.

4.6 **Single contrast staining – Silver Stains**

4.6.1 *Bodian silver technic [15, 16]*

Formula:
Reducing Solution:

Hydroquinone	0.5 pt
Sodium sulfite	2.5 pt
Distilled Water	50 pt

Copper Sheet:

Copper Sheet	5 pt (1/2″)
Hydrochloric Acid	1N

Preparation:
1. Wash copper sheet in 1N hydrochloric acid, rinse in tap water and distilled water.

Formula:
Protargol-S Solution:

Protargol-S	0.5 pt
Distilled Water	50 pt
Copper sheet, acid cleaned	

Preparation:
1. Add copper sheet and distilled water in a coplin jar. Sprinkle Protargol-S on top of water. Do not mix; allow solution to sit undisturbed until dissolved.

Formula:
0.5% Oxalic Acid:

Oxalic acid	1 pt
Distilled Water	200 pt

Formula:
5% Hypo:

Sodium thiosulfite	5 pt
Distilled Water	100 pt

Procedure:

1. Deparaffanize and hydrate to distilled water.
2. Immerse in Protargol solution at 37°C overnight.
3. Rinse quickly in distilled water.
4. Immerse in reducing solution, 5 min.
5. Wash in tap water, rinse in distilled water.
6. Expose to 0.5% gold chloride, 3–5 min, or until gray.
7. Rinse in distilled water.
8. Expose to 0.5% oxalic acid, 10 min, observe under the microscope for black fibers.
9. Rinse in distilled water.
10. Immerse in 5% hypo, 3 min.
11. Wash in tap water.
12. Dehydrate, clear in xylene, coverslip.

4.6.2 *Cajal's block silver stain [17]*

Procedure:

1. Fix tissue in 70% ethanol.
2. Postfix in 5% acetic, 85% ethyl alcohol, 3.7% formaldehyde and 6.3% water solution for 2–4 hr.
3. Wash in 70% ethanol.
4. Immerse in ammoniacal alcohol (2 pt ammonium hydroxide with 98 pt 100% ethanol) for 24 hr.
5. Rinse in distilled water.
6. Transfer to 4% silver nitrate and keep in the dark for 3–5 days.
7. Develop stain in aqueous 4% pyrogallol solution for 6 hr.
8. Dehydrate, embed in resin and section.

4.6.3 *Holmes-blest technic [18]*

Formula:

Silver Nitrate I:

Silver Nitrate	2 pt
Distilled Water	8 pt

Silver Nitrate II:

M/5 boric acid	27.5 pt
M/20 borax	22.5 pt
1% silver nitrate solution	5 pt
Pyridine or derivative	2 pt
Distilled Water	250 pt
pH to 8.4	

Hydroquinone:

Hydroquinone	3 pt
Sodium sulphite	30 pt
Distilled water	300 pt

Procedure:

1. Fix insects in alcoholic Bouin's for 24 hr.
2. Wash in 50% alcohol, dehydrate through graded alcohols, and embed in wax.
3. Section, de-wax, and rehydrate in graded alcohols.
4. Immerse in Silver Nitrate I for 2 to 3 hr.
5. Rinse in distilled water 5 min.
6. Incubate in Silver Nitrate II at 37°C.
7. Immerse in Hydroquionone solution at 60°C.
8. Wash in running tap water 3 min.
9. Rinse in distilled water 1 min.
10. Tone in 0.2% sodium gold chloride 1 to 3 min.
11. Rinse in distilled water, 1 min.
12. Reduce in 2% oxalic acid, 5 min.
13. Wash in distilled water, 3 min.
14. Fix in 5% sodium thiosulphate, 1 min.
15. Wash thoroughly in distilled water, dehydrate, and mount.

4.6.4 *Uranyl acetate and lead citrate staining [19–21]*

Application:
Recommended as a general contrast stain for ultrathin sections.

Formula:

Lead Citrate	
$Pb(NO_3)_2$	1.33 pt
$Na_3(C_6H_5O_7)$	1.76 pt
Distilled Water	30 pt

Preparation:

1. Shake vigorously for 1 min.
2. Add 8 pt 1N NaOH and 50 pt distilled water and mix.

Procedure:

1. Float grid on a drop of 2% uranyl acetate for 30 min.
2. Incubate in lead citrate for 10 min.

4.6.5 *Wigglesworth's osmium-ethyl gallate method [17]*

Procedure:

1. Fix tissue in 4% formaldehyde in phosphate buffered saline.
2. Rinse in buffer.
3. Post-fix in 0.5% Osmium Tetroxide.
4. Rinse in buffer, then distilled water.
5. Treat with saturated ethyl gallate (e.g., a salt or ester of gallic acid).
6. Rinse, dehydrate and embed in resin.

4.7 **Polychrome staining techniques – General**

4.7.1 *Rae's modified masson trichrome [22]*

Formula:

Solution A: Groat's Hematoxylin

Water	50 pt
Ferric Ammonium Sulfate	0.8 pt
Sulfuric Acid	50 pt
Ethyl Alcohol (95%)	50 pt
Hematoxylin	0.5 pt

Solution B:

Xylidene Ponceau	0.25% in 1% aqueous acetic acid

Solution C:

Phosphomolybdic Acid	1% aqueous solution

Solution D:

Light Green SF	2% in 1% aqueous acetic acid

Preparation:
1. Dissolve ingredients in water in order listed.
2. Filter.

Note:
1. Peak performance is attained with 5 hr and maintained for 4 to 8 weeks.

Procedure:
1. Transfer to Groat's Hematoxylin for 12 min.
2. Transfer to solution B for 3 ½ min.
3. Rinse in distilled water.
4. Transfer to solution C for 4 min.
5. Rinse in distilled water.
6. Transfer to solution D for 3 min.
7. Rinse in distilled water.
8. Dehydrate twice with absolute alcohol (quickly).
9. Clear and mount (Canada balsam).

Result:
1. Nuclei appear brownish purple.
2. Connective tissue appears green.
3. Muscle displays green striations on reddish-pink or green background.
4. Nervous tissue appears pink and greenish gray.

4.7.2 *Kramer's technic [23]*

Application:
Recommended for differential staining of insect whole mount.

Procedure:
1. Fix in Kramer's Bouin's.
2. Transfer to alcohol (50%) for 10 min.
3. Transfer to alcohol (70%) for 1 hr.
4. Transfer to alcohol (95%) for 10 min.
5. Transfer to eosin for 6 to 8 hr or until uniform pink color is achieved.
6. Transfer to alcohol (95%) for 30 min.
7. Slowly add (by drops) methyl salicylate in alcohol.
8. Transfer to methyl salicylate.

Note:
1. Methyl salicylate (step 7) must be added at a rate that will prevent collapse of specimen.
2. Time periods suggested above are calculated for 4-day-old house fly larvae.

Result:

1. Muscle appears pink and other tissues appear green.

4.7.3 *Lower's trichrome stain [24]*

Application:

Recommended as a procedure providing optimal differentiation of insect tissues under bright field, dark field, and phase-contrast.

Formula:

Solution A:

Azocarmine G	1% (aqueous)
Glacial Acetic Acid	1 pt added drop by drop with constant stirring

Solution B:

Orange G	3 pt
Acetic Acid (1% aqueous)	100 pt

Preparation:

1. Add orange G to 100 pt of acetic acid.
2. Let stand one week with frequent agitation.
3. Let particles settle and decant.

Solution C:

Phosphotungstic Acid	5% (aqueous)

Solution D:

Methyl green	0.1 pt dissolved in 100 pt acetic acid (1% aq).

Solution E:

Camsal (equal parts camphor and phenol salicylate)	4 pt
Eucalyptol	8 pt
Dioxane	8 pt
Paraldehyde	4 pt

Procedure:

1. Transfer deparaffinized sections to water.
2. Rinse in 1% acetic acid.
3. Leave in solution A for 15 min or longer.
4. Rinse in 1% acetic acid.
5. Transfer to solution B for about 20 sec.
6. Wash in 1% acetic acid (examine under microscope).
7. Repeat procedures until azocarmine is removed from endocuticle and turns yellow.
8. Rinse in water.
9. Transfer to solution C for 3 min.
10. Rinse in water and then in 1% acetic acid.
11. Quickly drop in solution D (10 sec).

12. Rinse in water.
13. Repeat to desired effect: when endocuticle is green (about 30 sec) it is usually sufficient time.
14. Differentiate in 95% alcohol (two 2 min changes).
15. Clear in solution E.
16. Mount in Mohr and Wehrle's medium (Euparal type mountant).

4.7.4 *Modified gomori's [25]*

Application:
Recommended for mosquitoes and originally suggested for crustaceans.

Formula:
Solution A:

Azocarmine	0.1 pt
Glacial Acetic Acid	2 pt

Preparation:
1. Dissolve Azocarmine in 100 pt distilled water.
2. Boil for 5 min.
3. Allow to cool and add acetic acid.
4. Filter.

Solution B:

Distilled Water	1000 pt
Phosphotungstic Acid	10 pt
Aniline Blue	1.2 pt
Orange G	4.4 pt

Procedure:
1. Transfer from xylene through series of alcohol solutions to water.
2. Transfer to solution A for 15 min.
3. Rinse in distilled water.
4. Transfer Aniline (1% in 95% alcohol).
5. Transfer to distilled water.
6. Transfer to solution B for 15 min to counterstain.
7. Transfer through three changes of absolute alcohol for 1, 1 and 10 min.
8. Transfer through two changes of xylene.
9. Mount in Piccolyte (a beta-pinene polymer).

Note:
1. Bouin's picro-formalin or any alcoholic modification is recommended as a fixative.

Result:
1. In crustacean tissues, epicuticle appears red, endocuticle appears blue, epidermal cells appear yellowish, cytoplasm appears pink with red-orange nuclei, muscle appears bluish-purple, etc.

4.7.5 *Phosphotungstic hematoxylin [26]*

Application:
Recommended as a polychrome stain for connective tissue and muscle.

Formula:

Hematoxylin Crystals (dissolved in 80 pt water)	0.1 pt
Phosphotungstic Acid (10% solution)	20 pt
Hydrogen Peroxide	0.2 pt

Preparation:

1. Dissolve Hematoxylin under heat.
2. Add phosphotungstic acid then add hydrogen peroxide.
3. Let stand several weeks.

Procedure:

1. Transfer sections through an alcohol series to water.
2. Transfer to potassium permanganate (0.25%) for 4 min.
3. Wash in distilled water.
4. Transfer to oxalic acid (5%) for 5 to 10 min.
5. Transfer through five changes of distilled water for 1 min each.
6. Transfer to stain for 12 to 25 hr.
7. Wash in distilled water (rapidly).
8. Transfer to alcohol (95%) for 1 min.
9. Transfer to xylol for 3 min.
10. Mount in balsam.

4.7.6 *Baker staining [6]*

Formula:

Solution A:

Calcium dichromate	5 pt
Anhydrous calcium chloride	1 pt
Distilled Water	Up to 100 pt

Solution B: Acid Hematein

Crystallized hematoxylin	1 pt
1% aqueous sodium iodate	2 pt
Acetic Acid	2 pt
Distilled Water	95 pt

Procedure:

1. Stain with calcium dichromate solution for 18 hr.
2. Wash three times in distilled water.
3. Counterstain with acid hematein for 5 hr at room temperature.
4. Wash three times in distilled water.

4.7.7 *Periodic acid schiff's stain [6]*

Procedure:

1. Stain with alcian blue (pH 2.5) for 30 min.
2. Wash in distilled water.
3. Immerse in 1% periodic acid for 5 min.
4. Wash in distilled water.
5. Immerse in Schiffs reagent for 30 min.
6. Wash in sulfur water for 1 min.
7. Wash in tap water 10 min.

8. Counterstain with Harris hematoxylin for 2 min.
9. Wash in tap water.
10. Dehydrate, clear in xylol and mount in Canada balsam.

4.7.8 *Toluidine blue-azur II [27]*

Application:
Recommended for staining mast cells.

Formula:

Toluidine Blue	1 pt
Azur Blue	1 pt
Distilled Water	98 pt

4.7.9 *Methylene blue-azur [28]*

Formula:

1% Methylene Blue	1 pt
1% Azur Blue	11 pt
Sodium Tetraborate	1 pt

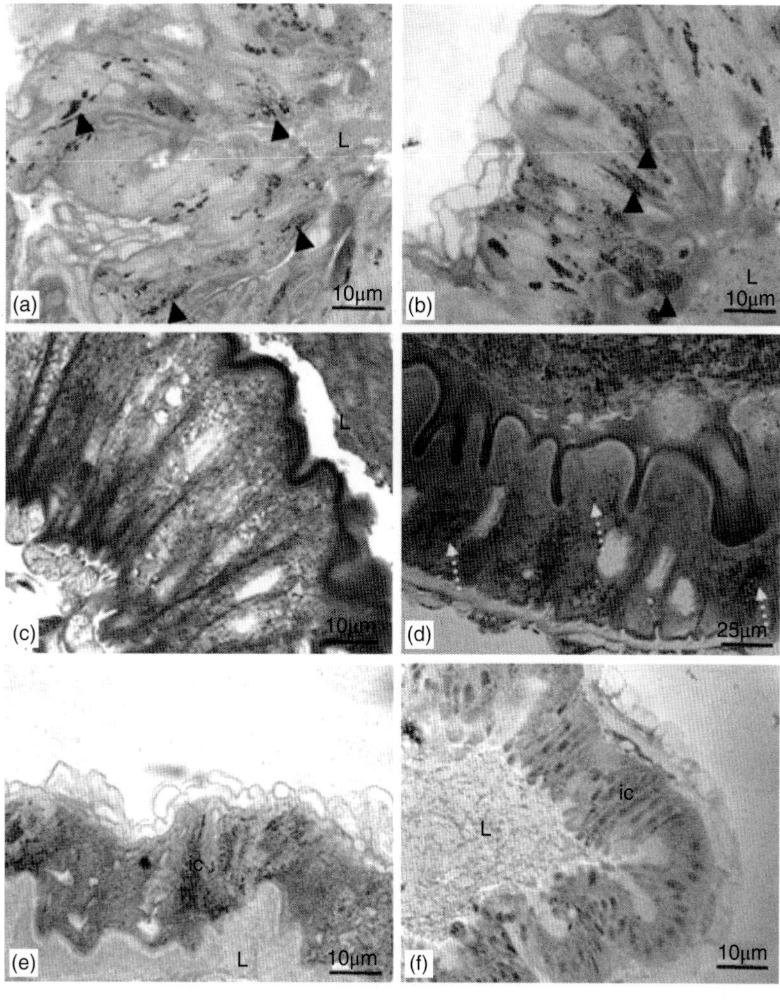

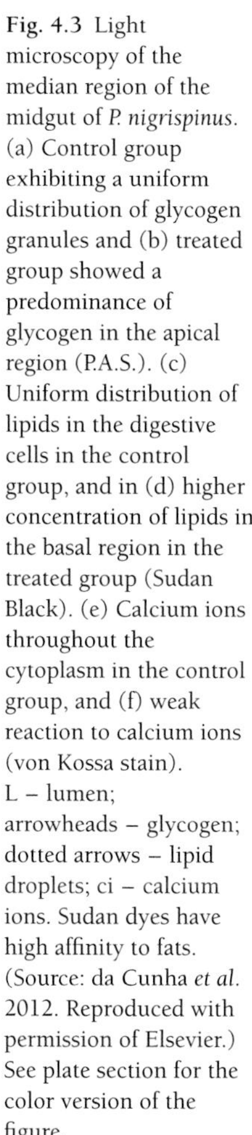
Fig. 4.3 Light microscopy of the median region of the midgut of *P. nigrispinus*. (a) Control group exhibiting a uniform distribution of glycogen granules and (b) treated group showed a predominance of glycogen in the apical region (P.A.S.). (c) Uniform distribution of lipids in the digestive cells in the control group, and in (d) higher concentration of lipids in the basal region in the treated group (Sudan Black). (e) Calcium ions throughout the cytoplasm in the control group, and (f) weak reaction to calcium ions (von Kossa stain). L – lumen; arrowheads – glycogen; dotted arrows – lipid droplets; ci – calcium ions. Sudan dyes have high affinity to fats. (Source: da Cunha *et al.* 2012. Reproduced with permission of Elsevier.) See plate section for the color version of the figure.

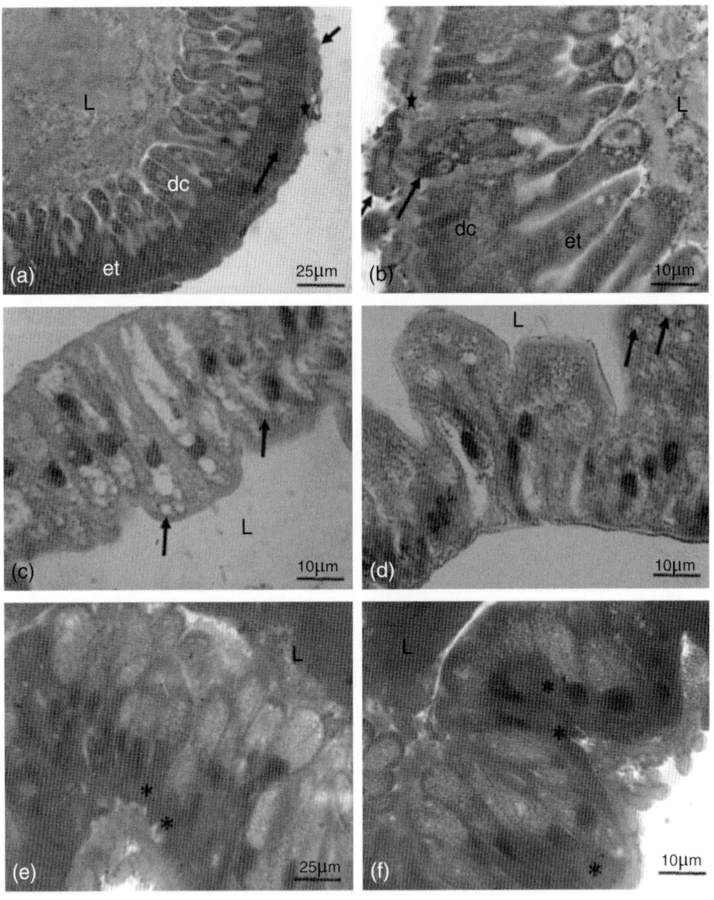

Fig. 4.4 Light microscopy of the median region of the midgut of *P. nigrispinus* control group (a) and treated (b). Toluidine blue: take a positive reaction for carboxylated glycosaminoglycans in the apical region of cells for both control (c) and treated (d). Alcian Blue: protein compounds in the basal region of digestive cells in the control group and treated group (e and f, respectively). Bromophenol Blue: Ep – epithelium simple; dc – digestive cell; long arrow – cell regenerative; short arrow – longitudinal muscle; star – circular muscle; L – lumen; dotted arrows – droplets of glycosaminoglycans; arrowhead – nucleus; asterisks – protein compounds. (Source: da Cunha *et al.* 2012. Reproduced with permission of Elsevier.) See plate section for the color version of the figure.

Procedure:

1. Stain slides for 1 min.

4.7.10 *Methylene blue-azur II [29]*

Formula:

Methylene Blue	1 pt
Azur Blue	1 pt
Borax	1 pt
Distilled Water	97 pt

4.7.11 *Kluver and Barrera Luxol Fast Blue [30]*

Formula:

Solution A: Luxol Fast Blue

Luxol fast blue	1 pt
Methanol	1000 pt
10% Acetic Acid	5 pt

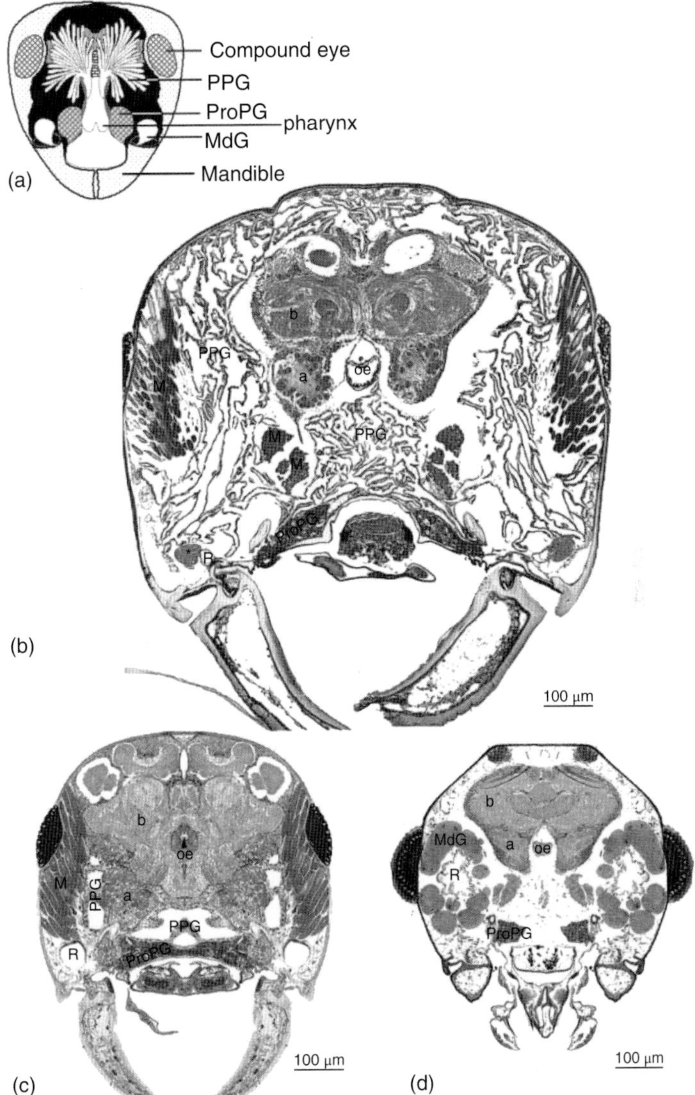

Fig. 4.5 Histological aspects of the head glands of *Lasius niger*. (a) Schematic drawing of an ant head showing the position of the mandibular gland (MdG), the propharyngeal gland (ProPG) and the postpharyngeal gland (PPG) inside the head capsule. (b–d) Frontal sections of the head showing consistent differences in gland size between castes. The postpharyngeal glands (PPG) occupy half of the entire head in queens (b) and the mandibular glands (MdG) are particularly hypertrophied in males (d). The propharyngeal glands (ProPG) touch each other on the axial region of the head only in workers (c). Abbreviations: a = antennal lobe, b = brain, M = muscle, oe = oesophagus, R = reservoir of the mandibular gland, * = mandibular gland in queens, MdG = mandibular gland. Double-staining with 1% azur blue and 1% methylene blue (11:1) in 1% sodium tetraborate for 1 min and washing with distilled water. (Source: Niculita *et al.* 2007. Reproduced with permission of the authors.) See plate section for the color version of the figure.

Preparation:
1. Mix and filter before use.

Solution B: Cresyl Violet

Cresyl Violet	0.5 pt
Distilled Water	100 pt

Preparation:
1. Mix and filter before use.

Solution C: Cresyl Violet Differentiator

Alcohol	100 pt
Glacial Acetic Acid	0.25 pt

Procedure:
1. Rehydrate sections to 95% alcohol.
2. Stain in solution A for 2 hr at 60°C or overnight at 37°C.
3. Wash in 70% alcohol.
4. Wash in tap water.
5. Differentiate in saturated lithium carbonate solution.
6. Wash in tap water.
7. Stain with solution B for 10–20 min.
8. Wash in tap water.
9. Differentiate in solution C for 4–8 sec.
10. Dehydrate, clear and mount.

4.7.12 *Eosin and methylene blue [26]*

Procedure:
1. Transfer deparaffinized sections to water for 10 min.
2. Transfer to 3 to 5% eosin (aq. solution) for 20 min.
3. Transfer to distilled water for 1 min.
4. Transfer to polychrome methylene blue for 15 min.
5. Remove excess stain in distilled water.
6. Differentiate in 95% alcohol (with several drops of 10% solution of eosin added).
7. Quickly dip in absolute alcohol when nuclei are blue and pink color returns.
8. Transfer to xylol and mount in balsam.

4.7.13 *Delafield-eosin technic [2]*

Application:
Recommended for staining cytoplasm of glandular, embryonic and rapidly growing tissues of insects.

Procedure:
1. Transfer slides to xylol for 15 to 20 min to remove paraffin.
2. Transfer to fresh xylol for 1 to 2 min.
3. Transfer to absolute alcohol for 3 to 6 min.
4. Transfer through 95, 70, and 30% alcohol for 1 min each.
5. Transfer to solution A for 5 min.
6. Rinse in water.
7. Transfer to alcohol (35%) for 2 min.
8. Transfer to alcohol (70%) for 1 min.

9. Transfer to alcohol (70%) with 0.1% to 1.0% HCL (acid alcohol) until sections are reddish.
10. Rinse in alcohol (70%) with a few drops of 0.1% bicarbonate of soda (alkaline alcohol) until sections are blue.
11. Transfer to alcohol (95%) for 3 min.
12. Transfer to eosin in alcohol (95%) for 5 min.
13. Dip into alcohol (95%).
14. Transfer to absolute alcohol for 5 min.
15. Transfer to xylol for 10 min or until clear.
16. Mount in balsam.

Note:
1. Changes in steps 5, 9, and 10 may be observed under a microscope.
2. If eosin does not adhere to sections in steps 13 and 14 add 2 to 3 drops of (5%) acetic acid.

4.7.14 *Levy's technic [31]*

Application:
1. Recommended for tissues that have been in 70% alcohol for long periods.
2. Recommended originally for mayfly tissue.

Formula:
Solution A: Methylene blue saturated solution in dioxane (95%)

Solution B: Erythrosin, saturated solution in dioxane (95%)

Procedure:
1. Transfer to xylene for 2 to 3 min.
2. Transfer to xylene for 1 min.
3. Transfer to dioxane (100%) for 2 to 3 min.
4. Transfer to dioxane (100%) for 1 min.
5. Transfer to solution A until tissues turn blue-green.
6. Wash in dioxane (100%) for 5 sec.
7. Transfer to solution B until desired contrast is achieved.
8. Wash in dioxane (100%) for 15 sec.
9. Transfer to dioxane (100%) for 2 to 3 min.
10. Transfer to xylene for 2 to 3 min.
11. Transfer to xylene for 1 min.
12. Mount in gum damar.

Note:
1. For best results, solution A should be at least a week old.
2. All 100% dioxane solutions should be changed frequently.

Result:
1. Cell nuclei of connective, epithelial, fatty and glandular tissues appear blue.
2. Cytoplasm, muscle fibers, nerve cord and hypodermal layer appear red-pink.
3. Epicuticle appears black.
4. Exocuticle appears yellow to green.

4.8 **Polychrome staining – Brain/Nerve**

4.8.1 *Kingsbury and Johannsen technic [32]*

Application:
Recommended for staining of insect brain tissue.

Formula:
Solution A: Stock

Phosphomolybdic Acid (10% solution)	1 pt
Hematoxylin Crystals	1 pt
Chloral Hydrate	6–10 pt
Water	100 pt

Procedure:
1. Fix in formalin (10%) for an average of 6 hr (range of 4 to 20 hr).
2. Dehydrate, clear in xylene and embed in paraffin.
3. Stain in eosin for 20 to 30 sec.
4. Stain in Delafield's Hematoxylin for 30 sec.
5. Wash in water.
6. Transfer to phosphomolybdic acid (1%) for 2 to 3 min.
7. Wash in water.
8. Dehydrate, clear, and mount.

Modification A:
1. Fix in formalin (10%) for an average of 6 hr (range of 4 to 20 hr).
2. Wash in water.
3. Transfer to copper sulfate (5%) for 24 hr.
4. Wash, dehydrate, and embed.
5. Stain in solution A.
6. Dehydrate, clear, and mount.

Note:
1. Other fixatives may be used, but those containing alcohol should be avoided.

4.8.2 *Buxton's picro-nigrosin [33]*

Application:
Recommended for the histological display of tracts of nerve fibers in insect brains.

Formula:

Nigrosin (1% aqueous)	1 pt
Picric Acid (saturated aqueous)	9 pt

Procedure:
1. Fixation with osmic acid or fixatives containing osmic acid is suggested for best results.
2. Counter-staining and over-staining should be avoided, because washing stain out with acid alcohol is difficult.

4.8.3 *Mann's stain [33]*

Application:
Recommended for the histological display of insect nerve cells.

Formula:

Methyl Blue (1% aqueous)	35 pt
Eosin	35 pt
Water	100 pt

Procedure:
1. Fix material in Bouin's, Gilson's or a picro-chloracetic mixture.
2. Proceed with appropriate post-fixation procedure.
3. Stain for 10 min or more.
4. Wash in alcohol (with 1% pyridine), in which it will destain slowly.

Note:
1. This stain is not proper for materials fixed in osmic fixatives.
2. Length of staining is not critical since over staining does not occur.

4.9 Polychrome staining – blood

The following staining techniques have been recommended for blood cells of insects.

4.9.1 *Dubuscq [25]*

Formula:

Water	250 pt
Copper Chloride	0.6 pt
Copper Acetate	0.6 pt
Osmic Acid	0.6 pt
Acetic Acid	0.6 pt
Thionin	0.6 pt

Procedure:
1. Mix equal parts of stain and blood.
2. Let stand for 5 min.
3. Make smear.

4.9.2 *Yeager's Wright-nicotine oxalic acid technic*

Formula:

Solution A:

Wright's Blood Stain	Undiluted

Solution B:

Wright's Blood Stain	Diluted with equal volumes of next solution
Nicotine Solution	2.5% (aqueous)

Solution C:

Oxalic acid	0.0006% (in distilled water)

Preparation:

1. Wright's stain is prepared as usual by dissolving dye powder in pure methyl (acetone free) alcohol but twice the usual concentration (generally 0.2–0.3 pt) is used. Solution is refluxed (6 to 7 hr) and allowed to cool and sediment to settle.

Procedure:

1. Immerse insect in hot water (55–60°C) for 5 min.
2. Dry insect thoroughly, cut leg and collect a large drop of blood on a slide.
3. Make a smear of the blood and let dry.
4. Flood smear with Wright's stain (undiluted) for 1.5 min.
5. Flood with Wright's stain (diluted) and wash with four medicine droppers full of distilled water.
6. Blot smear carefully and dry over small flame.
7. Differentiate with ethyl alcohol (70%).
8. Blot smear and dry over small flame.
9. Flood with oxalic acid solution for 1 to 2 min.
10. Blot smear and dry over small flame.
11. Mount in balsam.

Note:

1. Technique may require slight adaptation to particular insect blood or Wright's blood stain.

4.9.3 *HEMA3 staining of mosquito hemocytes [34]*

Formula:

Solution A: Citrate buffer with anticoagulant

1M NaOH	98 pt
1M NaCl	186 pt
0.5M EDTA	3.4 pt
1M Citric acid, pH 4.5	41 pt
dH_2O	Up to 1000 pt

Solution B: Hemolymph diluent

Schneider's Medium	60 pt
Fetal Bovine Serum	10 pt
Solution A	30 pt

Procedure:

1. Prepare microinjector system and pull needles according to manufacturer's protocol.
2. Encircle a small area (1 cm diameter) on a clean microscope slide to denote area where hemolymph will be collected.
3. Cold-anesthetize adult female mosquitoes for about 8 min, or until they are no longer moving.
4. Place one mosquito onto a depression slide for microinjection.
5. Gently holding the mosquito in place, inject 3–3.5 µl of the hemolymph diluent into the mosquito between the seventh and eighth abdominal segments. Following injection, the abdomen should look "filled" or bloated.
6. Place injected mosquitoes into a separate clean container on ice to recover for 5 min.
7. After the recovery period, with the mosquito still on ice, snip off legs, wings and the tip of the abdomen (at the eighth segment) using microscissors.
8. Place clipped mosquito onto the marked microscope slide and inject an additional 8–10 µl of hemolymph diluent into the mosquito on the lateral side of its mesothorax.

9. Position the abdomen so that it is just inside the edge of the encircled area and gently squeeze the abdomen with forceps to direct the hemolymph into the marked area.
10. Allow hemolymph to dry on slide for about 10 min or until visibly dry.
11. Stain hemocytes with HEMA3 staining kit from Fisher® (Cat. 123–869).
12. Dip slide in fixative five times 1 sec each.
13. Dip slide in solution A three times 1 sec each.
14. Dip slide in solution B three times 1 sec each.
15. Rinse the back of the slide with dH_2O, do not rinse the front of the slide.
16. Mount in Shur/Mount or Polymount or similar mounting medium, cover with cover glass.

Note:
1. When cutting off the legs and wings, be sure to not cut too close to the thoracic attachment sites to avoid introducing additional openings for hemolymph to escape.

4.10 Single contrast procedures for chitinous material

The following staining procedures have been recommended for the staining of chitin and the insect exoskeleton.

Material that has been treated with KOH can be stained with (a) acid fuchsin, (b) one-quarter per cent picric acid solution (aq.) or (c) gentian violet. Occasionally chitinous materials will also be stained by (a) Hematoxylin and (b) acid anilin dyes.

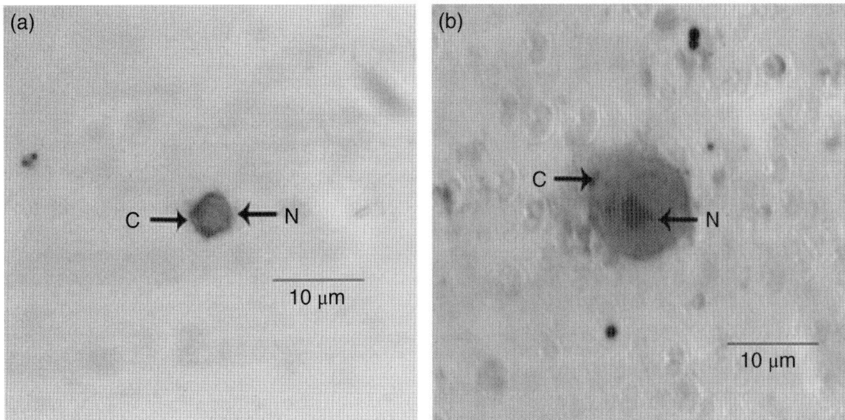

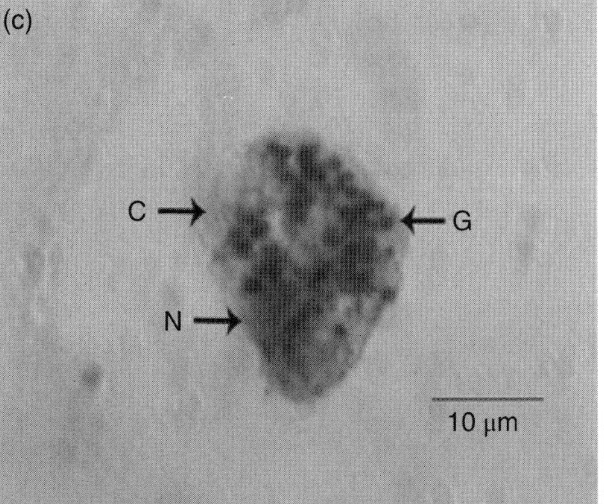

Fig. 4.6 Light microscope images of a representative HEMA 3™ (which is similar to Giemsa stain) fixed, stained prohemocyte (a), oenocytoid (b), and granulocyte (c) from an *Ae. aegypti* adult female. C = cytoplasm, N = nucleus G = granules. (Source: Qayum & Telang 2011. Reproduced with permission of the authors.) See plate section for the color version of the figure.

4.10.1 *Kingsbury and Johannsen's pyrogallic technic [32]*

Procedure:
1. Treat material with caustic potash.
2. Stain with alcoholic or glycerine solution of pyrogallic acid.
3. Differentiate by bleaching with weak acid.

4.10.2 *Nuttall's [23]*

Application:
Recommended for potash-cleared specimens.

Formula:

Xylene	100 pt
Picric Acid	To saturation (about 10%)

4.10.3 *Azariah's modified lugol's iodine technic [35]*

Application:
Recommended for demonstration of chitin.

Formula:
Lugol's Iodine

Iodine	1 pt
Potassium Iodine	2 pt
Distilled Water	300 pt
pH	1.4 to 4.0

Procedure:
1. Heat in saturated potassium hydroxide at 160°C for 45 min.
2. Wash in running water.
3. Immerse in Lugol's Iodine.

4.11 Polychrome staining procedures for chitinous material

4.11.1 *Chatton's eosin Y-light green [23]*

Formula:
Solution A:

Alcohol (95%)	100 pt
Light Green	1 pt
Eosin Y	2 pt

Preparation:
1. Dissolved with occasional shaking for several days.
2. Filter.

Solution B:

Acetic acid (in absolute alcohol)	5%

Procedure:
1. Transfer to alcohol (95%).
2. Transfer to solution A for 5 min.
3. Transfer to solution B (until connective tissue is distinctly green).
4. Clear and mount (balsam).

Result:
1. Green chitin on a red background.

4.11.2 *Trichrome stain [36]*

Procedure:

1. Fix aphids for a minimum of 24 hr at 4°C in Duboscq-Brazil's fluid (see Chapter 2).
2. Embed specimens in agar–agar, dehydrate through a graded series of 95 and 100% ethyl alcohol (two times, 10 min in each), and in butylic alcohol (>24 h).
3. Embed specimens in paraffin wax and section.
4. Melt sections onto cleaned slides, de-wax in xylene and rehydrate in 100% and 95% ethyl alcohol (two times each) to water.
5. Stain sections with 0.1% toluidine blue for 5 min at 0°C, followed by Ponceau red solution for 2–4 min.
6. Rinse in dH_2O.
7. Incubate in 5% phosphotungstic acid for 15–20 min, then in Heidenham blue solution for 2–6 min.
8. Dehydrate sections in 95% and 100% ethyl alcohol 2 min each, clear in toluene and mount in xylene-dammar.

4.11.3 *Semichon methyl blue-Eosin-Victoria yellow [23]*

Formula:

Water	100 pt
Methyl Blue	0.04 pt
Eosin Y	0.2 pt
Victoria Yellow	0.1 pt

Procedure:

1. Transfer to water.
2. Transfer to stain overnight.
3. Drain and transfer to absolute alcohol for differentiation.
4. Transfer to xylene.
5. Mount in balsam.

Result:

1. Chitin appears yellow and other tissues orange and blue.

4.11.4 *Mallory's triple [23]*

Formula:

Solution A:

Acid Fuschin (aqueous)	1/5% solution

Solution B:

Grubler's water soluble aniline-blue	0.5 pt
Orange G	2 pt
Phosphomolybdic acid (1% aqueous solution)	100 pt

Procedure:

1. Fix in Bouin's or Zenker's.
2. Transfer to solution A for 5 min.
3. Without washing, transfer to solution B for 20 min.
4. Wash, differentiate, and dehydrate with 95% and absolute alcohol.
5. Clear in xylene and mount in balsam.

4.11.5 *Gier's Giemsa [37]*

Application:

1. Recommended for detecting early chitin formation, chitinous linings, and egg capsules.
2. Recommended for differential staining of tissues associated with chitin.

Formula:

Solution A: Stock

Giemsa (powder)	1 pt
Glycerine	66 pt
Methyl Alcohol (acetone-free)	66 pt

Preparation:

1. Dissolve powdered stain in glycerine at 60°C for 24 hr.
2. Add methyl alcohol.

Solution B: Working Fluid

Distilled water	50 pt
$NaHCO_3$ (0.1% solution)	0.5 pt

Preparation:

1. Mix 1 pt solution A with water.
2. Add $NaHCO_3$.

Procedure:

1. Wash slide in water.
2. Place in working solution for about 1 hr.
3. Rinse in water.
4. Differentiate in acetone (about 30 sec).
5. Dip in xylol to stop further differentiation.
6. Mount as temporary or permanent preparation.

Notes:

1. Commercially available Giemsa blood stain or Azure II-eosin can be substituted for the stock solution.
2. General fixatives (except osmic acid fixatives) are compatible with this staining regime.
3. Tap water (pH 7–8) can be substituted for distilled water plus $NaHCO_3$ (see preparation).
4. Giemsa's stain (or substitutes, e.g., Azure I) degrade; fresh solutions must be used periodically.
5. Several drops of methyl alcohol will delay precipitation and aid stain penetration.
6. At 40°C staining time may be reduced to 30 min.
7. Neutral to slightly alkaline solutions must be used for proper staining.
8. Stain may fade in 2 to 3 years.

Result:

1. Chitinous material appears light green.
2. Nuclei are blue to black.

Fig. 4.7 Comparison of Different Stains. Light microscopy images of *Thelohania solenopsae* spores in smears of tissue from *Solenopsis invicta*. a. – Giemsa stain, b. – Modified trichrome stain, c. – calcofluor stain. Bar = 5 μm. (Source: Milks *et al.* 2004. Reproduced with permission of John Wiley & Sons.) See plate section for the color version of the figure.

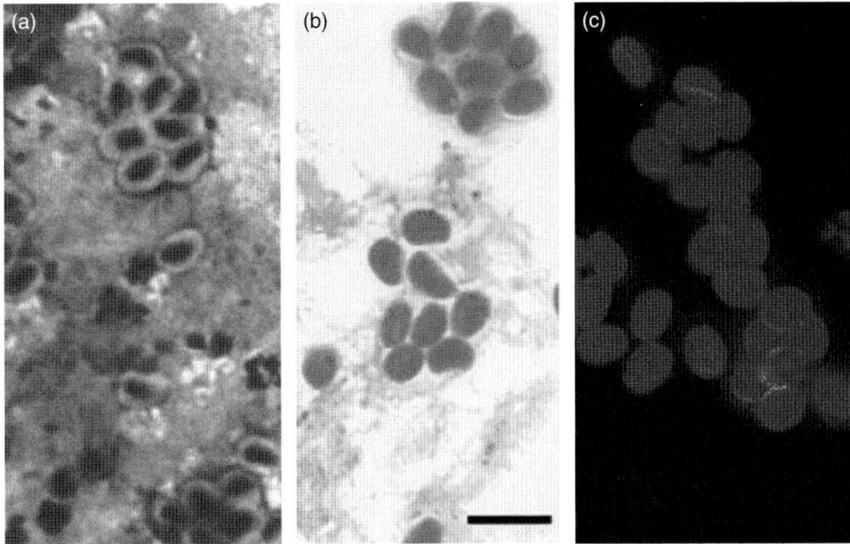

3. Cytoplasm appears pink to red.
4. Mustard-green indicates understaining, while blue-green indicates overstaining of chitinous material.

4.11.6 *Wismar's quadrachrome [38]*

Formula:

Solution A:

Saffron du Gatinais	6 pt
Absolute Alcohol	100 pt

Preparation:

1. Extract saffron in alcohol at 58–60°C for 48 hr, decant and store.
2. A second extraction may be made at 58–60°C for 15 days.

Solution B:

Woodstain Scarlet (0.1% in 0.5% aq. acetic acid)	2 pt
Acid Fuchsin (0.1% in 0.1% aq. acetic acid)	1 pt

Solution C: Lugol's Iodine

Potassium Iodide (in 70% alcohol)	15%

Solution D: Verhoeff's Hematoxylin

Hematoxylin (5% in absolute alcohol)	5 pt
Ferric Chloride (10%)	2 pt
Lugol's Iodine Solution	2 pt

Procedure:

1. Fix in Movat's FSA.
2. Take sections through alcohol changes up to 80%.

3. Place in solution C for 5 min.
4. Decolorize in $Na_2S_2O_3$ (5% aq.) for 5 min.
5. Wash in running water for 5 min.
6. Stain in 1% Alcian blue (in 1% acetic acid) for 30 min.
7. Wash in running water for 1 min.
8. Stain in solution D for 4 to 6 hr.
9. Place in ethanol (95%) for 3 min.
10. Stain in solution B for 3 min.
11. Place in aqueous acetic acid (1%) for 1 min.
12. Differentiate in aqueous phosphotungstic acid (5%) for 20 min.
13. Place in aqueous acetic acid (1%) for 1 min.
14. Complete Hematoxylin differentiation in aqueous $FeCl_3$ (10%) by checking resolution of nuclear chromatin pattern.
15. Dehydrate in acetone and in two changes of absolute alcohol.
16. Stain in solution A for 5 to 8 min.
17. Dehydrate quickly with absolute alcohol (two changes).
18. Clear in xylene and mount.

Note:
1. When double embedding is used, the total removal of celloidin is essential.
2. Optimal section thickness is 4 to 7 microns. Thicker sections require longer staining time.
3. Pooled blood serum diluted with distilled water (1:1) must be used to attach sections. Gelatin or albumen are not appropriate.
4. If Alcian blue solution is over 24 hr old add a few thymol crystals (as a preservative).
5. Staining time should be decreased by 1 hr for each day solution D is kept. Do not use after 3 days.
6. Change solution B when it changes from clear red to deep purple.
7. Absolute alcohol used for dehydration (steps 15 and 17) should contain a dehydrant to remove all water.

Result:
1. Chitin and cytoplasm appear bright red to lavender.
2. Nucleic acids appear purple to black.
3. Ground substances appear blue-green.

4.12 Polychrome staining for chitinous material – KOH

The following staining procedures are recommended for the staining of insect whole mount material which has been cleared in KOH for too long a period and thus is too transparent.

4.12.1 *Gage's [23]*

Formula:

Water	100 pt
Hydrochloric Acid	1 pt
Acid Fuchsin	0.2 pt

Note:
1. Levereault (1934) indicates stain washes out with alcohol and is of no value for alcohol-preserved specimens.

4.12.2 *Levereault's hematoxylin [39]*

Formula:

Heidenhain's Iron Hematoxylin	0.5%
Iron Alum	as needed

Procedure:

1. Treat specimen in potassium hydroxide (10%).
2. Rinse well in distilled water.
3. Neutralize with acetic acid.
4. Rinse in water.
5. Transfer to iron alum for 1 hr.
6. Transfer to iron Hematoxylin for 1 hr.
7. Rinse in distilled water.
8. Differentiate in iron alum until membranous areas are colorless.
9. Wash in distilled water for 1 hr.
10. Dehydrate through 35, 80% alcohol solutions (10 min each) up to 80 or 85% for preservation.

4.12.3 *Racovitza's [23]*

Formula:

Solution A:

Pyrogallic Acid	1%

Solution B:

Hydrochloric Acid	0.5%

Procedure:

1. Treat exoskeletons with KOH (19%).
2. Transfer to water (wash thoroughly).
3. Transfer to solution A for ½ to 1 hr.
4. Transfer to 70% alcohol (in strong light) until material darkens.
5. Transfer to solution B for differentiation.
6. Clear and mount (balsam).

4.12.4 *Smith's [23]*

Formula:

Phenol	100 pt
Ethyl eosin (sat. alkaline solution, about 1.5%)	12.5 pt

Procedure:

1. Treat with potassium hydroxide (10%).
2. Stain at 60°C for 15 min.
3. Clear in clove oil.
4. Mount in balsam.

4.13 **Polychrome staining for chitinous material – Differential staining of Individual Organs**

Stains can also be used to visually differentiate the internal organs of insects. Procedures for these purposes are usually designated as intravitam staining (staining of the whole body by intravenous injection) techniques. Such processes can be very helpful aids in general dissections and for classroom demonstrations. The following procedure has been recommended for general use with any dye of choice.

4.13.1 *Hershberger's general technic [40]*

Procedure:

1. Anesthetize specimen.
2. Insert hypodermic needle at the base of the third abdominal segment on the ventral side and introduce dye solution (0.1 pt).
3. Dyes can be prepared in a 1% solution by dissolving one pt of dye powder in 100 pt of normal salt solution (insect saline).
4. Stained internal organs can be displayed after an interval of from ½ hr to 2 or 3 days.

The following dyes (at 1% concentration) were reported to give outstanding results in the staining of various tissues.

4.13.2 *Aniline blue*

Staining Time:
Half an hour.

Result:

1. Portions of fat bodies appear light blue.
2. Thoracic muscles appear gray.
3. Dorsal wall material appears white.
4. Caeca and tracheae appear in blue outlines.
5. Gizzard is tinged blue.
6. Salivary glands appear gray-blue.

4.13.3 *Azur II*

Staining Time:
Half an hour.

Result:

1. Fat bodies appear light lavender.
2. Heart and tracheae appear in lavender outline.
3. Thoracic muscles appear blue.
4. Malpighian tubules appear greenish blue.

4.13.4 *Brilliant cresyl blue*

Staining Time:
Two days.

Result:

1. Gizzard, caecae, salivary glands, large intestine appear lavender.
2. Malpighian tubules appear yellowish.
3. Dorsal wall material appears dark blue.
4. Fat bodies appear bright blue with lavender marks.
5. Heart is tinged blue.
6. Femurs appear green.
7. Thoracic muscles appear bluish green.

4.13.5 *Rose Bengal*

Staining Time:
Three days.

Result:

1. Crop and gizzard appear deep red.
2. Caeca appears orange red.
3. Thoracic muscles appear streaked red.

4. Salivary glands appear deep pink.
5. Dorsal wall material appears light scarlet.
6. Heart appears deep wine.

4.14 Staining of specific tissues

The following recommendations were made for staining of specific tissues.

<u>Toluidin Blue</u>
Tissue:
Heart
Staining Time:
Half an hour.

<u>Trypan Blue</u>
Tissue:
Heart
Staining Time:
1–3 days.

<u>Aldehyde Green</u>
Tissue:
Caecae
Staining Time:
One day.

<u>Acid Fuchsin</u>
Tissue:
Caecae
Staining Time:
One day.

<u>Methyl Violet B</u>
Tissue:
Fat bodies
Staining Time:
Three days.

<u>Azur II</u>
Tissue:
Fat bodies
Staining Time:
Half an hour to two days.

<u>Safranin O</u>
Tissue:
Thoracic Muscles
Staining Time:
Two days.

<u>Congo Red</u>
Tissue:
Digestive Tract
Staining Time:
Half an hour.

<u>Aldehyde Green</u>
Tissue:
Digestive Tract
Staining Time:
Half an hour.

<u>Basic Fuchsin</u>
Tissue:
Tracheae
Staining Time:
Three days.

<u>Borax Carmine</u>
Tissue:
Ventral Nerve Cord
Staining Time:
Half an hour.

<u>Brilliant Cresyl Blue</u>
Tissue:
Salivary Glands
Staining Time:
Half an hour to two days.

<u>Trypan Blue</u>
Tissue:
Salivary glands and ovary
Staining Time:
Half an hour.

<u>Toluidin Blue</u>
Tissue:
Thoracic Muscles
Staining Time:
One to two days.

<u>Azur II</u>
Tissue:
Malpighian Tubules
Staining Time:
Two days.

<u>Light Green SF Yellowish</u>
Tissue:
Malphighian Tubules
Staining Time:
Half an hour to two days.

<u>Aldehyde Green</u>
Tissue:
Ovaries
Staining Time:
Three days.

4.15 **Two dye combinations**

Staining Time:
Two days.

4.15.1 *Bordeaux red-toluidin blue [41]*

Result:
1. Gizzard and intestine appear deep purple.
2. Gastric caeca appears dark lavender.
3. Malpighian tubules appear deep wine red.
4. Heart appears purple.
5. Dorsal wall material (fat) appears greenish blue.
6. Fat bodies appear pale blue.
7. Leg muscles appear dull green.
8. Ovaries appear yellowish green.
9. Ventral nerve cord appears blue.

4.15.2 *Bordeaux red-azur-II [41]*

Staining Time:
One day.

Result:
1. Gastric caeca appear tinged light green.
2. Intestine appears purple.
3. Malpighian tubules appear faint pink.
4. Heart appears in brown outline.
5. Dorsal aorta appears orange-pink.
6. Heart wing muscle appears light tan.
7. Fat bodies appear blue and pink.

4.15.3 *Biebrich scarlet-azure II [41]*

Staining Time:
Half an hour.

Result:
1. Salivary glands appear light to bright blue.
2. Malpighian tubules appear blue and red.
3. Heart and wing muscles appear purple.
4. Leg muscles and testes appear light blue.

4.15.4 *Dusham's technic [2]*

Application:
Recommended as a procedure to display the tracheae of insect.

Procedure:
1. Place the specimen in a vacuum flask with India ink covering the specimen.
2. Evacuate the flask.
3. Allow air to re-enter vessel.
4. Tracheae are filled as air enters.

Note:
1. It is suggested that wings should be removed from Coleoptera and Hemiptera.

4.15.5 *Hagmann's technic [2]*

Application:

1. Recommended for the permanent demonstration of insect tracheae.
2. The method is restricted to insects with holopneustic or hemipneutic tracheal systems.

Formula:

Solution A: Dye

Trypan Blue	2 pt
Santomerse No. 3	1 pt
Glacial Acetic Acid	10 pt
Distilled Water	90 pt

Solution B: Fixative

Formaldehyde (40%)	15 pt
Glacial Acetic Acid	10 pt
Barium Chloride (saturated aqueous solution)	75 pt

Procedure:

1. A glass container with a two-hole stopper is needed. Also required are a stop-cock with a straight tube inserted and a stop-cock with L-shaped glass tube inserted.
2. Assemble apparatus.
3. Place specimen in a small basket and suspend above solution A with fine wire passed through opening of stop-cock.
4. Stop-cock is maintained half-closed and air is exhausted with a vacuum pump.
5. After 15 to 20 min open stop-cock fully dropping basket in the stain.
6. Maintain vacuum for 5 min.
7. Allow air to re-enter slowly through other stop-cock.
8. Maintain specimen in dye for 15 min.
9. Place specimen in solution B for 3 hr to overnight (as determined by specimen size).
10. Wash in water and then dehydrate up to 70% alcohol.
11. Counter stain with safranin if needed.
12. Sections may be cut using either the celloidin or paraffin method.

Note:

1. In step 4 the reduced pressure should be at least 29 inches Hg.
2. Larger insects may be fixed by direct injection into the body cavity.
3. Penetration of dye may extend to tracheae 0.2 u. in diameter.
4. Santomerse No. 3 was reported to be a Monsanto product (detergent).

4.15.6 *Noland's [42]*

Application:

1. Recommended as a stain for the observation of live sperm.
2. Recommended as particularly useful for house flies.

Formula:

Phenol (saturated aqueous solution)	80 pt

Formalin (40%)	20 pt
Glycerol	4 pt
Gentian Violet	20 pt

The following procedures are designed for the counting of ovarioles and their differentiation from fat bodies in general dissections.

4.15.7 *Weaver and Thomas' ovarian technic [43]*

Formula:
Solution A: Herxheimer's
1. Saturated solution of Sudan IV in alkalinized alcohol (70%)

Procedure:
General A
1. Tease apart ovarioles (slightly).
2. Stain with aqueous crystal violet solution (0.01%) for 5 to 10 sec.
3. Wash immediately with distilled water.

General B
1. Stain abdominal contents for 30 sec with Herxheimers.
2. Wash with water.
3. Stain with crystal violet (0.01%) for 30 sec.

Note:
1. Alcohol (70%) may be alkalinized with 1 to 2% NaOH.

The following procedure is designed to highlight areas of cuticular permeability with particular emphasis on sense organs.

4.15.8 *Slifer's technic [44]*

Application:
Recommended as an aid in finding the distal tips of chemoreceptor neurons.

Procedure:
1. Anesthetize specimen with CO_2 or KCN.
2. Drop into Bouin's or formalin (10% solution) for at least 24 hr.
3. Wash well.
4. Place in crystal violet solution (0.5%) if insect is small.
5. If insect parts (antenna, leg, etc.) are being treated place between layers of cotton gauze or glass cloth.
6. Saturate cloth by adding dye slowly.
7. Stain for 5 min to 2 hr.
8. Quickly dip in two changes of distilled water.
9. Place on absorbent paper.
10. When dry clean in xylol.
11. Mount.

Notes:
1. Specimens must be clean and uninjured.
2. Dye will enter through any damaged area of cuticle.

3. It is suggested that specimens be reared in isolation before use.
4. When insect parts are treated, the cut end must protrude from between cloth layers and dye must not reach the cut end.
5. Dyed layers of cloth should be kept in moist chamber.
6. Clearing of small parts may require 1 to 2 hr.

References

1. Comstock J.H. & Kellogg V.L. *The Elements of Insect Anatomy*. Comstock Pub. Co., Ithaca, NY (1899).
2. Kennedy C.H. *Methods for the Study of the Internal Anatomy of Insects*. Department of Entomology, Ohio State University (Mineographed) (1932).
3. Peterson A. *Entomological Techniques. How to work with Insects*. Edwards Brothers Inc., Ann Arbor, Michigan (1964).
4. Galigher A.E. & Kozloff E.N. *Essentials of Practical Microtechnique*. Lea & Febiger, Philadelphia (1964).
5. Davenport H.A. *Histological and Histochemical Technics*. W.B. Saunders Company, Philadelphia (1960).
6. Vieira A.S., Bueno O.C. & Camargo-Mathias M.I. The functional morphology of the metapleural gland of the leaf-cutting ant *Atta laevigata* (Formicidae: Attini). *Micron* **41**, 149–157 (2010).
7. Costa-Filho A. *et al.* Identification and tissue-specific distribution of sulfated glycosaminoglycans in the blood-sucking bug *Rhodnius prolixus* (Linnaeus). *Ins. Biochem. and Mol. Biol.* **34**, 251–260 (2004).
8. Dallai R. *et al.* Fine structure of the Nassonow's gland in the neotenic endoparasitic of female *Xenos vesparum* (Rossi) (Streptsiptera, Insecta). *Tissue & Cell* **36**, 211–220 (2004).
9. Luna L.G.(ed.). *Manual of Histologic Staining Methods of the Armed Forces Institute of Pathology*. McGraw Hill Co., New York (1968).
10. Funk C.J. Alkaline phosphatase activity in whitefly salivary glands and saliva. *Arch. Ins. Biochem. and Phys.* **46**, 165–174 (2001).
11. Li Y.S. & Strausfeld N.J. Morphology and sensory modality of mushroom body extrinsic neurons in the brain of the cockroach, *Periplaneta americana*. *J. Comp. Neurol.* **387**, 631–650 (1997).
12. Jones T.A., Donlan N.A. & O'Donnell S. Growth and pruning of mushroom body Kenyon cell dendrites during worker behavioral development in the paper wasp, *Polybia aequatorialis* (Hymenoptera: Vespidae). *Neurobiol. Learn. Mem.* **92**, 485–495 (2009).
13. Sukontason K. *et al.* Identification of forensically important fly eggs using a potassium permanganate staining technique. *Micron* **35**, 391–395 (2004).
14. Makka T., Seino A., Tomita S., Fujiwara H. & Sonobe H. A possible role of 20-hydroxyecdysone in embryonic development of the silkworm *Bombyx mori*. *Arch. Insect Biochem. Physiol.* **51**, 111–120 (2002).
15. Rospars J.P. & Hildebrand J.G. Sexually dimorphic and isomorphic glomeruli in the antennal lobes of the sphinx moth *Manduca sexta*. *Chem. Senses* **25**, 119–129 (2000).
16. WebPath. Bodian's Method – Nerve Fibers. Staining Manual. Surgical Pathology – Histology. 2011.
17. Buschbeck E.K. & Hauser M. The visual system of male scale insects. *Naturwissenschaften* **96**, 365–374 (2009).
18. Blest A.D. Some modifications of Holmes' silver method for insect central nervous systems. *Quart J. Microsc. Sci.* **102**, 413–417 (1961).
19. Alves S.N., Serrao J.E. & Melo A.L. Alterations in the fat body and midgut of *Culex quinquefasciatus* larvae following exposure to different insecticides. *Micron* **41**, 592–597 (2010).
20. Reynolds E.S. The use of lead citrate at high pH as an electron-opaque stain in electron microscopy. *J. Cell. Biol.* **17**, 208–212 (1963).

21. Watson M.L. Staining of tissue sections for electron microscopy with heavy metals. II. Application of solutions containing lead and barium. *J Biophys. Biochem Cytol.* **4**, 727–730 (1958).
22. Rae C.A. Masson's Trichrome stain after Petrunkewitch or Susa fixation. *Stain Technol.* **30**, 147 (1955).
23. Gray P. *The Microtomist's Formulary and Guide*. The Blakiston Company, Inc., New York (1954).
24. Lower H.F. A trichrome stain for insect material. *Stain Technol.* **30**, 209–212 (1955).
25. Hubschman J.H. A simplified azan process well suited for crustacean tissue. *Stain Technol.* **37**, 379–380 (1962).
26. Becker E.R. & Roudabush R.L. *Brief Directions in Histological Technique*. Collegiate Press, Inc., Ames, Iowa (1935).
27. Courrent A. *et al.* The fine structure of colleterial glands in two cockroaches and three termites, including a detailed study of *Cryptocercus punctulatus* (Blattaria, Cryptocercidae) and *Mastotermes darwiniensis* (Isoptera, Mastotermitidae). *Arthropod Struct. & Dev.* **37**, 55–66 (2008).
28. Niculita H., Billen J. & Keller L. Comparative morphology of cephalic exocrine glands among castes of the black ant *Lasius niger*. *Arthropod Struct. & Dev.* **36**, 135–141 (2007).
29. Eberhard M.J.B. *et al. Structure and function of the arolium of Mantophasmatodea (insecta)*. *J. Morph.* **270**, 1247–1261 (2009).
30. Tolbert L.P., Matsumoto S.G. & Hildebrand J.G. Development of synapses in the antennal lobes of the moth *Manduca-Sexta* during metamorphosis. *J. Neurosci.* **3**, 1158–1175 (1983).
31. Levy H.A. Dioxane as an aid in staining insect cuticle. *Stain Technol.* **18**, 181–182 (1943).
32. Kingsbury B.F. & Johannsen O.A. *Histological Technique*. John Wiley & Sons, Inc., New York (1927).
33. Buxton B.A. On the protocerebrum of *Micropteryx* (Lepidoptera). *Trans. Entomol. Soc. Lond. Part I* **65**, 112–153 (1917).
34. Qayum A.A & Telang A. A protocol for collecting and staining hemocytes from the yellow fever mosquito *Aedes aegypti*. *J. Vis. Exp.* **51**, (2013).
35. Azariah J. An optimum pH for the demonstration of chitin in *Periplanata americana* using Lugol's iodine. *Acta Histochem.* **53**, 238–240 (1975).
36. Sauvion N., Nardon C., Febvay G., Gatehouse A.M.R. & Rahbe Y. Binding of the insecticidal lectin Concanavalin A in pea aphid, *Acyrthosiphon pisum* (Harris) and induced effects on the structure of midgut epithelial cells. *J. Insect Phys.* **50**, 1137–1150 (2004).
37. Gier H.T. Differential stains for insect exoskeleton. *J. Kansas Entomol. Soc.* **22**, 79–80 (1949).
38. Wismar B.L. Quad-type stain for the simultaneous demonstration of intracellular and extracellular and tissue components. *Stain Technol.* **41**, 309–313 (1966).
39. Levereault P. Iron hematoxylin stain for differentiation of sclerites from membranous areas. *Ann. Entomol. Soc. Amer.* **27**, 313–314 (1934).
40. Hershberger R.V. Differential stains of insect tissues. *Ohio J. Sci.* **46**, 152–162 (1946).
41. Hershberger R.V. Stain combinations in living insects. *Ohio J. Sci.* **48**, 161–168 (1948).
42. Michelsen A. Experiments on the period of maturation of the male house fly, *Musca domestica L. Oikos* **11**, 250–264 (1960).
43. Weaver N. & Thomas R.C. Jr. A fixative for use in dissecting insects. *Stain Technol.* **31**, 25–26 (1956).
44. Slifer E.H. A rapid and sensitive method for identifying permeable areas in the body wall of insects. *Entomol. News* **71**, 179–182 (1960).

5 Immunohistochemical techniques

5.1 Introduction

Whereas the dye-based stains noted in Chapter 4 are powerful tools for identification of specific structures and organelles or biochemical properties of the cell or tissues, Immunostaining (or Immunohistochemistry (IHC)) refines this further through the use of protein-specific antibodies to localize and quantitate individual proteins. IHC can be performed on tissue sections or on whole (small) animals or tissues for *in vivo*, three-dimensional localization. Bound primary antibodies can be detected in a number of ways, including direct labeling methods, indirect labeling methods that allow for amplification of signal, and chromogenic methods for routine pathology. Advanced and highly informative techniques also include fluorogenic methods (see Chapter 7 Fluorescence for Additional Protocols) that allow tight and specific labeling as well as quantification of intensity, and electron dense particle labeling methods for combination of IHC with electron microscopy for unparalleled, sub-cellular localization of proteins. Individual antibodies vary in their ability to bind fixed tissue and tissues can vary in their ability to allow penetration of antibodies generally. As such, a variety of IHC procedures are presented to cover the complexity of labeling approaches and variations in antibodies and tissues to be studied. Where one protocol may work beautifully for a specific antibody on a particular tissue, a different antibody or a different tissue may require an entirely different protocol. As such, the reader is encouraged to consider multiple approaches if an initial technique proves unsuccessful for his/her antibody and tissue combination.

5.1.1 *Formaldehyde fixation of* Drosophila *testes [1]*

Application:
For good chromosome morphology, for visualization of the chromosome condensation–decondensation cycle, and for immunostaining of microtubules (i.e., components of the cytoskeleton, found throughout the cytoplasm).

Procedure:
1. Dissect *Drosophila* larval or adult testes in saline.
2. Fix testes in 3.7% formaldehyde in phosphate buffered saline (PBS) at room temperature for 30 min.
3. Transfer testes to 45% acetic acid for 30 sec.
4. Transfer testes into a drop (4 µl) of 60% acetic acid on a 20 × 20 mm coverslip for 2–3 min at room temperature.

Insect Histology: Practical Laboratory Techniques, First Edition.
Pedro Barbosa, Deborah L. Berry and Christina S. Kary.

Companion Website: www.wiley.com/go/barbosa/insecthistology

5. Place a clean slide over the coverslip, invert the sandwich and squash testes using moderate pressure.
6. Freeze the slide in liquid nitrogen.
7. Remove the coverslips with a razor blade, immediately immerse the slides into ethanol at −20°C for 15 min.
8. Transfer the slides into PBS with 0.1% Triton X-100 for 10 min at room temperature.
9. Wash the slides twice for 5 min each in PBS at room temperature.

5.1.2 *Methanol-acetone fixation of* Drosophila *testes I [2]*

Application:
Recommended for preparation of testes slides for immunostaining.

Formula:
Testis Buffer:

1M EDTA	1 pt
1M KCl	183 pt
1M NaCl	47 pt
1M phenylmethylsulfonyl fluoride (PMSF)	1 pt
1M Tris-HCl	10 pt
Add distilled water to 1 L, adjust pH to 6.8	

Procedure:
1. Dissect testes in testis buffer from *Drosophila* larvae, pupae, or young adults.
2. Transfer testes into 2 μl drop of testis buffer on coverslip. Tear open the adult testes using either very thin forceps or tungsten needles.
3. Place a clean slide over the coverslip without pressing and invert the sandwich.
4. Freeze the slides in liquid nitrogen.
5. Remove the coverslips with a razor blade and incubate in methanol at −20°C for 5 min.
6. Transfer the slides to acetone at −20°C for 1–2 min.
7. Transfer slides to PBS with 0.5% acetic acid and 1% Triton X-100 for 10 min at room temperature.
8. Wash slides in PBS two times 5 min each.

5.1.3 *Methanol-acetone fixation of* Drosophila *testes II [2]*

Application:
Recommended for preparation of slides of testes, for immunostaining and chromosome analysis.

Formula:
Testis Buffer:

1M EDTA	1 pt
1M KCl	183 pt
1M NaCl	47 pt
1M phenylmethylsulfonyl fluoride (PMSF)	1 pt
1M Tris-HCl	10 pt
Add distilled water to 1 L, adjust pH to 6.8	

Procedure:

1. Dissect testes in testis buffer from *Drosophila* larvae, pupae, or young adults.
2. Transfer testes into 2 μl drop of testis buffer on coverslip. Tear open the adult testes using either very thin forceps or tungsten needles.
3. Place a clean slide over the coverslip without pressing and invert the sandwich.
4. Freeze the slides in liquid nitrogen.
5. Remove the coverslips with a razor blade and incubate in methanol at −20°C for 15 min.
6. Transfer the slides to acetone at −20°C for 30 sec.
7. Transfer slides to PBS with 0.1% acetic acid and 0.1% Tween-20 for 10 min at room temperature.
8. Wash slides in PBS with 0.1% Tween-20, three times 10 min each.
9. Wash in PBS for 5 min.

5.1.4 *Preparation of* Drosophila *brains for IHC staining [3]*

Larval Brains

Procedure:

1. Prepare egg laying chamber and collect embryos on a fruit agar plate.
2. Wait until embryos start to hatch (~21 hr at 25°C) and remove all hatched larvae from plate.
3. After 2 hr collect all newly hatched larvae and allow them to develop in a vial to the necessary stage.
4. Collect the larvae with a paint brush dampened with phosphate buffered saline with 0.3% Triton-X (PBTx).
5. Place larvae PBTx in a dissecting dish.
6. Hold the larval body with one pair of forceps and the mouth hook with a second pair. Pull the forceps apart gently to pull the mouth hook with the brain attached to it.
7. While still holding the mouth hook, gently remove excess tissue surrounding the brain.
8. Still holding the mouth hook, transfer the brain to a tube containing 500 μl freshly prepared 4% paraformaldehyde, on ice.
9. Continue to dissect brains and add to the tube of fixative until sufficient numbers for staining have been obtained.
10. Proceed with staining protocol for brains detailed below.

Pupal Brains

1. Collect white pre-pupae with a paint brush dampened with phosphate buffered saline (PBS).
2. Place 0 hr pupae in a small Petri dish in a humidified chamber at 25°C to age the pupae to the appropriate developmental stage.
3. Place Petri dish on ice to stop development.
4. Transfer a pupa to PBS in a dissecting dish.
5. Hold the puparium (i.e., pupal case) dorsal slide up with forceps. Squeeze a second pair of forceps closed and use the sharp tips to puncture a hole in the abdominal end of the pupa.
6. Open the forceps to tear the puparium apart and pull the pupa out.
7. Use forceps to gently remove abdominal tissue, up to the level of the thorax.

8. Gently hold the pupa at the very end of the head case and use the side of the second pair of forceps to gently push the brain from the head and out through the thorax.
9. Gently remove excess fat particles.
10. Transfer the brain to a tube containing 500 µl freshly prepared 4% paraformaldehyde, on ice.
11. Continue to dissect brains and add to the tube of fixative until sufficient specimen numbers for staining have been obtained.
12. Proceed with staining protocol for brains detailed below.

Adult Brains

1. Anesthetize adult flies and place into a dish on ice.
2. Transfer one fly to PBTx in a dissection dish.
3. Hold the fly belly up with forceps and insert a second pair of forceps into the cavity just below the eye to obtain a grip of the eye. Be careful to avoid internal structures, notably the brain.
4. Gently pull the head off the fly and discard the body.
5. Use the free forceps to obtain a grip of the other eye from the underside and gently pull the forceps apart to separate the head cuticle.
6. Remove the brain from the head cuticle and carefully remove the surrounding trachea.
7. Transfer the brain to a tube containing 500 µl freshly prepared 4% paraformaldehyde, on ice.
8. Continue to dissect brains and add to the tube of fixative until sufficient numbers for staining have been obtained.
9. Proceed with staining protocol for brains detailed below.

IHC Staining of Brains

1. Place the tube with brains in fixative on a nutator ™ mixer and fix for 20 min at room temperature.
2. Place the tube into a tube rack and allow the brains to settle to the bottom of the tube so that the fixative can be removed.
3. Add 0.5 ml PBTx to the tube, close and gently invert tube. Allow brains to settle, remove the PBT and repeat once more for a total of two quick washes at room temperature.
4. Add fresh PBTx and rock samples on a nutator for 20 min at room temperature. Repeat washes for a total of three 20 min washes.
5. Following the last wash, replace PBTx with 0.5 ml of 5% normal goat serum in PBTx (blocking solution) and rock samples on a nutator for 30 min at room temperature.
6. Remove the blocking solution and replace with the primary antibody diluted in PBTx. Incubate on nutator for two nights at 4°C.
7. Remove primary antibody and perform two quick washes with PBTx, as in step 3, followed by three, 20 min washes, as in step 4.
8. Remove the PBTx and add an appropriate fluorescently conjugated secondary antibody diluted in PBTx. Incubate on nutator for two nights at 4°C.
9. Remove the secondary antibody and perform two quick washes with PBTx, as in step 3, followed by three 20 min washes, as in step 4. DAPI can be added into one of the 20 min washes to stain nuclei.

10. Replace PBTx with 200 μl SlowFade® (Invitrogen™) and allow the brains to settle at 4°C.
11. Use a cut pipette tip, transfer the brains to a mounting slide minimizing the amount of SlowFade® transferred (remove any excess SlowFade®).
12. Orient the brain(s) so that the anterior side of the antennal lobes is facing upward.
13. Arrange two broken coverslips on the microscope to form a bridge around the brains.
14. Cover the bridge with another coverslip and image.

5.1.5 *Preparation of* Drosophila *embryos for immunostaining – embryo fixation with methanol [4]*

Application:
Recommended for preparation of embryos for immunostaining.

Formula:
Fixative:

10X PBS	1 pt
4% freshly prepared formaldehyde	5 pt
dH_2O	4 pt

Procedure:
1. Prepare an egg laying chamber with a fruit juice agar plate and allow flies to lay eggs.
2. Cover the plate with 100% bleach and brush the embryos with a fine paintbrush to detach them from the plate.
3. Pour the bleach with embryos into an egg basket and place the basket on an empty Petri dish. Fill the dish with enough bleach to cover the embryos for 2 min.
4. Rinse embryos extensively in tap water and transfer the dechorionated embryos to a glass tube containing 1 ml heptane.
5. Add 1 ml of fixative. Wait until two phases form with the embryos in between the two layers, then fix for an additional 20 min at room temperature.
6. Completely remove the lower fixative layer and add 1 ml methanol. Mix vigorously with a vortex mixer for 15 sec.
7. Let mixture settle and the devitellinized embryos will fall to the bottom of the tube whereas the embryos that do not lose their vitelline membranes will remain floating in between the two layers. Remove the upper layer and intervening embryos, then remove the lower layer.
8. Transfer the embryos to a clean 0.5 ml microfuge tube and wash with 400 μl of the following graded series of methanol for 5 min each: 80%, 50%, and 30% methanol.
9. Wash two more times in phosphate buffered saline with 0.4% Triton X-10 (PBTx).
10. Incubate with primary antibody overnight at 4°C with gentle rocking.
11. Wash five times for 5 min each in PBSTx.
12. Incubate with labeled secondary antibody for 6 hr at 4°C with gentle rocking.
13. Wash five times for 5 min each in PBSTx.
14. Remove as much PBSTx as possible and replace with 50% glycerol.
15. To avoid crushing the embryos, use nail polish to mount a small coverslip on each end of a slide to form a bridge. Place the embryos in glycerol in the middle of the slide and cover with a coverslip so that the coverslip rests on top of the other two coverslips.
16. Image embryos using appropriate fluorescent filters.

5.1.6 *Miller and Albert's embryo technic [5]*

Application:
For immunohistochemistry of antigens not well preserved by the methods of Karr and Albert, this modification of the technique is recommended.

Formula:
Triton-NaCl Wash Solution:

1M NaCl	150 pt
Triton X-100	0.5 pt
Distilled Water	850 pt

Procedure:
1. Place embryos in Triton-NaCl wash at 90°C in a boiling water bath for less than 5 sec.
2. Vortex embryos in solution.
3. Add 4 pt more ice cold Triton-NaCl solution and plunge into ice-water bath.
4. Replace Triton-NaCl solution with a mixture of 1 pt heptane and 1 pt 25 mM Na_3EGTA in 95% methanol and shake for 30 sec.
5. The devitelinized embryos will sink and can be collected from the bottom of the tube and transferred to fresh 25 mM Na_3EGTA/95% methanol.
6. Wash embryos two more times in 25 mM Na_3EGTA/95% methanol and incubate for 1 hr.
7. Embryos are rehydrated through graded methanol in phosphate buffered saline and processed using standard IHC (immunohistochemistry) techniques.

5.1.7 *Swevers immunohistochemistry for ovarioles [6]*

Procedure:
1. Fixed ovarioles are made permeable by placing them in phosphate buffered saline with 0.3% Triton X-100.
2. Ovarioles are treated with 0.5 mg/ml collagenase (Type IV) followed by 1 μg/ml proteinase K.
3. Ganglia are rinsed in phosphate buffered saline, post-fixed in 4% paraformaldehyde, and stained by standard IHC techniques.

Note:
1. Additional proteinase treatments may be necessary for adequate penetration of antibodies into ovarioles.

5.1.8 *Preparation of eye imaginal discs for immunostaining [4]*

Procedure:
1. Remove L3 larvae from the wall of a culture vial with a paintbrush and place into a drop of phosphate buffered saline (PBS) on a slide.
2. Under a stereomicroscope, hold the middle of the larval body with one set of forceps and the mouth hooks with a second set. Firmly pull the mouth parts away from the rest of the body.
3. Look for the eye-antennal imaginal discs which will be attached to the mouth hooks, the optic lobes, and ventral nerve cord.
4. Remove all surrounding structures including salivary glands and fat body, but leave the discs attached to the mouth hooks and optic lobes.
5. Using a cut 200 μl pipette tip, transfer the discs to 50 μl PBS in a microfuge tube.

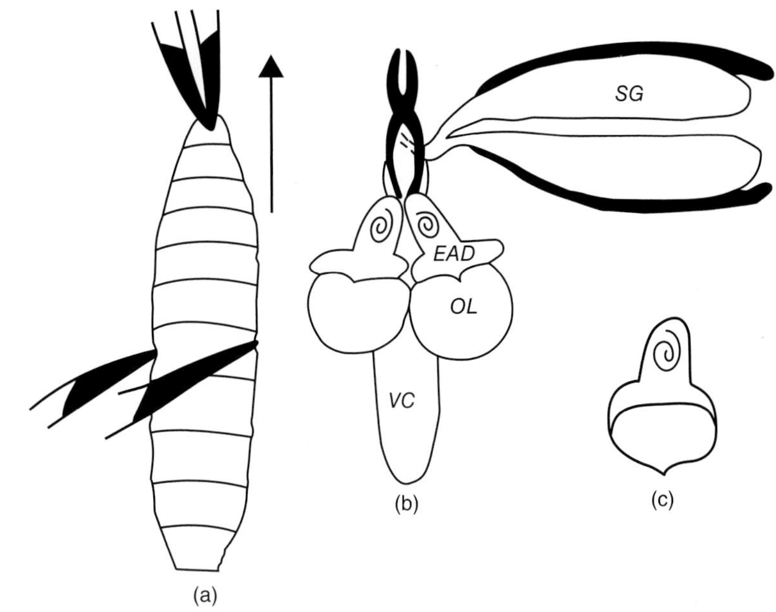

Fig. 5.1 (a) Schematic drawing illustrating the larval dissection procedure (For details *see* "Larval Eye Imaginal Discs" and "Salivary Glands"). (b) Drawing of eye-antennal imaginal discs (*EAD*) attached to the larval brain (comprising the optic lobes [*OL*] and the ventral cord [*VC*]. Salivary glands [*SG*], lined with fat bodies, (*in dark*) and attached to the larval mouth hooks are also shown. (c) Drawing of a single isolated eye-antennal imaginal disc. (Source: Ramos *et al.* 2010. Reproduced with permission of Springer.)

6. Continue dissections, adding into the 50 μl of PBS from step 5 until sufficient discs have been collected for analysis.
7. Carefully remove the PBS and fix discs in 4% paraformaldehyde for 30 min.
8. Remove fixative and add 2% glycine in PBS with 0.1% Triton X-100 (PBTx) for 10 min.
9. Wash five times, for 5 min each, in PBTx.
10. Incubate with primary antibody for 1 hr at room temperature or overnight at 4°C with gentle rocking.
11. Wash five times, for 5 min each, in PBTx.
12. Incubate with labeled secondary antibody for 0.5 to 2 hr at room temperature, or 4–6 hr at 4°C with gentle rocking.
13. Wash five times, for 5 min each, in PBTx.
14. Remove as much PBTx as possible and replace with 50% glycerol in PBTx.
15. With a cut pipette, quickly transfer the tissues to a slide.
16. Dissect out the optic lobe with needles and mount the slide with only the eye disc on it.
17. Image slide with appropriate microscope settings.

5.1.9 *Preparation of pupal retina for immunostaining [4]*

Application:
Recommended for the staining of pupal retinas between 12 and 50 hr after puparium formation.

Procedure:
1. Under a stereomicroscope, hold the posterior end of a puparium with forceps. Use a second pair of forceps to remove the puparium from the anterior end to expose the pupal head.

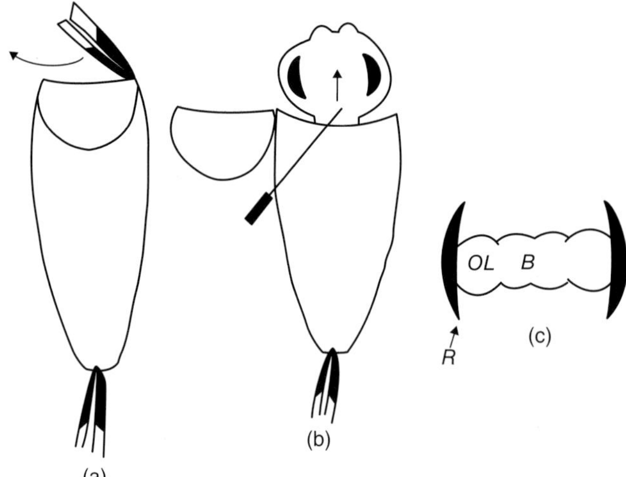

Fig. 5.2 Schematic drawings depicting the procedure for dissection of pupal retinas. (a) Opening of the pupal case. (b) Removal of the pupal head from the case. (c) The retinae (*R*) attached to the optic lobe (*OL*) and brain (*B*). (Source: Ramos *et al.* 2010. Reproduced with permission of Springer.)

2. While holding the posterior end, put the head into a drop of phosphate buffered saline (PBS) on a slide. Put a needle in the neck of the pupa and pull to the anterior end of the head.
3. Look for the pupal retinae attached to the optic lab and brain inside the PBS drop.
4. Clean the retina/optic lobe structures and use a trimmed pipette tip to transfer the retina/optic lobes to 50 µl PBS in a microfuge tube.
5. Continue dissections, adding into the 50 µl of PBS from step 4 until sufficient retinas have been collected for analysis.
6. Carefully remove the PBS and fix discs in 4% paraformaldehyde for 30 min.
7. Remove fixative and add 2% glycine in PBS for 10 min.
8. Wash five times for 5 min each in PBS.
9. Incubate with primary antibody for 1 hr at room temperature or overnight at 4°C with gentle rocking.
10. Wash five times, for 5 min each, in PBS.
11. Incubate with labeled secondary antibody for 0.5 to 2 hr at room temperature, or 4–6 hr at 4°C with gentle rocking.
12. Wash five times, for 5 min each, in PBS.
13. Remove as much PBS as possible and replace with 50% glycerol in PBS.
14. With a cut pipette, quickly transfer the tissues to a slide.
15. Dissect out the optic lobe with needles and mount the slide with only the retina on it.
16. Image slide with appropriate microscope settings.

5.1.10 *Preparation of* Drosophila *salivary glands for immunostaining [4]*

1. Remove L3 larvae from the wall of a culture vial with a paintbrush and place into a drop of phosphate buffered saline (PBS) on a slide.
2. Under a stereomicroscope, hold the middle of the larval body with one set of forceps and the mouth hooks with a second set. Firmly pull the mouth parts away from the rest of the body.
3. The bi-lobed salivary glands are transparent and attached at their anterior ends. Use forceps to hold this attachment point and transfer the glands away from the mouth parts and imaginal discs to a fresh drop of PBS.

4. Use a needle to carefully pull the fat away from each lobe and transfer the clean salivary glands to 50 µl PBS in a microfuge tube.
5. Continue dissections, adding into the 50 µl of PBS from step 4 until sufficient glands have been collected for analysis.
6. Carefully remove the PBS and fix glands in 500 µl 4% paraformaldehyde and 500 µl heptane for 20 min with agitation.
7. Carefully remove the lower phase (paraformaldehyde) and add 500 µl methanol, leave for 1 min.
8. Remove the upper phase (heptane) and the interphase.
9. Rinse three times with methanol.
10. Rinse with 1:1 methanol and PBS with 0.1% Triton (PBTx).
11. Wash three times with PBTx, then five times with PBTx with 1% BSA (PBSBT).
12. Block in PBSBT for 1 hr at room temperature.
13. Incubate with primary antibody overnight at 4°C with gentle rocking.
14. Wash six times, for 10 min each, in PBSBT.
15. Incubate with labeled secondary antibody for 2 hr at room temperature.
16. Wash five times for 10 min each in PBSBT.
17. Remove as much PBSBT as possible and replace with 50% glycerol in PBS.
18. With a cut pipette, carefully transfer glands to a slide.
19. Spread and separate the salivary glands so that they can be imaged individually.

5.2 General immunostaining techniques

5.2.1 *IHC on the central nervous system in adult* Drosophila *[7]*

Formula:

Permeabilizing Buffer:

Phosphate buffered saline (PBS)	1X
Paraformaldehyde	4%
Triton X-100	0.1%
Deoxycholate	0.1%

Blocking Solution:

Phosphate buffered saline (PBS)	1X
Triton X-100	0.1%
Bovine serum albumin	0.5%
Normal serum	5%
Sodium azide	1 mM

Protocol:

1. Following dissection, place the central nervous system (CNS) in permeabilizing buffer for 30 min.
2. Wash CNS three times, for 5 min each, in PBS with 0.1% Triton X-100 (PBTx) at room temperature on a nutator.
3. Incubate in blocking solution for 1 hr at room temperature.
4. Add diluted primary antibody and incubate for 48 hr (or up to 3 weeks) at 4°C on a nutator.
5. Wash CNS in PBTx five times, over a 10–12 hr period, at room temperature on a nutator.
6. Incubate in blocking solution for 1 hr at room temperature.
7. Add diluted secondary antibody for 2 hr at room temperature.
8. Wash CNS in PBTx, five times, over a 3 hr period.

9. If a fluorescent secondary antibody was used, mount in Vectashield® and image.
10. If an HRP conjugated secondary antibody was used, place CNS in DAB (3,3′-Diaminobenzidine) and watch development of the stain under a microscope. Develop until appropriate staining is achieved.
11. Dehydrate the CNS through graded ethanol, wash three times, for 20 min each, in 100% ethanol, clear in Histo-Clear® for 30 min and mount with Araldite epoxy resin.

5.2.2 *Chiang brain technic [8]*

Application:
Recommended for preserving original size and shape of the brain.

Protocol:
1. Dissect brain samples in phosphate buffered saline (PBS) and fix in 4% paraformaldehyde on ice with microwave irradiation for 90 sec with continuous rotation, three times.
2. Wash in PBS with 1% Triton-X and 10% normal goat serum (PBS-T) for 30 min at room temperature.
3. In PBS-T, degas in a vacuum chamber to expel tracheal air with six cycles (depressurize to ~70 mmHg and hold for 10 min).
4. Block samples overnight in PBS-T at 4°C.
5. Expose to primary antibody in PBS-T at 4°C for 2 days.
6. Wash in PBS-T three times.
7. Expose to secondary antibody conjugated with biotin diluted in PBS-T at 4°C for 2 days.
8. Wash in PBS-T, three times.
9. Expose to tertiary streptavidin conjugated to Alexa Fluor® 635 at 4°C overnight.
10. Wash extensively in PBS-T and clear brains in FocusClear ™ (CelExplorer®) for 5 min.
11. Mount in a drop of MountClear ™ between two coverslips separated by a spacer ring of 200-μm thickness to prevent flattening of the brain.

5.2.3 *Antibodies recommended for IHC staining of* Drosophila *brains [9]*

I. To detect green fluorescent protein or β-gal in fixed specimen:
   1. Rabbit anti-GFP, Molecular Probes® A11122, 1:1000 dilution.
   2. Sheep anti-GFP, Serotec® 4745–1051, 1:1000 dilution.
   3. Rabbit anti-β-gal, ICN/Cappel®, 1:20,000 dilution.
   4. Mouse anti-β-gal, Promega® Z3781, 1:500 dilution.

II. To label relevant architectural features:
   1. Mouse anti-nc82 (neuropil), Developmental Studies Hybridoma Bank, University of Iowa. (DSHB), 1:50 dilution.
   2. Rat anti-DE-cadherin (DCAD2) (larval neuroepithelium and neuropile in larvae and early pupae), DSHB, 1:50 dilution.
   3. Rat anti-DN-cadherin (DN-EX 8) (neuropil) DSHB, 1:50 dilution.

III. To label neuronal populations in larvae:
   1. Mouse anti-24B10 (photoreceptors and their projections), DSHB, 1:50 dilution.
   2. Mouse anti-Dachshund (mAbdac 2–3) (lamina precursors and lobula cells) DSHB, 1:100 dilution.
   3. Mouse anti-Elav (9F8A9) (neurons), DSHB, 1:25 dilution.
   4. Rat anti-Elav (7E8A10) (photoreceptors and their projections) DSHB, 1:25 dilution.

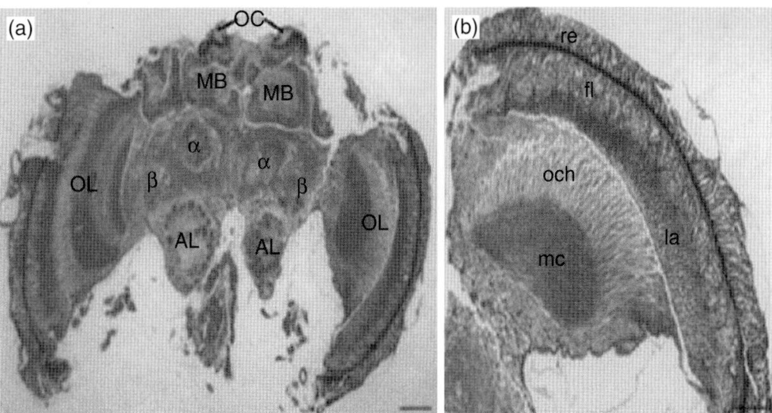

Fig. 5.3 Panoramic view of the honeybee *Apis mellifera* brain stained with hematoxylin eosin. Horizontal section of the honeybee brain shows the regions: ocelli (OC), mushroom bodies (MB), antennal lobe (AL), alpha lobe (a), beta lobe (b) and optical lobe (OL); Scale bar = 200 mm. (b) Detail of the optical lobe with its subregions: retina (re), fenestrated layer (fl), lamina (la), outer chiasm (och) and medula (me); Scale bar = 80 mm. (Source: Calábria *et al.* 2010.)

5.2.4 *Low pH antigen retrieval [10]*

Application:
Many antibodies require antigen retrieval in order to facilitate appropriate staining. This procedure describes retrieval using a low pH Citrate antigen retrieval procedure.

Procedure:
1. Fix tissues in 4% paraformaldehyde, dehydrate in ethylic alcohol, clear, infiltrate with paraffin, and embed in paraffin wax.
2. Cut 5 µm sections and mount onto slides.
3. Perform antigen retrieval by incubating slides in pre-heated 10 mM sodium citrate, pH 6.0, for 30 min at 100°C in a steam cooker (Tefal Steam Cuisine®, mod. 364688, power 100%).
4. Block sections with 4.5% hydrogen peroxide in phosphate-saline buffer for 15 min.
5. After rinsing with dH_2O, treat sections with 0.1 M Tris-glycine buffer, pH 7.4, for 30 min.
6. Block sections in phosphate buffered saline (PBS) with 0.3% Triton X-100, 5% dried milk and goat inactive serum 1:6 (v/v) for 4 hr.
7. Incubate slides in diluted primary antibody for 10 hr at 4°C.
8. Wash slides in PBS.
9. Incubate slides in diluted secondary antibody conjugated with peroxidase for 1 hr at 25°C.
10. Develop the stain in 10 mg/mL DAB (3, 3'- diaminobenzidine).
11. Rinse slides in dH_2O, dehydrate through a graded ethanol series, clear and mount the slides with Permount.

5.2.5 *Wicklein and Strausfeld [11]*

Application:
Recommended for improved antibody penetration of tissues for IHC.

Procedure:
1. Fix tissues in 4% formaldehyde in phosphate buffer overnight.
2. Rinse in buffer and dehydrate tissue through graded ethanol series.
3. Rinse in propylene oxide for 10 min.
4. Rehydrate through graded ethanol series.
5. Proceed with IHC using standard techniques.

5.2.6 *Wildman immunohistochemistry for ganglion Cells [12]*

Application:
Additional proteinase treatments may be necessary for adequate penetration of antibodies due to ganglion sheath.

Procedure:
1. Dehydrate fixed ganglia through ethanol series to 70% ethanol and store.
2. Place ganglia in 90% ethanol for 1 hr then rehydrate through graded series of ethanol and wash in phosphate buffered saline.
3. Treat ganglia with 0.1% collagenase (Type IV) followed by 0.1% hyaluronidase for 1 hr at 37°C.
4. Rinse ganglia in phosphate buffered saline, and follow with standard IHC techniques.

5.2.7 *Truman's immunohistochemistry for insect tissues [13]*

Application:
Many insect tissues require additional permeabilization techniques to allow for penetration of antibodies.

Procedure:
1. Treat fixed tissues with collagenase (Type IV) at 0.5 mg/ml for 20 min at room temperature.
2. Post-fix in 4% paraformaldehyde.

5.2.8 *Immunostaining on frozen sections [14]*

Application:
Recommended as a routine IHC protocol on frozen tissue sections.

Procedure:
1. Open the local surface cuticle of a jumping plant bug permit fluid entry, and fix in paraformaldehyde in phosphate buffered saline (PBS) and then in 4% formaldehyde for 4 hr.
2. Leave specimen, overnight, in 25% sucrose solution in PBS.
3. Submerge tissues in Tissue-Tek® OCT and freeze in liquid nitrogen.
4. Cut the metathorax sections at 10–50 μm on a cryostat.
5. Rehydrate the sections for 25 min in a 0.25% solution of Triton-X in PBS (PBTx), and permeabilize in 2% PBTx for 10 min, then in 0.25% PBTx for 10 min.
6. Block in 5% normal goat serum (NGS) for 30 min and incubate overnight at 4°C in a primary antibody diluted in 2.5% NGS.
7. Warm the sections to 25°C, wash in 0.25% PBT for 30 min, then incubate overnight at 4°C with an appropriate secondary antibody conjugated to Cy3.
8. Wash sections for 30 min in 0.25% PBT and mount in a drop of Vectashield® mounting medium.

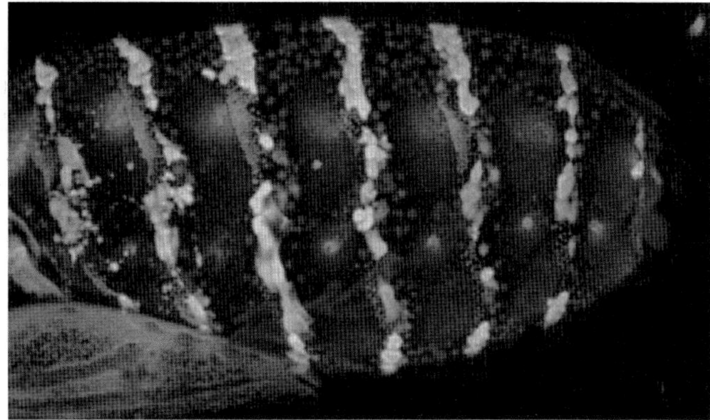

Fig. 5.4 Segmentation gene products in *Drosophila* pupae. Wingless protein and Engrailed expression (En-gal4:uas-GFP) were visualized at 26 hr after puparium formation with mouse anti-Wg (4D4: Iowa Hybridoma Bank) and anti-GFP. 10× magnification, anterior is left and dorsal is up. Nuclei were counterstained with DAPI. (Source: Wang & Yoder 2011. Reproduced with permission of the authors.) See plate section for the color version of the figure.

5.2.9 *Immunohistochemistry for Agarose blocks [15]*

Procedure:

1. Dissect brains from adults or heads from staged embryos into ice cold phosphate buffered saline (PBS).
   a. For heads from staged embryos, dissect heads of pharate larvae out of the egg and free them from embryonic membranes.
2. Fix brains or heads in 3.7% paraformaldehyde overnight at 4°C.
3. Wash for 1 hr in PBS with 0.025% Triton X-100 (PBTx).
4. Embed in 5% agarose/PBS at 55–60°C and allow block to cool.
5. Serial section the block on a Vibratome at 40 μm thickness.
6. Collect sections in PBS, free them from the agarose and position onto Superfrost Plus® microscope slides.
7. Cover slides with tris buffered saline (TBS) with 1% normal goat serum and 0.25 mg BSA, and 0.05% Triton X-100 for 1 hr.
8. For antigen retrieval, incubate in 10 mM Sodium Citrate buffer pH 6.0 for 30 min at room temperature.
9. Incubate with primary antibody diluted in TBS for 72 hr at 4°C.
10. Wash in TBS for 3 hr at room temperature, then expose to appropriate secondary antibody conjugated to Cy3 diluted 1:150 in TBS with 1% normal goat serum and 0.25 mg BSA, and 0.05% Triton-X for 3 hr at 4°C in a humid chamber in the dark.
11. Wash sections overnight in TBS at 4°C in the dark.
12. Mount in Vectashield® (Vector Labs) and coverslip.
13. Image using appropriate fluorescent microscope settings.

5.2.10 *Immunohistochemistry of* Drosophila *pupal abdomen [16]*

Application:

Recommended preparation and treatment of *Drosophila* pupal abdomen for IHC.

Procedure:

1. Collect 0 hr white prepupae and age to necessary stage in a humid chamber at 25°C.
2. Create a dissection platform by placing a strip of double sided tape to a microscope slide. Remove protective paper so that the tape is exposed.

3. Place staged pupa onto the dissection platform with the anterior end of the pupa facing the broadest width of the tape. Use a paintbrush to position the pupa before it has completely adhered to the tape.
4. Air dry pupa and allow to adhere to tape for 5–10 min.
5. Dissect each pupa bilaterally with a surgical scalpel.
6. Dampen a paint brush with phosphate buffered saline (PBS) and use it to gently dislodge the pupa from the tape.
7. Holding the anterior end of the pupa with forceps, transfer the pupa half to a dish of PBS and gently wash away the internal tissue of the abdomen, being careful to leave the epithelia cells intact.
8. Transfer the clean pupa half to PBS with 4% paraformaldehyde and 0.2% deoxycholic acid (for permeabilization), and fix for 1 hr at room temperature.
9. Rinse fixed pupa in PBS three times for 5 min each.
10. Block pupa in PBS with 2% BSA for 1 hr.
11. Incubate with primary antibody in PBS overnight at 4°C without rocking.
12. Wash pupa in PBS three times for 10 min each.
13. Block pupa in PBS with 2% BSA for 1 hr.
14. Incubate with secondary antibody in PBS for 3 hr at room temperature without rocking.
15. Wash pupa in PBS three times 10 min each.
16. Counterstain with nuclear stain such as DAPI, if preferred.
17. Place 75–100 μl mounting media in a depression well slide.
18. Remove the pupal case by grasping the anterior end of the pupal case with forceps and gently removing the internal pupal membrane from the pupal case with forceps.
19. Transfer to the depression well and continue to add additional samples.
20. Orient the pupa with the lateral side facing up and remove any air bubbles.
21. Coverslip and image samples.

Note:
1. Dissect ten or fewer pupae at a time to prevent proteolysis.

5.2.11 *Immunostaining of* Aedes aegypti *embryos [17, 18]*

Formula:
PEM:

1M PIPES	100 pt
0.5M EGTA	4 pt
1M $MgSO_4$	1 pt
dH_2O	To 1000 pt

Preparation:
1. Mix above reagents, adjust pH to 7.0.
2. Filter through 0.22 mm filter.
3. Add 1 ml 37% formaldehyde to 9 ml PEM to make PEM-F fresh before use.

Protocol:
1. Preparation of eggs and embryos.
   a. Use a squirt bottle to wet paper containing *Aedes aegypti* eggs. Use a paint brush with the squirt bottle to gently dislodge the eggs from the paper.
   b. Rinse eggs in water and dechorionate in 50% bleach for 3 min.
   c. Rinse eggs thoroughly with water.
   d. Use a squirt bottle to transfer eggs to a scintillation vial.

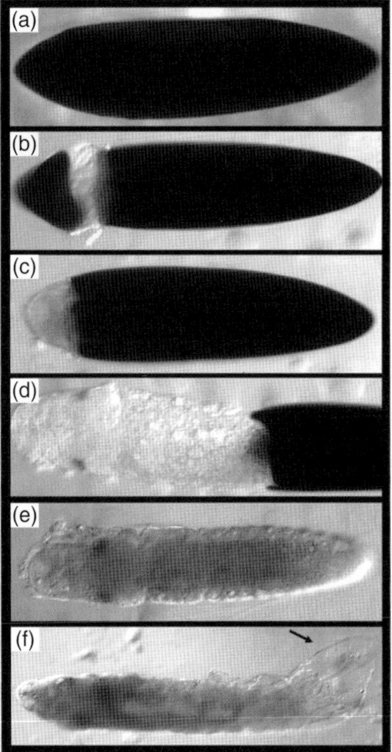

Fig. 5.5 Dissection methodology for *Ae. aegypti* embryos. The process of dissecting a fixed embryo is shown in a–f. It is critical to remove the black endochorion (a–d) and transparent serosal cuticle (arrowhead in f) in order to successfully perform immunohistochemical or *in situ* assays. (Source: Clemons *et al.* 2010. Reproduced with permission of Cold Spring Harbor Protocols.)

e. Remove water and add 10 ml PEM-F and incubate for 30 min at 60°C.
f. Replace PEM-F with chilled heptane for 5 min at −80°C.
g. Remove vial from freezer and add 10 ml methanol.
h. Hold vial under warm running water while shaking vigorously for 30 sec.
i. Remove the upper layer (heptane) and most of the methanol.
j. Wash three to four times with methanol to remove any remaining heptane.
k. Add 10 ml methanol. Embryos can be stored in methanol at −20°C.
l. Transfer eggs from methanol to a microfuge tube.
m. While keeping eggs submerged, remove the majority of the methanol and wash with 50% methanol in phosphate buffered saline (PBS), then with PBS and finally with PBS containing 0.1%Triton X-100 (PBTx).
n. Use a pipette to transfer ~15–30 eggs to a PBS dampened filter paper in a Petri dish.
o. Under a dissection microscope, use a tungsten needle or paintbrush to align the eggs such that the anterior (fattest) ends point towards you.
p. Prepare a depression slide with a strip of double sided sticky tape and lay the slide, tape-side down, over the eggs to stick the eggs to the tape.
q. Cover the eggs with PBS.
r. Under a dissection microscope, crack each egg by poking the anterior end with a tungsten needle. Insert the needle into the crack and remove the anterior tip (the tip will look like a cap when removed).
s. Use the blunt slide of the dissection needle to gently press against the outside of the egg to extrude the embryo from the endochorion.
t. Check to be sure the serosal cuticle is complete removed, if not, gently tease it away with a dissection needle.
u. Use a pipette to transfer the eggs to a microfuge tube containing 0.5 ml PBS.

2. Preparation of larvae and pupae.
   a. Fix larvae and pupae in PEM-F for 20 min and 1.5 hr, respectively.
   b. Wash two times in PBS, for 10 min each.
   c. Transfer to microfuge tubes with 0.5 ml PBS.
   d. Sonicate the tissues with an ultrasonic homogenizer outfitted with a microtip using an amplitude of 10% on a constant duty cycle.
   e. Wash pupae only in PBTx, then incubate in 0.3% SDS in PBTx for 30 min at room temperature.
3. Wash eggs/embryos/larvae/pupae three times in PBTx, for 10 min each.
4. Block in PBTx with 5% normal goat serum (blocking solution) for 30 min at room temperature.
5. Expose to 100 µl primary antibody diluted in blocking solution overnight at 4°C.
6. Wash three times for 5 min each, and then four times for 30 min each, in PBTx with rotation.
7. Block in blocking solution for 30 min at room temperature.
8. Expose to 200 µl secondary antibody diluted in blocking solution overnight at 4°C.
9. Wash three times for 5 min each, then four times for 30 min each, in PBTx with rotation.
10. For HRP-conjugated secondary antibodies, add 300 µl DAB solution, mix and incubate for 5 min. Add 30 µl of 0.3% H_2O_2 and develop to appropriate intensity of staining. Rinse three times for 5 min each in PBTx.
11. For fluorescently labeled secondary antibodies and developed DAB stains, wash two times in PBS for 2 min each.
12. Clear tissues in 50% glycerol at least overnight.
13. Replace 50% glycerol with 70% glycerol, then mount in fresh 70% glycerol.

5.2.12 *Immunostaining of mitotic chromosomes from* Drosophila *larval brain [19]*

Application:
Recommended as a modified preparation of chromosomes to preserve chromosomal proteins for immunostaining.

Formula:
Fixative:

Methanol	10 pt
Acetic Acid	2 pt
dH_2O	1 pt

Procedure:
1. Dissect larval brains
   a. Place larvae in a drop of saline and use two pairs of forceps to grab the larval mouth parts and the larval body midway toward the posterior part and pull apart. Remove brain with forceps.
2. Transfer to a drop of hypotonic solution on a siliconized slide for 2 min at room temperature.
3. Transfer to a drop of freshly prepared fixative on an 18 × 18 mm siliconized coverslip for 2 min. During fixation, use a syringe or tungsten needle to fragment the brain.
4. Place a clean non-siliconized slide on the coverslip, invert the sandwich, and squash between two to three sheets of blotting paper.

5. Freeze the slide in liquid nitrogen and flip off the coverslip with a razor blade.
6. Place slides in phosphate buffered saline (PBS) for 5 min at room temperature.
7. Place slides in PBS with 1% Triton X-100 (PBTx) for 10 min at room temperature.
8. Block in PBS with 1% BSA (PBS/BSA) for 30 min at room temperature.
9. Dilute primary antibody in PBS/BSA, decant excess PBS/BSA from slide, and add 15 µl diluted primary antibody. Cover with 18 × 18 mm non-siliconized coverslip, and incubate in a humid chamber at 4°C.
10. Carefully remove coverslip and wash the slides three times for 5 min each in PBS at room temperature.
11. Dilute fluorescently conjugated secondary antibody in PBS/BSA and add 15 µl to specimen. Cover with 18 × 18 mm coverslip and incubate in the dark in a humid chamber for 1 hr at room temperature.
12. Carefully remove coverslip and wash the slides three times for 5 min each in PBS at room temperature.
13. Counterstain with 0.05 µg/ml DAPI in 2X SSC for 5 min at room temperature.
14. Wash in PBS for 5 min at room temperature.
15. Mount in Vectashield® or similar antifade mounting medium and seal edges of coverslip with nail polish or rubber cement.

5.3 Immunolabeling of samples for Transmission Electron Microscopy (TEM)

Immunolabeling tissue samples before or after preparation for TEM provides utility or creates complications depending on the downstream goals of the labeling experiment. The reader is referred to MacDonald *et al.* [20] for a discussion of the most appropriate technique to use under a variety of circumstances.

5.3.1 *Post-embedding immunolabeling for TEM [20]*

Formula:
Blocking Buffer for TEM:

Preparation:
1. Gradually add 0.2 g BSA to 25 ml phosphate buffered saline (PBS) while stirring vigorously.
2. Prepare 10 ml of a 1% Tween-20 in PBS solution.
3. Add 25 µl cold water fish skin gelatin to the BSA mixture from step 1.
4. Add 0.5 ml of 1% Tween solution to BSA/gelatin mix and continue stirring.

Protocol:
1. Prepare resin sections of *Drosophila* tissues of interest.
2. Prepare a moist chamber by placing wet paper in the bottom of a 60-mm Petri dish.
3. Place an immunostaining pad made from polytetrafluoroethylene (PFTE; Ted Pella® 10526-1) on the wet paper.
4. Fill the top row of the pad with blocking buffer for TEM (~20 µl).
5. Float grids on the drops, section-side down, for 30 min. During the blocking step, the top should be on the humid chamber and the solution should be agitated on a rocker.
6. Fill the wells of the next row with primary antibody diluted in blocking buffer and transfer the grids to the primary antibody with a small wire loop, 2.5 mm in diameter.
7. Cover and rock for 1 hr at room temperature.

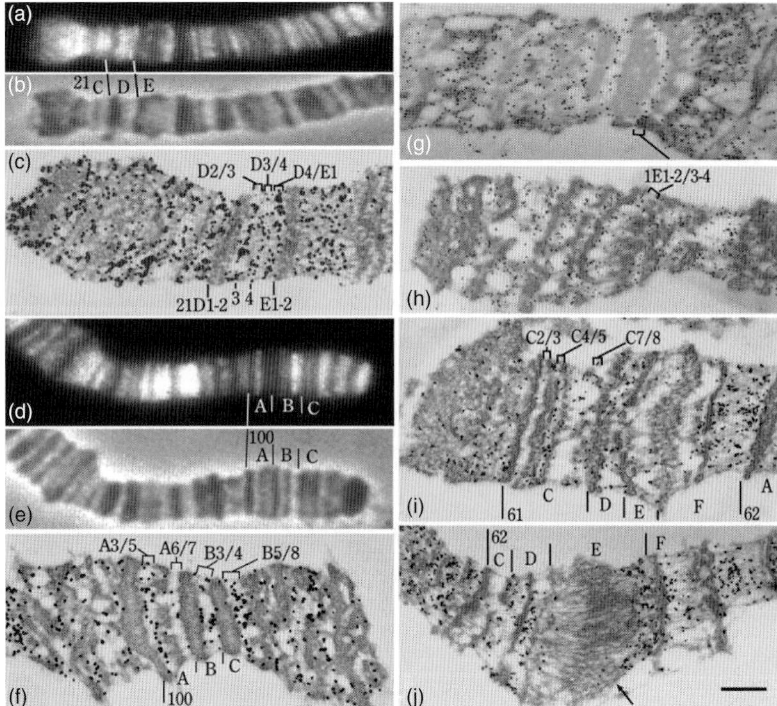

Fig. 5.6 Localization of the Z4 protein in the 21DE region of the 2L (a, b, c,), 100A–C region of the 3R (d, e, f,), 1E1-2–1E3-4 region of the X chromosome (g, h) and in the 61 (i) and 62 (j) regions of the 3L chromosomes. (a, d) immunofluorescent, (b, e) phase-contrast images; (c, f, g–j) immunogold EM images. The SuUR strain. (g) the larval stage; (h) 6-h prepupa. The arrow in j indicates the 62E puff; the region is almost unlabelled. Bar: 1 mm. (Source: Semeshin *et al.* 2001. Reproduced with permission of Springer.)

8. Rinse grids for 2 min in PBS with 2% Tween-20 (PBSTw).
9. Rinse four times for 2 min each in PBS.
10. Incubate in secondary antibody for 1 hr.
11. Rinse grids for 2 min in PBST, then four times for 2 min each, in PBS.
12. Fix in 0.5% glutaraldehyde in PBS for 5 min.
13. Rinse in PBS for 2 min.
14. Rinse four times for 2 min each, in H_2O.
15. Post-stain grids with 1% uranyl acetate in 70% methanol for 5 min.
16. Rinse in 70%, 50%, and 30% methanol, then two times with H_2O.
17. Post-stain grid with lead citrate for 3 min.
18. Allow the grids to air-dry and observe in microscope.

5.3.2 *Immunogold labeling of proteins in* Drosophila *polytene chromosomes [21]*

Protocol:

1. Dissect salivary glands in Ephurssi-Beadle's solution (see Appendix) supplemented with 0.1% Tween-20.
2. Pre-fix glands for 30 sec in PBS with 0.1% Tween-20 (PBSTw) with 1.8% formaldehyde, then fix in a drop of 45% acetic acid with 3.2% formaldehyde, and post-fix in a drop of 45% acetic acid for 1–2 min.
3. Squash glands in 45% acetic acid, freeze in liquid nitrogen, and pry off coverslips.
4. Fix in PBS with 1.5% formaldehyde for 1 hr and rinse in PBS.
5. Block in PBST with 0.1% BSA (10% non-fat dry milk or 2% fetal calf serum can substitute for the BSA).
6. Expose to diluted primary antibody for 2 hr at room temperature.

7. Wash in PBST four times for 5 min each.
8. Expose to secondary antibody labeled with either fluorochrome (Sigma®, 1:2000 dilution in any blocking solution) or 10 nm colloidal gold particles (Sigma®, 1:50 dilution in PBST + 0.1% BSA) for 1.5–2 hr.
9. Wash in PBST four times, for 5 min each.
10. For fluorochrome stains, mount slides in SlowFade® Antifade (Molecular Probes®) and image.
11. For colloidal gold stains, wash slides in ddH_2O six times for 3 min each and expose to Silver Enhancement Reagent (Boehringer Mannheim®) for 20–30 min to increase the size of the gold particles.
12. Wash again in ddH_2O six times 3 min each and dehydrate through a graded ethanol series (20%, 35%, 50%, 70%) for 5 min each.
13. Stain overnight in 1.5% uranyl acetate in 70% ethanol.
14. Proceed with standard dehydration and embedding for TEM.

5.3.3 *Immunogold TEM [22]*

Protocol:

1. Decapitate locusts, dissect the corpora cardiaca, and fix in a mixture of 2% glutaraldehyde and 2% paraformaldehyde in 0.1 M piperazinebis ethanesulphonic acid (PIPES) buffer pH 7.2 for 2 hr at 4°C.
2. Rinse glands in PIPES three times for 5 min.
3. Incubate in 0.1M PIPES with 1% paraformaldehyde and 30% glycerol for 30 min.
4. Freeze samples in liquid propane, transfer to liquid nitrogen.
5. Cryo-substitute at −90°C for 32 hr with 0.5% uranyl acetate in water-free methanol (refresh three times). During the last 9 hr of the 32-hr period, raise the temperature gradually from −90°C to −45°C in steps of 5°C/hr.
6. Immerse the tissues in Lowicryl HM20 embedding media (Polysciences®) at −45°C in steps of 1:1 methanol/Lowicryl for 2 hr, 1:2 methanol/Lowicryl overnight, and pure Lowicryl overnight.
7. Embed and polyermize at −45°C under UV light for at least 48 hr.
8. Cut ultrathin sections with an ultramicrotome and collect on 200-mesh nickel grids containing a Parlodion-supporting film.
9. Float grids on droplets of 0.5M glycine in Tris buffer (pH 8.2), two times for 10 min each.
10. Rinse three times for 5 min with Tris buffer (pH 8.2) containing 0.8% BSA and 0.1% teleostean gelatin.
11. Block in TBS with 0.8% BSA and 1% normal goat serum for 15 min.
12. Rinse three times for 5 min in Tris/BSA/gelatin buffer.
13. Float grids for 20 hr on primary antibody in Tris/BSA/gelatin buffer.
14. Rinse six times for 5 min in Tris/BSA/gelatin buffer.
15. Float grids on drops of colloidal 15-nm-gold-conjugated secondary antibody diluted in Tris/BSA/gelatin buffer.
16. Rinse three times for 5 min in Tris/BSA/gelatin buffer and six times for 5 min with Tris pH 8.2.
17. Fix for 10 min in 2% glutaraldehyde in Tris pH 8.2.
18. Rinse in Tris for 5 min, and five times for 2 min each with dH_2O.
19. Contrast sections with saturated uranyl acetate and lead citrate and visualize via TEM.

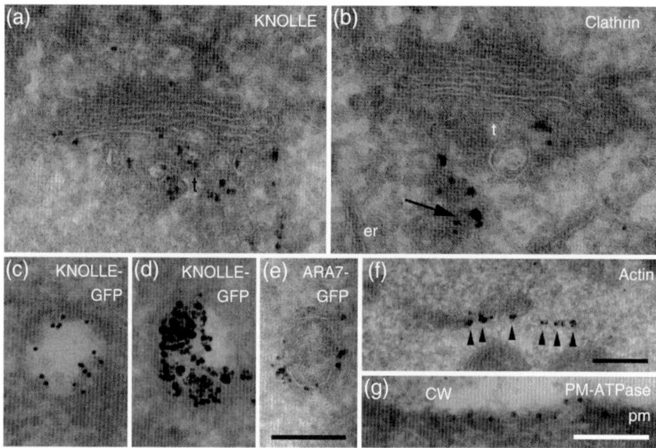

Fig. 5.7 Immunogold labelling of ultrathin cryo-sections of *A. thaliana* root tips after HPF, FS and rehydration. (a–g) Labelling of syntaxin-related protein KNOLLE (a), clathrin (b), GFP–KNOLLE (c, d), ARA7/RabF2b–GFP (e), actin (arrow heads, f) syntaxin-related protein KNOLLE and PM-ATPase (g) with Nanogold (a, b, d–g) or PAG-6 (silver-enhanced; c). Arrow in (b) indicates a coated vesicle, cw, cell wall; er, endoplasmic reticulum; pm, plasma membrane; t, TGN. Scale bars: (a–e, g) 250 nm; (f) 500 nm. Note: If FS was performed only with 0.5% GA and 0.5% UA (without OsO4), post-fixation has to be extended (2 hr; see Supplementary Table S1, variant 1). Without post-fixation, label density was very high, but at the expense of structural integrity, due to increased extraction (see Figure 2B in Dhonukshe *et al.*, 2007). Extraction happened most probably after 'opening' the cells by sectioning and during immunolabelling, because resin-embedded samples are well preserved after HPF and FS (without post-fixation). Nevertheless, it has to be kept in mind that optimum fixation conditions vary with samples and antigens, therefore fixative cocktail and fixative concentration should be adapted to the specimen used and to. (Source: Ripper *et al.* 2008. Reproduced with permission of John Wiley & Sons.)

5.3.4 *Immunolabeling of cryosections for TEM [23]*

Protocol:

1. Collect *Drosophila* embryos, dechorionate with 50% bleach and rinse.
2. Cryoprotect the embyros in 1-hexadecene in aluminium planchettes with 100–200 μm cavities.
3. Freeze embryos (using BAL-TEC High Pressure Freezing HPM 010) and free them from 1-hexadecene, below −100°C.
4. Transfer embryos to 2-ml cryotubes placed in a Leica FS unit at −90°C, filled with acetone containing 2–4% water, 0.1% OsO_4, 0.1–0.2% uranyl actetate and 0.5–2.5% methanol for 50–60 hr.
5. Warm samples to −60°C for 8 hr, then to −35°C. During this process, add 0.5% glutaraldehyde and incubate at −35°C for 8–10 hr.
6. Wash samples in acetone with 2–4% water and 0.5% glutaraldehyde, four times for 30–45 min each.
7. Warm samples to −20°C for 10–60 min, then warm to 0°C. During this warming process, perform the following series of incubations for 10 min each: 50% acetone with 0.38% glutaraldehyde at −20°C, 25% acetone with 0.25% glutaraldehyde at 0°C, 10% acetone with 0.25% glutaraldehdye at 0°C, and water with 0.25% glutaraldehyde at 0°C.
8. Fix for 90 min in water with 0.25% glutaraldehyde at 0°C.
9. Wash with water and then two times with water containing 50 mM glycine.
10. Embed samples in 10% gelatin in PBS.
11. Infiltrate step-wise with sucrose/PVP (1.8M sucrose/20% PVP).

12. Mount onto stubs, freeze in liquid nitrogen and section in a Leica Ultracut UCT cryro-ultramicrotome at −80°C for 300–500 nm sections for light microscopy, or at −110 to −115°C for 70–100 nm sections for electron microscopy.
13. Pick up cryrosections with a drop of 1:1 sucrose/PVP and transfer to coverslips for light microscopy or pioloform® (i.e., polyvinyl butyral), and carbon-coated grids for EM.
14. For immunolabeling in EM, block grids in PBS with 1% no fat dry milk and 0.5% BSA for 20 min.
15. Incubate with primary antibodies diluted in blocking buffer for 1 hr.
16. Wash five times, for 5 min each, in blocking buffer.
17. Detect primary antibodies with appropriate secondary antibodies conjugated to Nanogold (i.e. colloidal gold).
18. Wash two times, for 5 min each, in blocking buffer and three times, for 5 min each, in PBS.
19. Post-fix in PBS with 0.5% glutaraldehyde for 5 min.
20. Wash six times over 25 min with water.
21. Silver enhance gold particles with HQ Silver ™ (nanoprobes) for 8.5 min.
22. Wash six times, for 4 min each in water.
23. Stain with 1% uranyl acetate for 10 min and embed in 1.8% methyl cellulose containing 0.3–0.45% uranyl acetate.
24. For immunolabeling for light microscopy, mount cryosections onto coverslips and process the same as for EM, but use secondary antibodies conjugated to fluorescent markers. Counter stain in DAPI and embed in Mowiol® 4.88 (Calbiochem®) containing 25 mg/ml of the anti-fading reagent DABCO (Sigma®).

5.3.5 *Dual labeling for electron microscopy [24]*

Protocol:

1. Anesthetize flies expressing neuronal GFP on ice and restrain with wax and epoxy in a plastic chamber.
2. Open the tip of an antenna with forceps and inject dextran conjugated with tetramethylrhodamine and biotin (D-7162, Molecular Probes®).
3. Open a window on top of the head, remove the fat and air sacs above the brain, and gently remove the perineural sheathes around the antennal lobe.
4. Within 35 min after dye injection, dissect out the brain and fix in 0.5% glutaraldehyde, 4% formaldehyde in phosphate buffered saline (PBS) for 1 hr at room temperature.
5. Wash brains in PBS two times, for 5 min each and in PBS with 0.1% Triton X-100 (PBTx) for 5 min.
6. Incubate brains in a 1:100 dilution of rabbit anti-GFP antibody (A11122, Invitrogen™) in PBTx with 8% goat serum overnight at 4°C without shaking.
7. Wash thoroughly in PBTx.
8. Incubate brains in 0.5 mg/ml Alexa Fluor® 568-conjugated streptavidin (S-11226, Molecular Probes®) and a 1:50 dilution of goat anti-rabbit IgG conjugated with nanogold (#2004, Nanoprobes) in PBTx with 8% goat serum at 4°C overnight without shaking.
9. Wash brains in PBTx with 8% goat serum, then in PBS.
10. Post-fix brains in 2% glutaraldehyde for 30 min at room temperature.
11. Wash in PBS then in dH_2O.
12. Silver-enhance the brains (2012 HQ silver™, Nanoprobes) for 7 min at room temperature in the dark.

13. Rinse brains thoroughly in dH_2O for more than 5 min, then rinse in PBS.
14. Incubate in an avidin-biotin-peroxidase complex (ABC kit, Vectastain®; Vector Labs®) for 1 hr.
15. Wash the brains in PBS.
16. Incubate brains in 0.067% 3, 3'-diaminobenzidine tetrahydrochloride (DAB) for 30 min, then with 0.03% H_2O_2 in 0.067% DAB in Tris–HCl (pH 7.4) for 8 min.
17. Wash brains in Sorenson's buffer.
18. Post-fix in 0.2% osmium tetraoxide/Sorenson's buffer for 1 hr.
19. Wash brains in Sorenson's buffer.
20. Rinse in 0.1M acetate buffer and dehydrate through graded ethanol series.
21. Infiltrate brains with increasing concentrations of Epon™ resin.
22. Polymerize Epon in an oven at 60°C, overnight.
23. Cut ultrathin sections at 70 nm and mount on LUXFilm™ TEM supports (#12812-CU, Ted Pella®, Inc.).
24. Post stain with 2% uranyl acetate in 50% methanol and Reynolds lead citrate, and image.

5.4 **Proliferation assays**

Application:
Recommended for detection of proliferation in the moth *Manduca sexta*.

5.4.1 *BrdU labeling I [25]*

Protocol:
1. Inject *Manduca sexta* insects at the appropriate development stages with 50 μl of BrdU [5-bromo-2'-deoxyuridine] (50 mg/ml in dH_2O).
2. Maintain the animals for 12 hr at 26°C, then remove the nerve cords and wash for 10 min in 30% ethanol, followed by 10 min in 70% ethanol.
3. Fix the tissues for 2 hr in Carnoy's fixative (see Chapter 2).
4. Rehydrate through an ethanol series.
5. Wash in PBS.
6. Expose tissues to 0.5 mg/ml collagenase (Sigma®, type IV) in PBS with 1 mM $CaCl_2$ for 15 min at room temperature.
7. Rinse two times, for 5 min each, in PBS with 0.3% Triton X-100 (PBTx).
8. Denature the DNA in the tissue in 2N HCl in PBS for 1 hr at room temperature.
9. Wash three times in PBS for 10 min each.
10. Incubate tissue in a 1:100 dilution mouse anti-BrdU in PBS with 5% normal goat serum for 24 hr at 4°C.
11. Wash in PBTx then PBS for 1 hr each.
12. Expose to a 1:200 dilution of goat anti-mouse secondary conjugated to Cy3 in PBS with 1% normal goat serum for 24 hr at 4°C.
13. Wash in PBS, mount in 80% glycerol and image under appropriate fluorescent microscope settings.

5.4.2 *Brdu labeling II [26]*

Application:
Recommended for detection of proliferation in the cricket *Archeta domesticus*.

Protocol:
1. Administer BrdU as appropriate for the age/stage of the insect.
   a. Inject newly emerged adult females with 10 μl of BrdU (40 mg/ml in saline).
   b. Inject 9th or 10th instar larvae with 5 μl of BrdU (40 mg/ml in saline), respectively.

c. For hatching and mid-larval animals, incorporate BrdU into their drinking water at 10 mg/ml over 4 days.

d. For embryos, remove chorion in 50% bleach and permeabilize the vitelline membrane using discrete incisions. Incubate embryos in 20 mg/ml BrdU in saline for 2 hr at 30°C.

2. Dissect cerebral ganglia in saline and fix dissected ganglia; dechorionated embyros in Carnoy's fixative for 2 hr (see Chapter 2).
3. Rehydrate tissues in phosphate buffered saline (PBS) three times for 15 min each, then rinse in PBS with 0.5% Triton X-100 (PBTx) two times for 20 min each.
4. To hydrolyze the DNA, incubate tissues in 2N HCl diluted in PBS with 0.3% Triton X-100 for 1 hr at room temperature.
5. Wash three times for 15 min each in PBTx.
6. Treat tissues with 0.01% trypsin in saline for 25 min at 37°C.
7. Rinse tissues thoroughly in PBS.
8. IHC to detect BrdU can be performed with either whole mounts or paraffin embedded tissue sections.

a. For whole mounts, proceed to step 9.

b. For tissue sections, rinse the tissues in dH_2O, dehydrate through a graded series of ethanol, clear, infiltrate with paraffin and embed in paraffin.

c. Deparaffanize 5 µm sections in toluene, rehydrate through a graded ethanol series to dH_2O, and rinse in PBTx.

9. Block whole mounts or tissue sections in 5% normal goat serum in PBS for 1 hr at 4°C.
10. Expose to a 1:100 dilution of anti-BrdU antibody in PBS overnight at 4°C.
11. Wash thoroughly in PBTx.
12. Expose to peroxidase conjugated goat anti-mouse IgG antibody for 3 hr at 4°C.
13. Wash thoroughly in PBTx.
14. Detect peroxidase using DAB chromogen and rinse in dH_2O.
15. Counterstain with 0.4% indigo carmine in a saturated solution of picric acid.
16. Dehydrate through graded ethanol, clear in methyl salicylate and mount as appropriate for whole mounts or tissue sections.

5.4.3 *BrdU labeling III [27]*

Application:
Recommended for the detection of proliferation in the legs and larvae of *Manduca sexta*.

Protocol:

1. Label appropriate tissues with BrdU and fix:

a. For larvae, mix 10 mg/ml BrdU stock solution into the larval food mixture to a concentration of 0.2–0.3 mg/ml prior to hardening of the food. Allow larvae to feed on BrdU for an appropriate period for the experiment. Dissect and fix larval tissues in 3.7% formaldehyde in phosphate buffered saline (PBS) for 2 hr at room temperature, or overnight at 4°C.

b. Anesthetize *Manduca sexta* on ice and inject the abdomen with 20 µg/µl BrdU with a 100-µl Hamilton syringe. After 6 hr, spread and pin down the ventral thorax and the thoracic legs. Remove muscle, fat body, and trachea and fix in 3.7% formaldehyde in phosphate buffered saline (PBS) for 2 hr at room temperature, or overnight at 4°C. Use a sharp razor blade to split open individual legs into either dorsal/ventral or anterior/poster halves. Remove the nerves, tracheae, muscles, and fat body to expose the epidermis.

2. Rinse in PBS with 1% Triton X-100 (PBTx).
3. Acid-hydrolyze the DNA in 2N HCl in PBTx for 1 hr.
4. Wash thoroughly in PBTx.
5. Block in 5% normal donkey serum in PBTx for 25 min.
6. Expose to a 1:200 dilution of anti-BrdU (Becton Dickinson Immunocytometry Systems®) in PBTx.
7. Wash thoroughly in PBTx.
8. Expose to a 1:1000 dilution of Alexa Fluor® 488 goat anti-mouse antibody (Molecular Probes®) in PBTx.
9. Wash thoroughly in PBTx.
10. To detect nuclei, treat tissues with 100 μg/ml RNAse for 30 min and counterstain with a 1:200 dilution of 0.4 mg/ml propidium iodide.
11. Dehydrate tissues through a graded alcohol series, clear in xylene and mount.

5.4.4 *EdU incorporation assay [28]*

Application:
Recommended for the detection of proliferating cells.

Protocol:
1. Prepare reagents from Click-It reaction kit from Invitrogen™ (Cat. #C10337).
2. Dissect *Drosophila* brains in fresh Grace's unsupplemented cell culture medium.
3. Add an equal volume of 200 μM EdU (5-ethynyl-2′-deoxyuridine) solution in DMSO and incubate brains in the dark for 30 min at room temperature.
4. Remove the EdU and rinse brains in phosphate buffered saline (PBS).
5. Fix brains in fresh 3.7% formaldehyde in PBS in the dark for 15 min at room temperature.
6. Rinse brains two times in PBS.
7. Permeabilize the tissues in PBS with 0.1% Triton X-100 in the dark for 15 min at room temperature.
8. Rinse brains two times in PBS.
9. Incubate brains in the Click-iT™ Imaging Kit (Invitrogen®) for 30 min.
10. Rinse brains two times in the reaction rinse buffer (Invitrogen®).
11. Counterstain with Hoescht 33342 (a specific stain for AT-rich regions of double-stranded DNA, for 10 min).
12. Wash brains four times in PBS.
13. Mount on polylysine coated slides with Vectashield®.

5.5 Methods to detect specific proteins

Application:
Recommended as an alternative fixative and protocol to allow detection of octopamine by IHC.

5.5.1 *Octopamine immunolabeling [29]*

Protocol:
1. Deeply anesthetize adult spiders with CO_2 and fix the whole animal by perfusion through the heart using 4% 1-ethyl-3-(3-dimethylaminopropyl) carbodiimide hydrochloride (EDAC) in phosphate buffered saline (PBS).
2. Remove the legs, slit ventrally along the middle of each segment and fix for an additional 2 hr at 4°C, in 4% EDAC.
3. Wash four times in PBS and permeabilize in 0.5% Triton X-100 in PBS (PBTx) for 30 min.

4. Block in PBTx with 5% normal goat serum, 3% non-fat dried milk, and 1% BSA (blocking solution) for 1 hr.
5. Expose to a 1:500–1:1000 dilution of primary rabbit anti-octopamine antibody (AB1799, (Chemicon ®)) in the blocking solution overnight at 4°C.
6. Wash four times in PBTx and expose to 10 μg/ml secondary goat anti-rabbit Alexa Fluor® 488 antibody (Molecular Probes®) in blocking solution overnight at 4°C.
7. Wash four times in PBTx and five times in PBS.
8. Mount in a medium containing Mowiol® dissolved in 0.3 M Tris buffer enriched with 2.5% 1,4-diazabicyclo(2,2,2,) octane and image under a fluorescent microscope.

5.5.2 *Double labeling with biocytin and anti-octopamine [30]*

Formula:

Fixative:

Paraformaldehyde	2.5%
Glutaraldehyde	1.5%
Sodium cacodylate	0.1M
Sodium metabisulfite (SMB)	1%

Protocol:

1. Crudely dissect honeybee or cockroach brains (leaving connective tissue intact) in appropriate physiological saline solution containing 0.5% SMB at pH 7.0–7.2 (see Appendix).
2. Pin the brain's connective tissue onto Sylgard® in a small dish with saline solution.
3. Cut a tritocerebral connection between the brain and subesophageal ganglion, place the anterior end in a pool of 1% biocytin, isolated by a wall of Vaseline and incubate for 16–20 hr at 4°C.
4. Complete dissection of brain in fixative and fix in fresh fixative for 20–24 hr at room temperature.
5. Saturate double bonds with 0.5% sodium borohydride in 0.05 M Tris–HCl buffer, with 0.45% SMB, pH 7.4 for 20 min.
6. Wash in Tris–HCl SMB buffer four times for 20 min each.
7. Dehydrate brains in a graded ethanol series and permeabilize in propylene oxide.
8. Re-hydrate in a reversed graded ethanol series and wash in Tris–HCl SMB buffer four times for 20 min each.
9. Embed brains in gelatin and cut gelatin sections at 40–50 μm with a Vibratome.
10. Wash sections in Tris–HCl and expose to a 1:1000 dilution of streptavidin Cy2 conjugate (Molecular Probes®) in Tris–HCl at room temperature.
11. Wash sections four times for 25 min each in Tris–HCl SMB buffer with 0.5% Triton X-100.
12. Block in 5% normal swine serum in Tris–HCl with 0.5% Triton X-100 for 1 hr.
13. Expose to a 1:100 dilution of primary rabbit anti-octopamine antibody in Tris–HCl for 48 hr at room temperature.
14. Wash four times 25 min each in Tris–HCl with 0.5% Triton X-100 and expose to a 1:500 dilution of goat anti-rabbit Texas red-conjugated IgG (Molecular Probes®) for 24 hr at room temperature.
15. Wash four times 25 min each in Tris–HCl, mount sections on slides and coverslip under 80% glycerol.

References

1. Bonaccorsi S., Giansanti M.G., Cenci G. & Gatti M. Formaldehyde fixation of *Drosophila* testes. *Cold Spring Harbor Protoc.* **8**, prot067355 (2012).
2. Bonaccorsi S., Giansanti M.G., Cenci G. & Gatti M. Methanol-acetone fixation of *Drosophila* testes. *Cold Spring Harbor Protoc.* **10**, 1270–1272 (2011).
3. Wu J.S. & Luo L. A protocol for dissecting *Drosophila melanogaster* brains for live imaging or immunostaining. *Nat. Protoc.* **1**, 2110–2115 (2006).
4. Ramos R.G., Machado L.C. & Moda L.M. Fluorescent visualization of macromolecules in *Drosophila* whole mounts. *Methods Mol Biol* **588**, 165–179 (2010).
5. Miller K.G., Field C.M. & Alberts B.M. Actin-binding proteins from *Drosophila* embryos – A complex network of interacting proteins detected by F-actin affinity-chromatography. *J. Cell Biol.* **109**, 2963–2975 (1989).
6. Swevers L. & Iatrou K. The orphan receptor BmHNF-4 of the silkmoth *Bombyx mori*: ovarian and zygotic expression of two mRNA isoforms encoding polypeptides with different activating domains. *Mech. Dev.* **72**, 3–13 (1998).
7. Sweeney S.T., Hidalgo A., de Belle J.S. & Keshishian H. Antibody staining of the central nervous system in adult *Drosophila*. *Cold Spring Harbor Protoc.* **2**, 235–238 (2012).
8. Chiang A.S. *et al.* Three-dimensional reconstruction of brain-wide wiring networks in *Drosophila* at single-cell resolution. *Curr. Biol.* **21**, 1–11 (2011).
9. Morante J. & Desplan C. Dissection and staining of *Drosophila* optic lobes at different stages of development. *Cold Spring Harbor Protoc.* **6**, 652–656 (2011).
10. Calabria L.K. *et al.* Comparative analysis of two immunohistochemical methods for antigen retrieval in the optical lobe of the honeybee *Apis mellifera*: Myosin-v assay. *Biol. Res.* **43**, 7–12 (2010).
11. Wicklein M. & Strausfeld N.J. Organization and significance of neurons that detect change of visual depth in the hawk moth *Manduca sexta*. *J. Comp. Neurol.* **424**, 356–376 (2000).
12. Wildman M., Ott S.R. & Burrows M. GABA-like immunoreactivity in nonspiking interneurons of the locust metathoracic ganglion. *J. Exp. Biol.* **205**, 3651–3659 (2002).
13. Eystathioy T., Swevers L. & Iatrou K. The orphan nuclear receptor BmHR3A of *Bombyx mori*: hormonal control, ovarian expression and functional properties. *Mech. Dev.* **103**, 107–115 (2001).
14. Burrows M., Borycz J.A., Shaw S.R., Elvin C.M. & Meinertzhagen I.A. Antibody labelling of resilin in energy stores for jumping in plant sucking insects. *PLOS One* **6**, e28456 (2011).
15. Boyan G., Herbert Z. & Williams L. Cell death shapes embryonic lineages of the central complex in the grasshopper *Schistocerca gregaria*. *J. Morph.* **271**, 949–959 (2010).
16. Wang W. *Drosophila* pupal abdomen immunohistochemistry. *J. Vis. Exp.* **2**, (2011).
17. Clemons A., Flannery E., Kast K., Severson D. & Duman-Scheel M. Immunohistochemical analysis of protein expression during *Aedes aegypti* development. *Cold Spring Harbor Protoc.* **10**, (2010).
18. Clemons A. *et al.* Fixation and preparation of developing tissues from *Aedes aegypti*. *Cold Spring Harbor Protoc.* **10**, (2010).
19. Pimpinelli S., Bonaccorsi S., Fanti L. & Gatti M. Immunostaining of mitotic chromosomes from *Drosophila* larval brain. *Cold Spring Harbor Protoc.* **9**, (2011).
20. McDonald K.L., Sharp D.J. & Rickoll W. Immunolabeling of thin sections of *Drosophila* tissues for transmission electron microscopy. *Cold Spring Harbor Protoc.* **7**, 838–841 (2012).
21. Semeshin V.F., Andreyeva E.N., Shloma V.V., Saumweber H. & Zhimulev I.F. Immunogold electron microscope localization of proteins in *Drosophila* polytene chromosomes: applications and limitations of the method. *Chromosome Res.* **10**, 429–433 (2002).
22. Harthoorn L.F., Diederen J.H.B., Oudejans R.C.H.M. & Van der Horst D.J. Differential location of peptide hormones in the secretory pathway of insect adipokinetic cells. *Cell Tissue Res.* **298**, 361–369 (1999).

23. Ripper D., Schwarz H. & Stierhof Y.D. Cryo-section immunolabelling of difficult to preserve specimens: advantages of cryofixation, freeze-substitution and rehydration. *Biol. of the Cell* **100**, 109–123 (2008).
24. Tanaka N.K., Dye L. & Stopfer M. Dual-labeling method for electron microscopy to characterize synaptic connectivity using genetically encoded fluorescent reporters in *Drosophila*. *J. Neuorsci. Methods* **194**, 312–315 (2011).
25. Hesterlee S. & Morton D.B. Identification of the cellular target for eclosion hormone in the abdominal transverse nerves of the tobacco hornworm, *Manduca sexta*. *J. Comp. Neurol.* **424**, 339–355 (2000).
26. Malaterre J. *et al.* Development of cricket mushroom bodies. *J. Comp. Neurol.* **452**, 215–227 (2002).
27. Tanaka K. & Truman J.W. Development of the adult leg epidermis in *Manduca sexta*: contribution of different larval cell populations. *Dev. Genes and Evol.* **215**, 78–89 (2005).
28. Gouge C.A. & Christensen T.W. *Drosophila* Sld5 is essential for normal cell cycle progression and maintenance of genomic integrity. *Biochem. Biophys. Res. Comm.* **400**, 145–150 (2010).
29. Widmer A., Hoger U., Meisner S., French A.S. & Torkkeli P.H. Spider peripheral mechanosensory neurons are directly innervated and modulated by octopaminergic efferents. *J. Neurosci.* **25**, 1588–1598 (2005).
30. Sinakevitch I., Niwa M. & Strausfeld N.J. Octopamine-like immunoreactivity in the honey bee and cockroach: Comparable organization in the brain and subesophageal ganglion. *J. Comp. Neurol.* **488**, 233–254 (2005).

6 Use of genetic markers in insect histology

6.1 Introduction

The use of genetic markers makes it possible to identify specific anatomical regions and cell types, follow gene expression patterns, or localize proteins of interest. This chapter outlines genetic markers that can be visualized in insect tissues using histological techniques. A number of the methods featured in this chapter make use of common genetic strategies used to drive the expression of detectable markers. A transcriptional activator (a protein that activates the transcription of one or more genes) can be linked to a gene promoter of choice to drive transgene expression in a specific cell type or during a particular developmental stage. Combining transgene expression systems with tools to induce genetic recombination events allows for exquisite spatiotemporal control of gene expression and provides a wide variety of markers that can be detected histologically.

6.1.1 *X-Gal labeling of* Drosophila *neural tissues [1]*

Application:
Recommended for staining neural tissues for the activity of the *E. coli LacZ* gene product, β-galactosidase, using the substrate X-gal, which produces a blue precipitate that can be visualized by light microscopy.

Formula:
Solution A: X-gal (5-bromo-4-chloro-3-indolyl-ß-D-galactopyranoside)

Preparation:
8% X-gal in DMSO, stored in the dark in small aliquots at −20°C).

Solution B: X-gal Staining Solution for CNS

1 M Sodium phosphate buffer	10 pt
1 M NaCl	150 pt
1M $MgCl_2$	1 pt
1M $K_4Fe_2(CN)_6$	10 pt
1M $K_3Fe_3(CN)_6$	10 pt
Triton X-100	0.1%
dH_2O	819 pt

Insect Histology: Practical Laboratory Techniques, First Edition.
Pedro Barbosa, Deborah L. Berry and Christina S. Kary.

Companion Website: www.wiley.com/go/barbosa/insecthistology

Procedure:

1. Dissect adult fly brains as described in Chapter 5 of this manual (Preparation of *Drosophila* brains for IHC Staining).
2. Prewarm 0.5–1 mL of X-gal staining solution for CNS to 37°C. Pre-warm an aliquot of 8% X-gal in DMSO to 37°C in the dark.
3. To 0.5 mL of X-gal staining solution for CNS, add 12.5 µL of 8% X-gal in DMSO. Maintain at 37°C. If precipitate forms, heat the solution to 65°C until it dissolves.
4. Place the dissected CNS in 150–200 µL of the solution from Step 3 in a 96-well microtiter plate. Cover plate with Parafilm®.
5. Place the plate in the heating block. Cover with a cardboard box or another lid that excludes light.
6. Incubate at 37°C until staining is complete.
7. Dehydrate the CNS through an ethanol wash series of increasing concentrations, 30%, 50%, and 70%, followed by three washes in absolute ethanol (10–15 min in each).
8. Incubate in Histo-Clear® (National Diagnostics) for 30 min.
9. Mount on a microscope slide in Araldite®.

Note:

1. During the procedure, keep all solutions containing X-Gal at 37°C. At lower temperatures, the X-Gal will crystallize out of solution.

6.1.2 *β-Galactosidase activity staining of frozen adult* Drosophila *retinas [2]*

Application:

Recommended for the identification of specific cell types in adult *Drosophila* retinas using β-galactosidase staining that can be visualized with light microscopy.

Formula:

Solution A: 10X PBS without potassium salt

5 M NaCl	260 pt
1 M $Na_2HPO_4 \bullet 2H_2O$	70 pt
1 M $NaH_2PO_4 \bullet 2H_2O$	30 pt
dH_2O	640 pt

Preparation:

1. Sterilize by autoclaving.

Solution B: X-gal staining solution for *Drosophila* eyes, pre-warmed to 37°C.

0.2 M Na_2HPO_4	1.8 pt
0.2 M NaH_2PO_4	0.7 pt
5 M NaCl	1.5 pt
1 M $MgCl_2$	0.05 pt
50 mM $K_3Fe(CN)_6$	3 pt
50 mM $K_4Fe(CN)_6 \bullet 3H_2O$	3 pt
dH_2O	40 pt

Note:

1. This solution is stable for months when stored at room temperature in a foil-wrapped container.

Solution C: X-gal

8% X-Gal in DMF [*N,N*-dimethylformamide].

Note:

1. Store in 50 µL aliquots at −20°C.

Preparation:

1. To prepare fresh X-gal staining solution, prewarm the staining solution buffer at 37°C for 5 min.
2. Add 8 µL of 8% X-gal stock solution per 300 µL of staining solution buffer. Prewarming decreases precipitation of X-gal crystals.

Procedure:

1. Anesthetize *Drosophila* adults on a CO_2 pad (which modifies the atmosphere surrounding the object in question by displacing oxygen and ethylene with carbon dioxide gas).
2. Cut off the heads using a #11 scalpel blade.
3. Embed up to 12 heads in partially frozen Optimal Cutting Temperature (OCT) compound, Tissue-Tek® (Ted Pella®) on a cryostat chuck and orient using fine forceps as follows:
   a. Freeze a layer of OCT compound onto the chuck by placing it in a dry ice/ethanol bath.
   b. When the first layer is almost frozen, remove it from the dry ice/ethanol bath and add another drop of OCT compound to the frozen layer.
   c. Place up to 12 eyes in the unfrozen OCT compound. Use forceps to quickly orient the eyes (do this under the microscope) for either horizontal or transverse sections.
   d. Once oriented, put the chuck back in the dry ice/ethanol bath to freeze.
4. Place the chuck in the cryostat chamber and pre-trim the block with a razor blade. Allow the block to equilibrate to cryostat temperature (-14°C to -18°C) for 20–30 min.
5. Cut sections (10–14-µm thick) and transfer to slides.
6. To dry the sections onto the slides, heat slides on a drying plate pre-set to 40°C for no more than 1 min or air-dry at room temperature for 30 min. Do not overheat or over-dry.
7. Fix the sections in 2% glutaraldehyde/PBS at room temperature for 15–20 min in a Coplin jar.
8. Wash the slides three times in PBS (5 min for each wash).
9. Blot off excess PBS and place slides in humidified chamber.
10. Leave slides at room temperature until dry (5–10 min). Slides can be left for several days at this point with no loss of enzymatic activity.
11. Cover sections with 50 µL of pre-warmed X-gal staining solution and cover with a coverslip. Incubate tissue in staining solution at 37°C from 15 min to overnight (the length of time depends on the strength of expression of the *lacZ* line being used).
12. Once staining is complete, rinse slides twice in PBS (5 min for each wash).
13. The coverslip will float off the slide in the first rinse.
14. Mount slides in one of the following ways:
   i. Mount slides in 80% glycerol in PBS.
   ii. Dehydrate slides through a graded ethanol series (once in 30%, 50%, 70%, 90%, and twice in 100%; for 10 min each) and mount in Glycerol (80% in 1X PBS).

6.2 Inducible genetic markers

Inducible systems can be used to drive the expression of a transgene of choice by linking a genetic marker to a gene promoter. [3, 4] In the GAL4/UAS system, a construct linked to an upstream activating sequence (UAS) can be induced by any other gene promoter driving expression of the yeast GAL4 transcriptional activator. Once expressed, GAL4 binds the UAS sequence and drives expression of a visible marker. Inducible genetic strategies akin to the GAL4/UAS system allow spatiotemporal control of transgene expression, combined with readily detectible markers.

Inducible genetic markers, used in combination, provide myriad opportunities for the detection of gene expression patterns or proteins of interest. Reporters that fluoresce under excitation, such as Green Fluorescent Protein (GFP), can be detected in live or fixed tissues using fluorescence microscopy. [5] When combined with the GAL4/UAS system, fluorescent reporters can be used to visualize gene expression patterns or localize tagged proteins in specific cell types, in fixed or live specimens. [4] Through the use of differently colored fluorescent tags and/or dyes that specifically mark organelles, multiple proteins can be co-localized in subcellular compartments. [6] Fluorescent markers with improved characteristics, such as increased stability or brightness, further allow for the quantification of molecules visualized within a cell. [7]

6.2.1 *A visible dominant marker for insect transgenesis [8]*

Application:
Recommended for transgene expression of a visible genetic marker in a variety of insect taxa, including the silkworm *Bombyx mori* (L.), the ladybird beetle *Harmonia axyridis (Pallas)*, and the fruit fly *Drosophila melanogaster*.

This technique makes use of the GAL4/UAS system to overexpress the arylalkylamine-N-acetyl transferase (aaNAT) gene, linked to enhanced GFP (eGFP), a GFP variant reporter with increased stability and brightness compared to standard GFP. eGFP can be driven by any specific gene promoter such as aaNAT. Expression of aaNAT-eGFP leads to melanin pigment color changes (detectable by light microscopy) as well as expression of eGFP (detectable by fluorescence microscopy) in newly hatched larvae of the silkworm, *Bombyx mori*. This technique can also be used to drive marker expression during different developmental stages; for example,

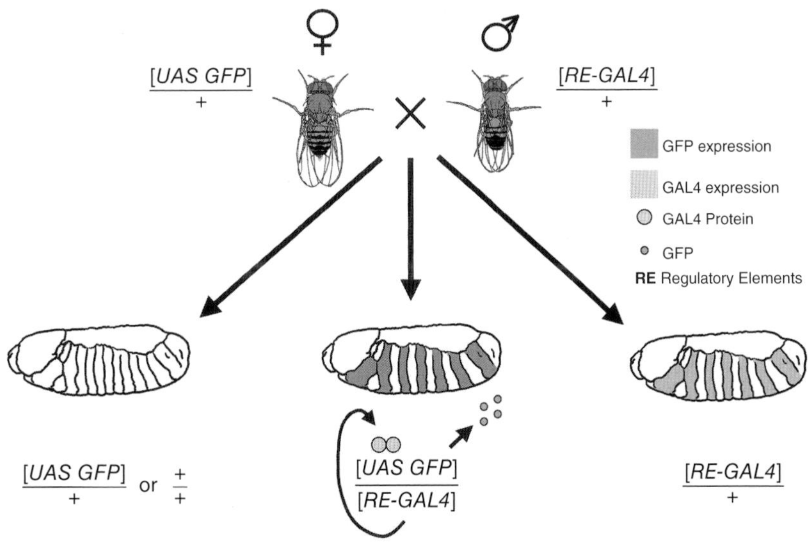

Fig. 6.1 The bipartite UAS/GAL4 system in *Drosophila*. When females carrying a UAS responder (*UAS-GFP*) are mated to males carrying a GAL4 driver progeny containing both elements of the system are produced. The presence of GAL4 in an alternating segmental pattern in the depicted embryos then drives expression of the UAS responder gene in a corresponding pattern. (Source: Duffy 2002. Reproduced with permission of John Wiley & Sons.) See plate section for the color version of the figure.

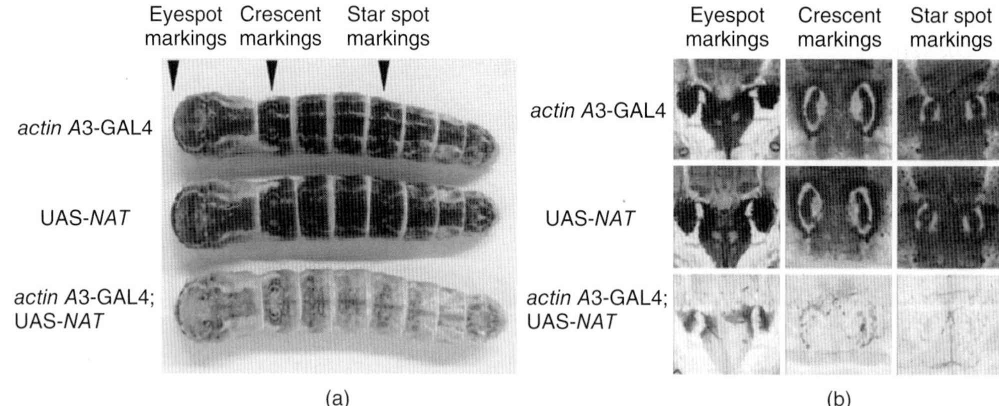

Fig. 6.2 The fifth-instar larvae shown here have the dominant *striped* (p^S) genetic background. (a) Dorsal view of the larvae. The positions of the markings magnified in (b) are indicated by the arrowheads. (b) Magnified images of the larval markings. (Source: Osanai-Futahashi *et al.* 2012. Reproduced with permission of Nature.) See plate section for the color version of the figure.

expression of the ß-alanyl-dopamine synthetase (Bm-ebony) gene leads to body color changes in later-stage silkworm larvae.

Procedure:

1. Generate transgenic silkworms expressing the desired pigment-altering gene linked to a florescent marker using established egg injection techniques for transgenesis.
2. Visually inspect transgenic silkworm larvae at the desired stage for larval coloration.
3. Use a combination of light microscopy, to image pigmentation differences, and fluorescent microscopy, to identify eGFP transgene expression in insect larvae.

6.2.2 *Targeted gene expression in the transgenic* Aedes aegypti *using the Gal4-UAS system [9]*

Application:

Recommended for the targeted expression of detectable genetic markers in *Aedes aegypti*. This method makes use of the GAL4/UAS system to express enhanced GFP (eGFP) in the previtollogenic fat body of the yellow fever mosquito, *Aedes aegypti*. This system can be used for the visual observation of adult females to detect eGFP expression in the fat body of transgenic mosquitoes after blood meal activation. This method thus allows visualization of transgenic markers in mosquitoes.

Formula:

Aedes physiological saline (APS) buffer solution:

1 M NaCl	150 pt
1 M $CaCl_2.2H_2O$	1.4 pt
1 M KCl	2 pt
1 M $NaHCO_3$	1.2 pt

Preparation:

1. Bring to 1 L with dH_2O.

Procedure:

1. Obtain mosquitoes expressing eGFP in the fat body using the driver construct pBac [3xP3-EGFP *afm,Vg-Gal4*] in transgenic mosquitoes.

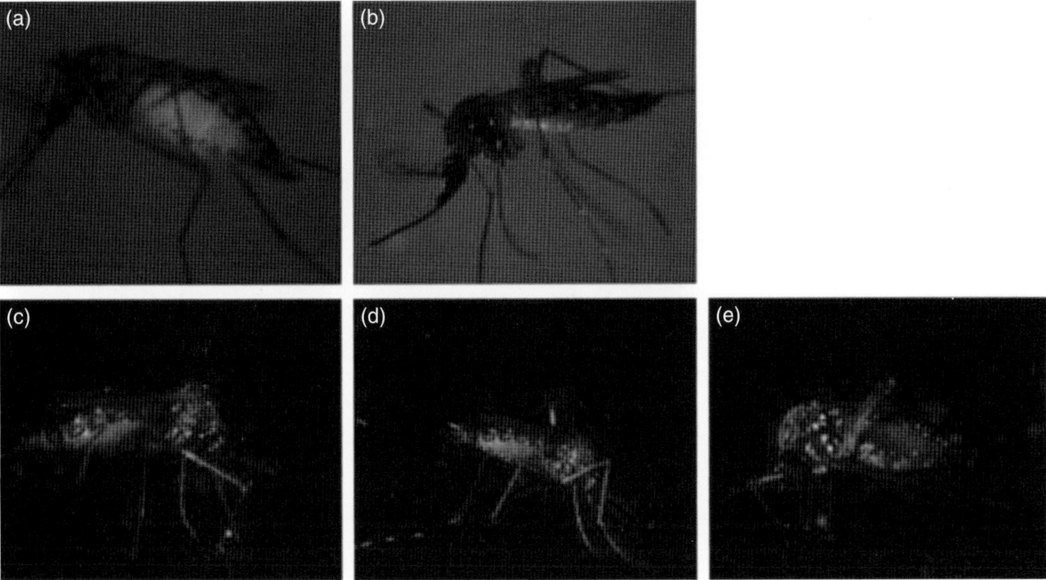

Fig. 6.3 Tissue- and stage-specific EGFP reporter expression in the *Vg-Gal4/UAS-EGFP* hybrid female mosquitoes after blood meal activation. Fluorescent images of adult vitellogenic females were captured using an EGFP filter set. Expression of EGFP reporter was detected only in the fat body of hybrid *Vg-Gal4/UAS-EGFP* females 24 (a) and 48 (b) h post-blood meal (PBM). No EGFP fluorescence was observed in the fat body of vitellogenic (24 h PBM) females of the *Vg-Gal4* driver (c) and *UAS-EGFP* responder (d) transgenics, or non-transgenic UGAL strain (e). (Source: Kokoza & Raikhel 2002. Reproduced with permission of John Wiley & Sons.) See plate section for the color version of the figure.

2. Dissect and fix fat bodies in 3.7% formaldehyde in APS buffer solution for 20 min at room temperature.
3. Wash fat bodies three times for 10 min in APS and transfer into a DNA-staining solution containing 5 μM Hoescht® 33342 (H1399, Invitrogen™) in APS for 10 min.
4. Mount the fat bodies in VectaShield-H1000® (Vector Laboratories) mounting media, and image using fluorescence microscopy.

6.2.3 *Live imaging of* Drosophila melanogaster *embryonic hemocyte migrations [10]*

Application:

Recommended for the live imaging of fluorescently labeled hemocytes, the embryonic macrophages in *Drosophila melanogaster*. Through the use of the GAL4/UAS system, this technique can be used to express a variety of fluorescently tagged markers, such as UAS-GFP, specifically in hemocytes in *Drosophila* embryos. Using florescence microscopy, the dispersal of hemocytes throughout the embryo can then be followed during development. This protocol provides a robust method to maintain embryos in a healthy condition for live imaging, and can be easily adapted to image other cell types by choosing different GAL4 drivers.

Procedure:

1. Obtain appropriate *Drosophila* lines containing a hemocyte-specific Gal4 driver (e.g. *srp-Gal4*[2]) and a genetically encoded fluorescent reporter under UAS control (e.g. *uas-GFP*). Flies homozygous for *crq-Gal4, uas-GFP* or *srp-Gal4, uas-GMA* (GFP fused to the actin-binding domain of the protein moesin) are particularly useful for imaging purposes.

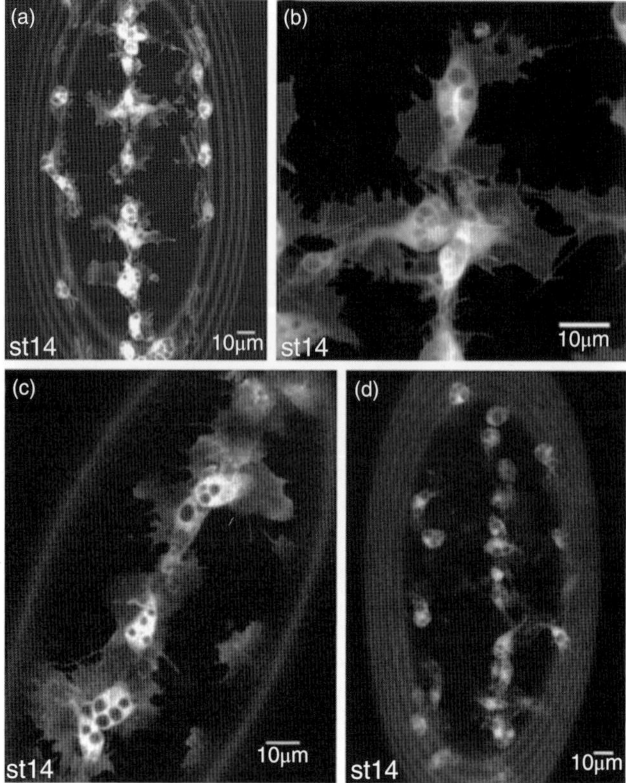

Fig. 6.4 Representative results from live imaging of GFP labeled hemocytes. Z-projections of hemocytes on the ventral side of a stage 14 *srp-Gal4,uas-GFP;crq-Gal4,uas-GFP* embryo (a-b). (a) is a lower magnification image such as used to monitor hemocyte developmental migrations in timelapse movies. (b) is a higher magnification still of hemocytes on the ventral midline, showing fine details of their morphology. (c) is a single 1 m slice of hemocytes on the ventral midline in a stage 14 *srp-Gal4,uas-GFP/+;crq-Gal4,uas-GFP/+* embryo, revealing that lower copy numbers of Gal4 drivers and uas constructs are also sufficient to generate good images. (d) displays a z-projection of hemocytes in a stage 14 *crq-Gal4,uas-GFP* embryo. Here hemocyte protrusions are less obvious due to lower expression of GFP but it is still possible to make movies and track hemocyte migration with this combination of Gal4 driver and uas construct. Images were taken on a Leica LSM510 confocal microscope; anterior is up in all images; the rings at the periphery of images are caused by vitelline membrane autofluorescence. (Source: Evans *et al.* 2010. Reproduced with permission of the authors.)

2. Collect embryos from an overnight apple juice agar plate maintained at 25°C or from a timed plate. Allow the flies to oviposit on a pre-warmed plate for 4 hr, before removing the plate and placing it at 18°C for 15–16 hr prior to mounting of embryos; this provides embryos from late stage 12 through to stage 15 of development. An overnight plate contains a greater diversity of stages but offers the advantage of higher levels of fluorescent reporter expression in hemocytes due to a longer period of time at 25°C, as the Gal4-UAS system is temperature sensitive.
3. Dislodge embryos from the apple juice agar plate using a small amount of water and a soft-tipped paintbrush. Dislodged embryos can be seen easily with the naked eye.
4. Transfer embryos to a cell strainer (Fisher Scientific®) or home-made basket by pouring water from the apple juice agar plate into the basket held over a beaker to collect waste water.

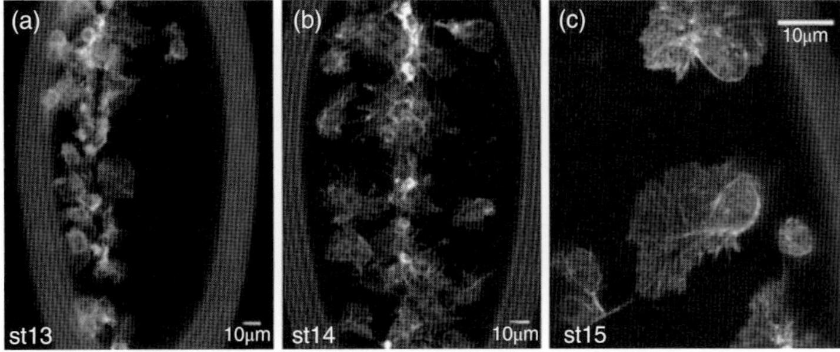

Fig. 6.5 Representative results from live imaging of GMA expressing hemocytes. Z-projections of hemocytes on the ventral midline of stage 13 (a) and stage 14 (b) *srp-Gal4,uas-GMA* embryos, taken from timelapse movies to show developmental migrations of hemocytes. Detailed information on actin dynamics can be obtained by higher magnification imaging of GMA expressing hemocytes (c). GMA consists of GFP fused to the actin-binding domain of moesin and labels actin filaments. Anterior is up in all images; images were taken on a confocal microscope. (Source: Evans *et al.* 2010. Reproduced with permission of the authors.)

5. Repeat step 2 until you are satisfied you have enough embryos transferred from the apple juice agar plate.
6. Wash embryos in cell strainer/basket using water.
7. Place cell strainer/basket in the Petri dish lid of the apple juice agar plate and add enough neat bleach to suspend embryos in the cell strainer/basket.
8. Follow dechorionation of the embryos on a dissection microscope under bright-field: dechorionation is complete when the dorsal appendages have dissolved, which should occur within 2 min.
9. Remove cell strainer/basket containing embryos from the bleach and wash off residual bleach using water. All traces of bleach should be removed before proceeding to step 8. One trick to assess whether all the bleach has been removed is to blot off residual water on blue-colored laboratory tissues, if there is residual bleach the blue color will be bleached white/pink.
10. Blot off remaining water using laboratory tissue applied to the underside of the cell strainer/basket.
11. Place a droplet of water in a Petri dish lid. With a fine paintbrush, collect all the dechorionated embryos from the embryo basket and re-suspend them in the droplet. Next dry the embryos by aspirating water using a micropipette or carefully absorbing it with a laboratory tissue.
12. Once the embryos have been dried, add a drop of voltalef® oil to cover all the embryos. Put a second small drop of oil adjacent to the droplet containing the embryos. As an alternative to voltalef® oil, halocarbon oil 700® (Sigma-Aldrich®) may be used.
13. Under a fluorescent dissection microscope select appropriately staged embryos of the desired genotype using a pair of watchmaker's forceps (number 5) from the oil droplet. These forceps should be bent inwards in order to scoop up the embryos without puncturing their vitelline membrane. Transfer selected embryos to the second oil droplet. It is important that you are able to see fluorescent hemocytes on the dissecting microscope in order to be able to collect good images on the confocal microscope. Stage 13/14 embryos can be mounted to image lateral migration of hemocytes on the ventral midline or stage 15 embryos to image the motility of hemocytes following dispersal over the embryo.

14. Stick two coverslips (18x18mm, thickness 1) to the underside of a Petriperm/ Lumox ™ dish (Sarstedt), using two small drops of voltalef® oil, leaving approximately 1cm between them; these will be used to support a coverslip placed over the embryos, so as not to crush them. Petriperm dishes (50mm diameter) contain a hydrophobic, gas-permeable membrane. The dishes become easier to use once they have been used several times (dishes can be wiped with 70% ethanol and reused).
15. Under brightfield on the dissection microscope, pick up selected embryos one by one with the bent forceps and line them ventral side up and parallel to the edge of the coverslips. It is possible to align up to 15 embryos in this way, depending on your dexterity and your patience. It is important to manipulate the embryos gently as both the embryos and Petriperm dish membrane are fragile and can be easily ruptured.
16. Once the embryos are aligned add a small drop of oil and let it spread to form a homogenous layer between the two coverslips. After the oil has spread (this may take a few minutes) check that the embryos are still ventral side up. If the embryos have rolled slightly, reposition them again with the forceps.
17. Finally, using number 3 tweezers, place a coverslip (18x18mm, thickness 1) over the embryos, resting upon the two previously adhered coverslips. Glue this coverslip to the coverslip supports using nail polish.
18. Take the Petriperm dish with mounted embryos to the confocal or wide-field microscope and mount the Petriperm dish on the stage using an appropriate adapter. Image embryos using fluorescence microscopy. Either an upright or inverted microscope may be used, with the objective lens focusing through the coverslip (as opposed to through the membrane).

6.2.4 In vivo *Visualization of Synaptic Vesicles within* Drosophila *Larval Segmental Axons [11]*

Application:
Recommended for the live imaging of fluorescently tagged synaptic vesicles within the larval segmental axons of *Drosophila*, in order to view axonal transport in real time. In this method, the UAS-GAL4 system can be used to express a single GFP-tagged axon in larval segmental nerves. Further, using different neuronal GAL4 drivers, one may mark vesicles within many axons. One may apply additional fluorescent dyes to mark organelles such as mitochondria (with MitoTracker®) or lysosomes (LysoTracker®) before imaging larvae using florescence microscopy. GFP-vesicle movement and particle movement can be viewed simultaneously using separate wavelengths on the fluorescent microscope.

Formula:
Dissection Buffer:

1 M NaCl	128 pt
1 M EGTA	1 pt
1 M $MgCl_2$	4 pt
1 M KCl	2 pt
1 M HEPES (4-(2-hydroxyethyl)-1-piperazineethanesulfonic acid)	5 pt
1 M Sucrose	36 pt
dH_2O	828 pt

Preparation:
1. Bring pH to 7.2 and filter sterilize.

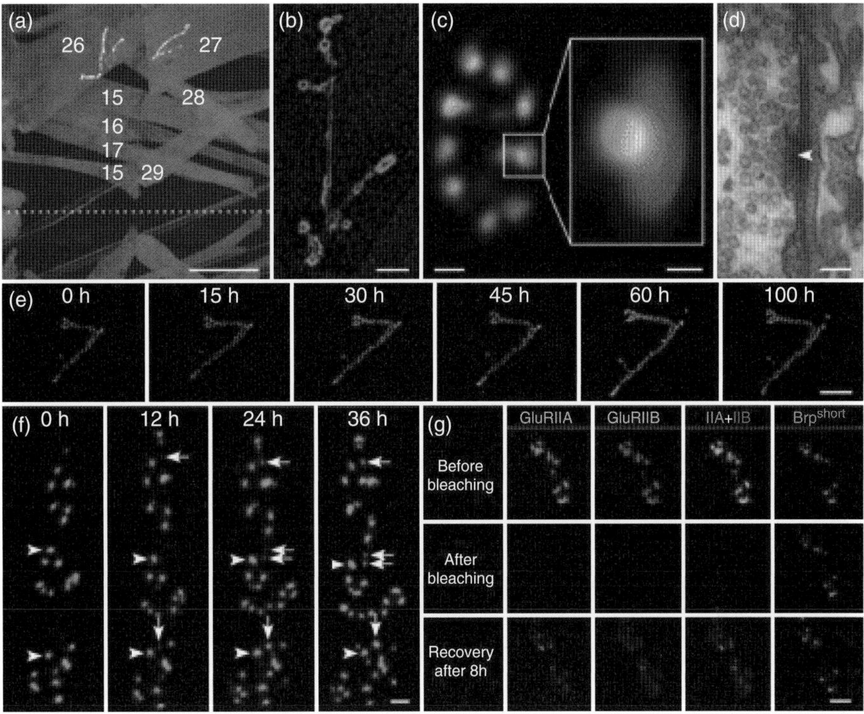

Fig. 6.6 Intravital imaging of neuromuscular junctions (NMJs). (a) Selected body wall muscle fibers in an abdominal segment as seen from the exterior (outside-out view). The dotted line indicates the ventral midline. The image was modified to highlight NMJs at muscles 26 (VA1) and 27 (VA2). Construct used for the visualization of muscle fibers and NMJs: *uas-dlgS97-gfp*, expressed using *c57-gal4*. Scale bar, 100 µm. (b) Morphological structure of a *Drosophila* larval NMJ at muscle 27 (VA2), outlined by the expression of *uas-dlgS97-gfp*, using *c57-gal4* as a driver. Scale bar, 10 µm. (c) Immunohistochemical stainings of a bouton of a larval NMJ and an individual synapse in lateral view (right box). Green, monoclonal antibody Bruchpilot^{Nc82} (Kittel et al. 2006); magenta, antibody against the glutamate receptor subunit DGluRIID (Qin et al. 2005). Scale bar (bouton), 1 µm; scale bar (synapse), 100 nm. (d) Ultrastructure of an active zone. The arrowhead points at the T-bar, synaptic vesicles cluster next to it. Scale bar, 100 nm. (e,f) In vivo imaging of synapse formation at an identified NMJ at muscle 27, tracked using the GFP-labeled glutamate receptor subunit DGluRIIA. (Adapted, with permission, from Rasse et al. 2005.) (e) The development of an identified NMJ at muscle 27 tracked >100 h at 16°C. Scale bar, 10 µm. (f) Higher magnification of the image shown in e: New synapses form de novo (arrows). Mature synapses (arrowheads) remain stable. Scale bar, 2 µm. (g) In vivo fluorescent bleach experiment (FRAP) of DGluRIIA-mRFP (green), DGluRIIB-GFP (magenta) and Brp^{short}-Cerulean (cyan). (*Top* row) Images taken directly before bleaching. (*Middle* row) Images taken directly after bleaching of DGluRIIA and DGluRIIB. (*Bottom* row) Images taken 8 h after bleaching. Recovery of both DGluRIIA and DGluRIIB can be observed. Scale bar: 2 µm. (Source: Kuznicki & Gunawardena 2010.) See plate section for the color version of the figure.

Procedure:

1. Prepare a Siligard® cube by cutting silicone gel to approximately one inch wide, 1 in long and about 1cm high. The gel cube is placed on a glass slide during dissection. Cubes may be used more than once.
2. Collect wandering 3rd instars that are expressing a GFP tagged vesicle protein onto a Petri dish.
3. Wash larvae in deionized water to get rid of all the food.

4. Transfer larvae to the dissection gel cube on a glass slide and immerse in dissection buffer.
5. Pin larvae at the anterior and posterior using two fine pins with the dorsal side up.
6. Using microdissection scissors, make a small cut at the midline near the posterior end. From this incision a cut is made through to the anterior. Using two pins, pin the cut cuticle on the lateral edges.
7. Add a drop of dissecting buffer. Carefully remove the gut, intestines, and fat bodies with forceps. Most of these tissues normally ooze out during the dorsal cutting.
8. Replace the buffer with fresh dissecting buffer. The larvae should never be dry and should constantly be in dissecting buffer.
9. Incubate larvae with organelle markers such as MitoTracker or LysoTracker by diluting the marker in dissecting buffer. After incubation with the marker, wash the dissected larvae several times (five quick washes) using fresh dissecting buffer.
10. After replacing the last wash, push pins into the Siligard® gel cube and immediately coverslip. The larvae are now ready for observation.
11. Place the coverslip and gel onto the microscope stage and image using an inverted microscope.
12. Use fluorescence microscopy to image tagged vesicles within larval segmental nerves. GFP expression in the cell bodies of the ventral ganglion is first evaluated before segmental nerves are imaged. If robust GFP staining is observed in the cell bodies in the ventral ganglion, the nerves are then imaged and used for evaluation.
13. Using a dual view system with filters for CFP/YFP or RFP/YFP and split view software, a dual-tagged protein can be imaged *in vivo*.

6.3 Mosaic gene expression

Genetic recombination methods can be used to produce subsets of cells that exhibit mosaic gene expression in a developing compartment. Methods to induce mitotic cell clones that express genetic markers provide a powerful set of tools for the analysis of gene function and cell autonomy. [12, 13] The FLP/FRT system exploits the yeast flippase (FLP) recombinase to generate mitotic recombination at FLP-recombinase target sequences (FRTs). Combining the FLP/FRT system for genetic recombination with the GAL4/UAS system for transgene expression [3] allows the generation of mutant clones in a given cell type that express detectable markers, thus providing spatial, temporal, and genetic information about individual cells. [14, 15]

6.3.1 *Domain specific genetic mosaic system in the* Drosophila *eye [15]*

Application:

This technique combines the flippase (FLP/FRT) system for genetic recombination with the GAL4/UAS system for domain-specific transgene expression to generate loss-of function or gain-of function cell clones in the *Drosophila* larval eye imaginal disc. [15] Here, use a GAL4 driver that is expressed in the developing eye imaginal disc, combined with a FLP/FRT cassette that allows the generation of homozygous loss-of-function clones that can be detected by the presence or absence of a GFP reporter. Once dissected, eye imaginal discs can be stained with antibodies with fluorescent epitopes to identify specific cell types. Discs can then be imaged with fluorescence microscopy to visualize the homozygous mutant clones that possess or lack the GFP reporter.

Procedure:

1. Use a genetic mosaic approach in *Drosophila* that combines the Gal4/UAS and FLP/FRT systems to generate loss-of function as well as gain-of function clones on the dorsal and ventral eye margins in eye imaginal discs. Use suitable genetic

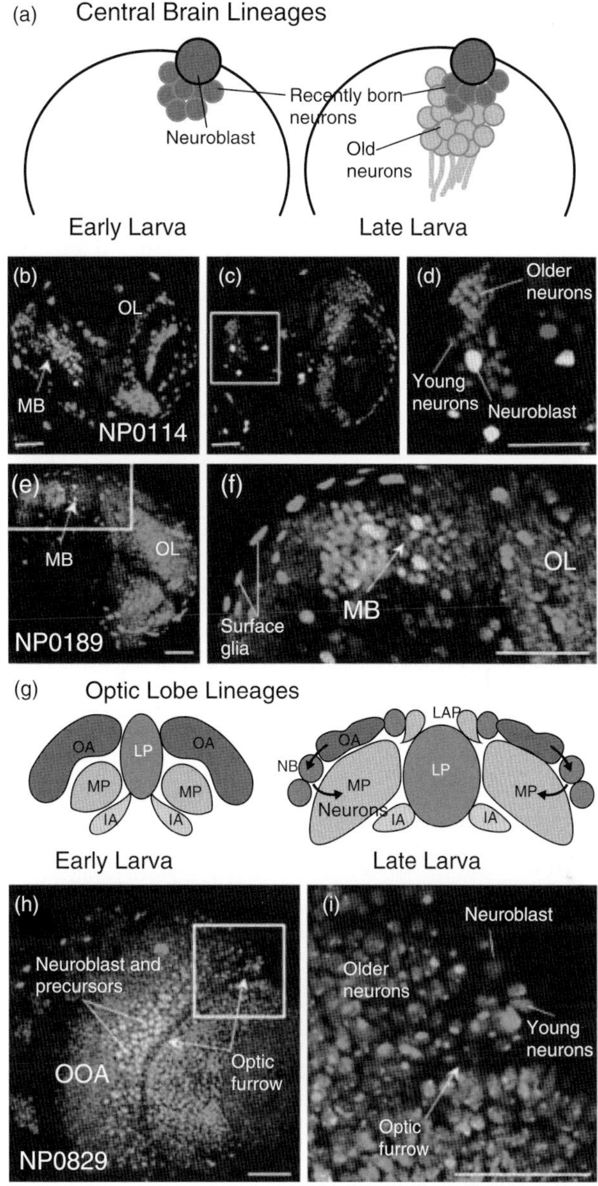

Fig. 6.7 G-TRACE reveals novel cell lineage markers a) Schematic of lineages from larval central brain neuroblasts. (b-f) G-TRACE in the Gal4-enhancer trap lines NP0114 (b-d) and NP0189 (e, f). d and f are close-ups of the boxed regions in c and e respectively. MB, mushroom body; OL, optic lobe g) Schematic depicting optic lobe development. Cross-section of a brain hemisphere at early and late stages showing the outer (OA; red) and inner (IA; gray) anlagen of the optic lobe. Neuroblasts (NB; green) at the lateral edge form neurons of the medullary primordium (MP) while medial edge neuroblasts form neurons of the laminar primordium (LAP). The inner optic anlage (IA) forms the lobula primordium (LP). (h, i) G-TRACE in the Gal4-enhancer trap line NP0829. OOA, outer optic anlage. i is a close-up of the boxed region in h. Scale: 50 μm. (Source: Evans *et al.* 2012. Reproduced with permission of Nature.) See plate section for the color version of the figure.

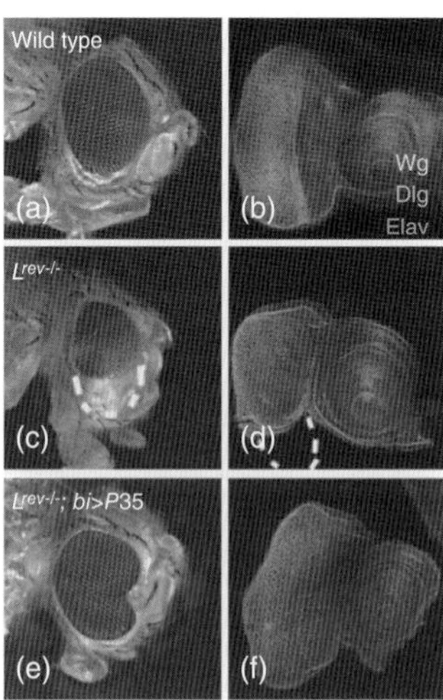

Fig. 6.8 Phenotypes of loss-of-function clones on dorsal and ventral eye margins. (a and b) Wild-type (a) adult eye and (b) eye imaginal disc. (b) Eye imaginal disc is stained for a membrane specific marker Disc large (Dlg: green channel), a signaling molecule Wingless (Wg: red channel), and a pan neural marker Elav (blue) that marks the photoreceptor neurons of the eye disc. (c and d) Loss-of-function clones of *L* on the dorsal and ventral margins of the developing eye imaginal disc by using cell-lethal strategy results in preferential loss of ventral eye phenotype as observed in the (c) adult eye and the (d) eye imaginal disc. The outline of ventral eye is marked by white dotted line both in the adult eye and the eye imaginal disc. Note that the dorsal eye margin does not exhibit any effect on the eye development and differentiation. (e and f) Misexpression of P35, to block caspase dependent cell death, in loss-of-function clones of *L* (using cell-lethal approach) in dorsal and ventral eye margins, result in suppression of loss-of-ventral eye phenotype as seen in the (e) adult eye and (f) eye imaginal disc. These reagents can be used for generating loss-of-function of a gene along with gain-of-function of another gene both in the dorsal and ventral eye margins. Magnification of all adult eye images (panel a, c, and e) is ×10 and all imaginal discs images (panel b, d, and f) are ×20. (Source: Tare *et al.* 2013. Reproduced with permission of John Wiley & Sons.) See plate section for the color version of the figure.

crosses to combine the FLP/FRT and Gal4/UAS systems to generate stocks where Gal4 can drive expression of FLP within its expression domain, and trigger homologous recombination at the site of the FRT cassettes to generate homozygous loss-of function clones. For example, using bifid-Gal4 selectively targets FLP on the dorso-ventral margins of the eye. The FLP will act on the DV margins to enable generation of genetic mosaic clones in a domain-specific manner. Establish fly stocks that allow generation of mutant clones for genetic mutations on particular arms (2R, 3R) of the second and third chromosomes. Set up crosses so that genetic mosaic clones can be detected by loss or gain of GFP expression.

2. Dissect eye imaginal discs from first-, second-, and wandering third-instars in PBS.

3. Fix and stain specimens using the immunostaining protocol for eye imaginal discs as described in Chapter 5 of this manual (Preparation of eye imaginal discs for Immunostaining). Stain specimens with desired primary antibodies to identify the desired cell types. For example, stain specimens with rat anti-ELAV (1:100) to identify photoreceptor neurons, mouse anti-Wg (1:50) for the signaling molecule Wg, and rabbit anti-Dlg (1:200) for the membrane-specific marker Dlg. Stain with appropriate secondary antibodies conjugated with florescent epitope tags, for example, goat anti-rat IgG conjugated with Cy5 (1:200), donkey anti-rabbit IgG conjugated to Cy3 (1:250), donkey anti-rabbit IgG conjugated to FITC, and donkey anti-mouse IgG conjugated to Cy3 (1:200).

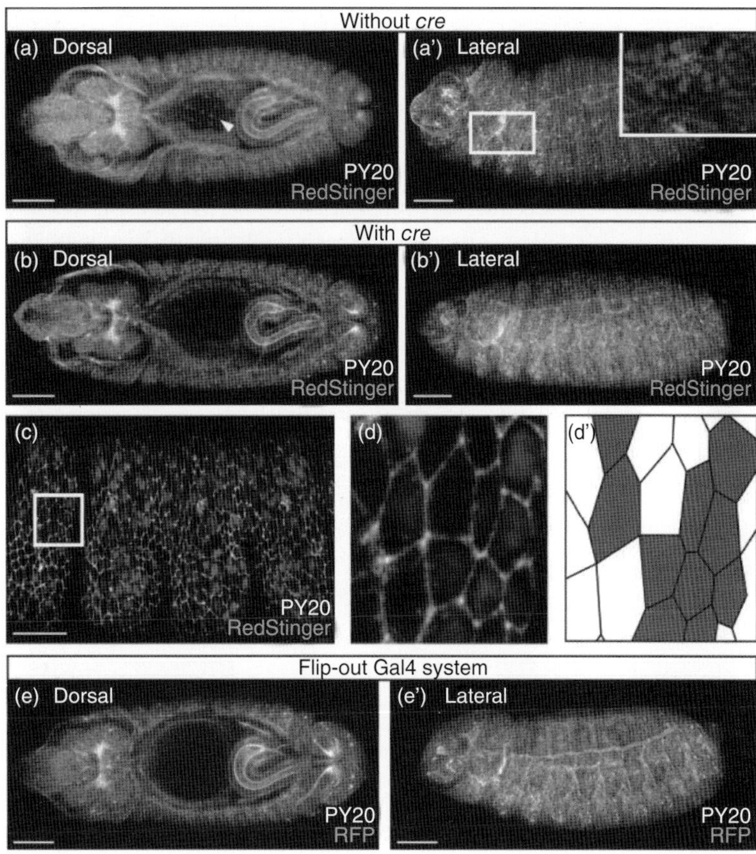

Fig. 6.9 *Gal4* mosaic expression is driven by *Aloxg-Gal4* in the presence of *cre*. a–e': Gal4 was detected by the UAS reporter *UAS-RedStinger* (magenta); plasma membranes were stained with the PY20 antibody (white). a, a': Dorsal (a) and lateral (a') views of a w^{1118} or w^{1118}/Y; *Aloxg-Gal4*; *UAS-RedStinger* embryo at stage 13. Arrowhead shows scattered signals in the yolk in A. Upper-right inset is a magnified view of the region in the white box in a'. b, b': Dorsal (b) and lateral (b') views of a *cre*/w^{1118} or *cre/Y*; *Aloxg-Gal4/+*; *UAS-RedStinger/+* embryo at stage 13. c: Left-side view of the epidermis in a *cre*/w^{1118} or *cre/Y*; *Aloxg-Gal4/+*; *UAS-RedStinger/+* embryo. d: The box in c, magnified. d': Schematic of the epidermal cells expressing *UAS-RedStinger* in d. e, e': Dorsal (e) and lateral (e') views of a *hsFLP*/w^{1118} or *hsFLP/Y*; +/*sna*^{Sco} or +/*Cyo, P{en1}wg*^{en11}; *P{GAL4-Act5C (FRT.CD2).P}S, P{UAS-RFP.W}3/+* embryo at stage 14. Scale bars: 50 µm (a, a', b, b' and e) and 20 µm (c). (Source: Nakazawa *et al.* 2012. Reproduced with permission of John Wiley & Sons.) See plate section for the color version of the figure.

4. Mount imaginal discs on slides in Vectashield®.
5. Image specimens using a florescent microscope for the presence of florescent markers identifying cell type, as well as cell genotype (via the presence or absence of GFP).
6. For bright field imaging of adult eyes, mount adult flies on a needle and image using a bright field scope.

6.3.2 *Mosaic gene expression in the* Drosophila *embryo using the Cre/loxP system [16]*

Application:

This method can be used to generate mosaic cells in *Drosophila* embryonic tissues. [16] In this system, Gal4 gene expression is suppressed by the presence of a gene insulator, gypsy, until the insulator is removed by induced recombination. Gal4 is expressed when site-specific recombination between loxP sites is induced by Cre recombinase. [17] Figure 6.1 illustrates GAL4 mosaic expression in *Drosophila* embryos with and without *cre* expression. The UAS-Red-Stinger reporter is expressed when Gal4 is present. The RedStinger protein can be detected by anti-Red fluorescent protein (RFP) antibody staining to mark cells in which Gal4 expression has been induced after Cre-mediated excision of the gypsy insulator.

Procedure:

1. Use a genetic approach in *Drosophila* that combines the GAL4/UAS and Cre/loxP systems to generate mosaic expression of Gal4 in embryonic tissues, leading to the Gal4-dependent activation of the desired gene under the control of the UAS sequence, and expression of the UAS-Red-Stinger reporter.
2. Collect embryos for 30 min at 25°C and incubate at 25°C for 1, 2, 3, 4, 5, and 6 hr after egg laying. Stage embryos based on the degree of germ-band retraction, gut morphology, degree of dorsal closure, and appearance of tracheal pits.
3. Methanol-fix and immunostain *Drosophila* embryos with appropriate antibodies as described in Chapter 5 of this manual.
4. To detect RedStinger, perform anti-RFP antibody staining along with antibody staining for desired embryonic tissue markers.
5. Mount immunostained embryos in 0.25% n-propyl gallate/50% glycerol/PBS.
6. Examine specimens with a fluorescence microscope.

Note:

1. This technique efficiently induces mosaic gene expression in most embryonic tissues, including the epidermis, amnioserosa (an extraembryonic epithelium of the embryo of higher flies), tracheal system, malpighian tubules, foregut, and midgut. However, it is inefficient in neuronal tissue and does not occur in the visceral muscles.

6.3.3 *Flybow: Genetic multicolor cell labeling for neural circuit analysis in* Drosophila *[18]*

Application:

This technique allows visualization of neural cell morphology at the single cell level using immunofluorescence. Here, the Gal4-UAS system to regulate transgene expression is combined with the heat-shock-inducible FLP/FRT system. The heat-shock-inducible system provides temporal control of the genetic recombination of sequence cassettes encoding different fluorescent protein pairs. This strategy allows for the labeling of specific cell populations in developing and adult *Drosophila* neural tissues with four different membrane-tethered fluorescent proteins. The sequences encoding different membrane-tethered fluorescent proteins are arranged in pairs within cassettes flanked by recombination sites. Stochastic recombination allows the expression of up to four reporters (eGFP, mCitrine, mCherry, and Cerulean-V5) within one sample, thus producing multicolor labeling in the same preparation.

Formula:

Fixative:

Paraformaldehyde	2%
1 M L-lysine monohydrochloride	10 pt
1 M Sodium phosphate buffer	5 pt
dH_2O	85 pt

Preparation:

1. Adjust pH to 7.4

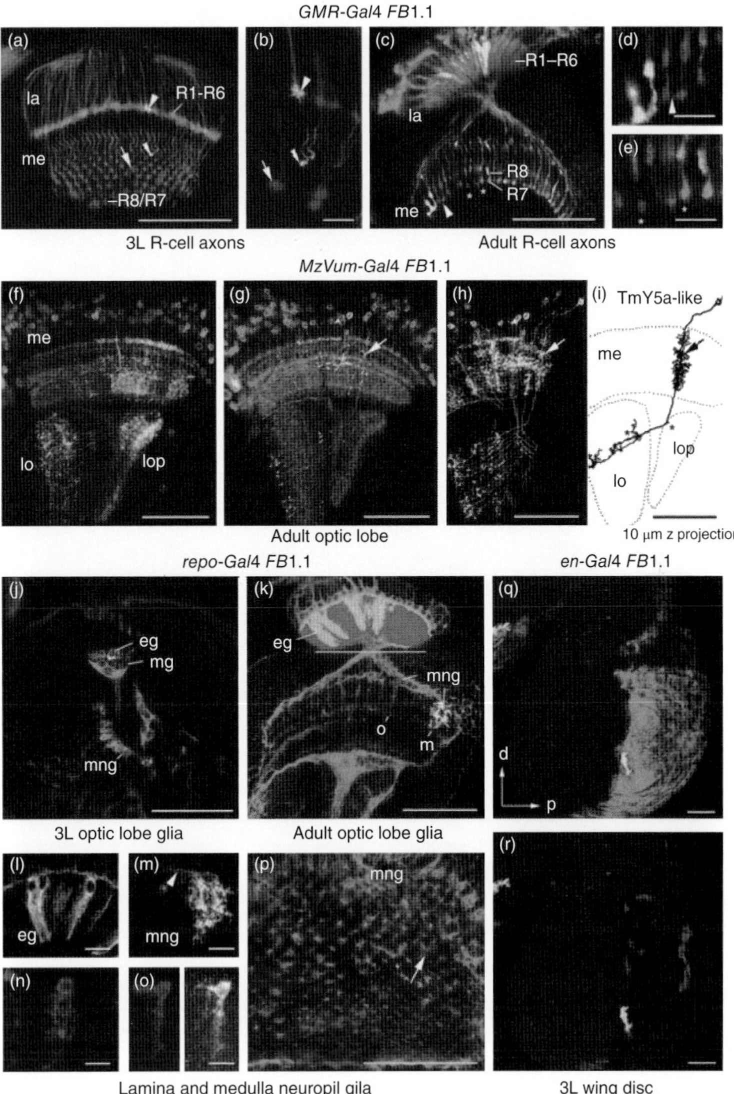

Fig. 6.10 (a–e) Confocal images showing third instar larval (3L) and adult R1–R8 photoreceptor axons labeled using *GMR-Gal4*. R1–R6 growth cones (arrowheads) in the lamina (la), and young (double arrowheads) and mature (arrows). R8 growth cones in the medulla (me) are highlighted in a,b. Adult R8 and R7 terminals in a column express the same fluorescent protein (arrowhead) or combinations of two fluorescent proteins (asterisks) (c–e). (f–i) Single optical sections (f,g) and a 10 μm *z*-stack projection of the mCitrine channel (h) of adult optic lobes in which *MzVum-Gal4* drives *FB1.1* expression in medulla neurons. One neuron (arrow; g,i) traced through a series of consecutive sections had TmY5a neuron-like features; asterisks indicate additional or absent branches compared to reported morphology[14]. (j,k) Epithelial (eg) and marginal (mg) glia in the lamina, and medulla neuropil glia (mng) at the distal medulla neuropil border were labeled with different fluorescent proteins in third instar larval (j) and adult (k) optic lobes using *repo-Gal4* and *FB1.1*. In k, fluorescence signals in the lamina above the white line were reduced relative to those in the medulla. (l–o) Higher magnification of the image in k, showing elaborate shapes of epithelial and medulla neuropil glia. (p) Area with overlapping medulla neuropil glial cell branches (arrowhead) in a medulla cross-section. (q,r) *en-Gal4*–driven expression of different fluorescent proteins in epithelial cell clones in the posterior (p) compartment of wing discs. d, dorsal. Scale bars, 50 μm (a,c,f–k,p–r) and 10 μm (b,d,e,l–o). (Source: Hadjieconomou *et al.* 2013. Reproduced with permission of Nature.) See plate section for the color version of the figure.

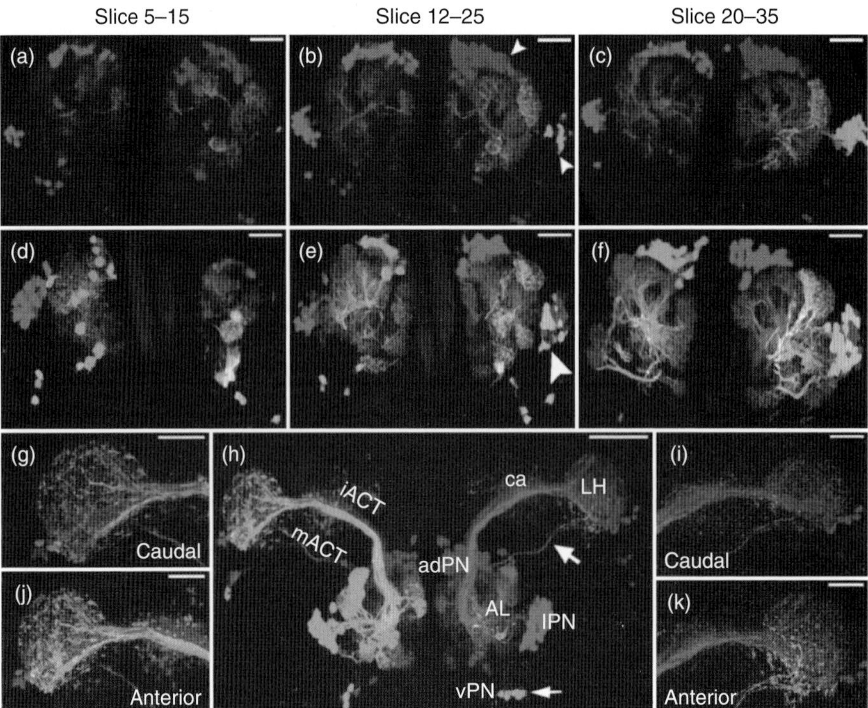

Fig. 6.11 (a–k) Single (a–c) and double (d–f) copies of *UAS-dBrainbow* were used along with *hs-Cre; GH146-GAL4* to label the three projection neuron lineages that express GH146-GAL4 (adPN, lPN and vPN) as well as the axon tracts (iACT and mACT) that connect the projection neuron cell bodies to the lateral horn (LH) (h) and calyx (ca) (g, i–k). Shown are maximum intensity projections of several 1-μm confocal sections (a–f) with each projection neuron lineage expressing a different fluorescent protein–epitope cassette and thus pseudocolored differently (arrowheads; b). The number and depth of confocal slices used to produce the merged images are indicated. In the right lPN lineage (arrowhead; e), recombination in the neuroblast occurred to select blue in one copy of *UAS-dBrainbow* but the recombinase did not act on the second copy until later to select green, so a subset of later-born neurons is labeled in cyan. In h, the antennal lobe (AL) and efferent neurons projecting to the lateral horn via the mACT and iACT; individual neurites can also be traced from the antennal lobe to the lateral horn via an alternative pathway (arrows). Higher-magnification views of the lateral horn from different orientations (g, i–k). Scale bars, 50 μm (a–f, h) and 20 μm (g, i–k). (Source: Hampel *et al.* 2011. Reproduced with permission of Nature.) See plate section for the color version of the figure.

Procedure:

1. Use hs-FLP/FRT recombination together with a GAL4/UAS system to obtain *Drosophila* adults expressing membrane-bound fluorescent proteins in developing and/or adult neural tissues.
2. Dissect larval or adult brains, or embryonic ventral nerve cords, in PBS.
3. Fix specimens for 1 hr at room temperature.
4. Wash in 0.5% Triton X-100 containing PBS to preserve endogenous fluorescence levels.
5. Incubate samples with appropriate primary and secondary antibodies as described in Chapter 5 of this manual.
6. Mount immunolabeled samples in Vectashield®.
7. Mount live embryos in PBS.
8. Image specimens using fluorescent microscopy to detect neural cells expressing different combinations of membrane-bound fluorescent proteins.

6.3.4 *Brainbow: Fluorescence labeling to subdivide neural expression patterns in* Drosophila *[19]*

Application:

1. This technique for multicolor neuron labeling combines the GAL4/UAS system to control cell-type-specific expression with the Cre/lox system for the stochastic recombination of epitope-tagged proteins. The *Drosophila* Brainbow (dBrainbow) construct contains a UAS sequence that allows cell-specific expression controlled by the presence of GAL4.
2. When stochastic recombination occurs with the Cre/lox system, different combinations of epitope-tagged proteins can be produced such that a single copy of the transgene can produce individual red, green, and blue colors; with two copies, cyan (blue and green), magenta (red and blue), and yellow (green and red) are also possible. Three different cytoplasmic, epitope-tagged proteins can be targeted in different combinations to particular groups of neurons, allowing for six color possibilities that can be detected by immunofluorescence. Thus, multiple neural cell lineages or individual neural projections can be visualized in the same preparation. Figure 6.2 demonstrates how expression of UAS-dBrainbow in three projection neuron lineages can be used along with hs-Cre and a GAL4 driver to label multiple projection neuron lineages. The projection of neuron lineages can be subdivided by including a second copy of the UAS-dBrainbow transgene, resulting in six possible colors.

Formula:

PAT Wash buffer:

1 M PBS	1 pt
dH_2O	99 pt
BSA	1%
Triton X-1000	0.5%

Procedure:

1. Combine hs-Cre/lox recombination with a GAL4/UAS system to obtain *Drosophila* adults containing neurons expressing cytoplasmic epitope-tagged proteins that can be detected by immunofluorescence.
2. Dissect adult brain and ventral nerve cords as described in Chapter 5 of this manual.
3. For endogenous fluorescence analysis, Dissect 3 to 7-day-old adult females in 1× PBS (pH 7).
4. Fix samples rotating at room temperature (~21°C) with 2% paraformaldehyde for 1 hr.
5. Wash samples three times for 20 min each in wash buffer.
6. Rinse tissue in 1× PBS.
7. Mount samples in Vectashield® on glass slides with two transparent reinforcement rings as spacers.
8. Before imaging, keep mounted samples at room temperature for 1 hr or store at −20°C.
9. For antibody staining, Dissect 3 to 7-day-old female flies in 1× PBS. (For proboscis staining, cut the proboscis segments into separate pieces to allow antibody penetration.)
10. Fix tissue with 2% paraformaldehyde overnight (~16 hr) at 4°C with gentle rocking.
11. Wash samples 3× for 20 min each with PAT wash buffer.

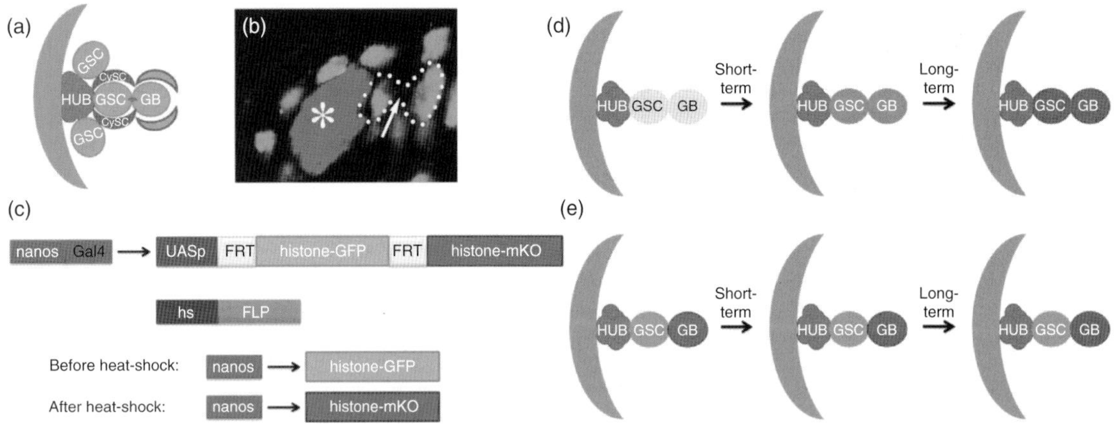

Fig. 6.12 Experimental design and potential results. (a) Diagram of the GSC niche. HUB, hub cells; CySC, cyst progenitor/somatic stem cell. (b) Immunofluorescent image of the niche: HUB (anti–Fas III, red, asterisk), GSC-GB pair expressing H3-GFP (green, dotted outline) connected by a spectrosome (anti-α-spectrin, red, arrow). (c) The *UASp-FRT-histone-GFP-PolyA-FRT-histone-mKO-PolyA* transgene. UAS, upstream activating sequence; FRT, FLP (flippase) recombination target; histone, H3, H2B, or histone variant H3.3; *nanos-Gal4*, a germline-specific driver; *hs-FLP*, the yeast FLP recombinase controlled by the heat shock (*hs*) promoter. (d and e) Two potential results. For simplicity, only one GSC-GB pair is shown, and each entire cell is colored according to histone fluorescence. (Source: Tran *et al.* 2012. Reproduced with permission of American Association for the Advancement of Science.) See plate section for the color version of the figure.

12. Block samples with 3% normal goat serum (NGS) in PAT wash buffer for 1 hr at room temperature.
13. Perform primary antibody incubations overnight at 4°C, nutating in 500 μl PAT wash buffer plus NGS with desired antibodies.
14. Wash samples three times for 20 min each with PAT.
15. Incubate samples overnight with secondary antibodies.
16. Wash samples three times for 20 min each with PAT wash buffer, rinse with 1× PBS.
17. Mounted samples in Vectashield®.
18. Before imaging, keep mounted samples at room temperature for 1 hr or store at −20°C.
19. Image specimens using fluorescence microscopy to detect the presence of different-colored neurons in the same preparation.

6.3.5 *Histone labeling in* Drosophila *male germline stem cells [20]*

Application:

This is a dual-color method to differentially label "old" versus "new" histones in *Drosophila* male germ line stem cells. This technique uses both spatial (via the GAL4/UAS system) and temporal (by heat shock induction) controls to switch fluorescently labeled histones from green (GFP) to red (monomeric Kusabira-Orange, mKO). Heat shock induces site-specific DNA recombination (via the FLP/FRT system) to shut down expression of GFP-labeled old histones and initiate expression of mKO (red)-labeled new histones.

Procedure:

1. Use a genetic mosaic approach in *Drosophila* that combines the Gal4/UAS and FLP/FRT systems, using hs-FLP, to generate switchable dual-color transgenic fly strains

with fluorescently labeled histone proteins. Use suitable genetic crosses to obtain fly stocks containing the UASp-FRT-histone-GFP-PolyA-FRT-histone-mKO-PolyA transgene. Using such a strain allows the generation of flies in which heat shock treatment induces site-specific DNA recombination to stop the expression of GFP-labeled histones and begin the expression of mKO-labeled histones in spermatogonia (an undifferentiated male germ cell).
2. Heat-shock males for 2 hr at 37°C in a circulating water bath to induce germ line stem cell DNA recombination.
3. After 16–20 hr, dissect testes and ovaries from young (0–2-day-old) flies and fix specimens as described in Chapter 5 of this manual.
4. Immunostain specimens using desired primary and secondary antibodies.
5. Mount specimens in 9:1 Citifluor:PBS containing 100 μg/ml p-phenylenediamine. (Citifluor is an antifadent mounting media).
6. Examine specimens using fluorescence microscopy.

6.4 Fluorescent markers for live imaging and kinetic microscopy

The availability of fluorescent markers that can be expressed in insects allows one to localize and follow the subcellular movements of proteins of interest through live imaging. Vector sets designed for the easy cloning of fusion proteins in different colors make it possible to co-localize proteins of interest with organelle-specific fluorescent markers. [6] Several kinetic microscopy tools that allow one to photoactivate or photobleach fluorescently tagged proteins make it possible to trace and quantify the kinetic movements and turnover dynamics of fusion proteins with high spatial and temporal resolution. [21] Fluorescent markers thus provide a wealth of opportunities for studies in living cells.

6.4.1 *Marking chromosomes in* Drosophila *with enhanced green fluorescent protein (eGFP) [22]*

Application:
Recommended for the localization of a gene of interest on chromosomes during mitosis. By tagging a protein of interest with an enhanced variant of GFP (eGFP) that possesses increased stability and brightness, and co-expressing a centromeric protein (Cenp-C) tagged with Red Fluorescent Protein (RFP), one may determine whether a given protein of interest co-localizes with a centromeric marker. Thus, using this technique, *Drosophila* embryos co-expressing two different fluorescent markers can be imaged to determine the sub-cellular localization of a protein of interest.

Procedure:
1. Collect *Drosophila* embryos co-expressing an eGFP-tagged protein of interest, and the centromeric protein Cenp-C-mRFP, on apple juice agar plates.
2. Age to the desired developmental stage.
3. For *in vivo* imaging, dechorionate embryos, immobilize on glass slides, and cover with halocarbon oil.
4. Image live embryos using fluorescence microscopy.
5. For fixed embryo imaging, dechorionate embryos for 2 min in 7% NaOCl.
6. Methanol-fix embryos following the embryo fixation protocol in Chapter 5 of this manual.
7. Immunostain embryos with the desired antibodies following an immunostaining method outlined in Chapter 5 of this manual.
8. Stain DNA with Hoechst 33258 at 1 μg/ml in phosphate-buffered saline.

9. Mount embryos in 70% glycerol, 50 mM Tris-HCl, pH 8.5, 5 mM *p*-phenylenediamine, and 50 mM *n*-propylgallate.
10. Image fixed embryos for GFP and RFP detection using fluorescence microscopy, in order to determine whether or not a protein of interest (marked with GFP) co-localizes with centromeric protein Cenp-C (marked with RFP).

6.4.2 *Fluorescence recovery After Photobleaching (FRAP) in live* Drosophila *embryos [7, 21]*

Application:
FRAP is a technique in which an area of the cell is photobleached with a high-intensity laser pulse. The movement of unbleached molecules into the bleached area is then recorded by time-lapse microscopy using low-intensity laser power. This allows the determination of the motility and kinetic behavior of fluorescently tagged proteins.

Procedure:
1. Obtain *Drosophila* embryos expressing a GFP-tagged protein of interest using standard genetic strategies.
2. Collect embryos for 1–4 hr on apple juice agar plates.
3. Dechorionate the embryos as described in Chapter 5 of this manual.
4. Mount embryos on their dorsolateral/ventrolateral sides on a chambered coverglass.
5. Slowly add enough PBS from the side of the chamber in order to avoid air bubbles.
6. The appropriate confocal instrument parameters for prebleach and postbleach image acquisition and photobleaching must be determined. Parameters will vary depending on the photobleaching characteristics of the fluorescent protein used, recovery kinetics of the process studied, the geometry of the region of interest, and image analysis.
7. Acquire a series of five to ten prebleach images to measure the total steady-state fluorescence intensity before photobleaching.
8. Selectively photobleach the region of interest using the established conditions.
9. Acquire a series of postbleach images until the fluorescence recovery process has reached a steady state.
10. FRAP data processing:
    a. Determine the raw fluorescence intensity measurements in the regions of interest in the cell.
    b. Subtract the background and correct for the bleached fraction.
    c. Correct for laser fluctuations and photobleaching during acquisition.
    d. Normalize the data to compare different experiments.

6.4.3 *Fluorescence Loss in Photobleaching (FLIP) in live* Drosophila *embryos [7, 21]*

Application:
The FLIP technique also provides information about a protein's kinetic properties. In FLIP, an area of the cell is repeatedly photobleached while images of the entire cell are collected. Monitoring fluorescence in the non-photobleached regions allows the mobility of a fluorescently tagged protein to be observed and quantified. The continuity between subcellular compartments can also be determined.

Procedure:
1. Obtain *Drosophila* embryos expressing a GFP-tagged protein of interest using standard genetic strategies.
2. Collect embryos for 1–4 hr on apple juice agar plates.

3. Dechorionate the embryos as described in Chapter 5 of this manual.
4. Slowly add enough PBS from the side of the chamber in order to avoid air bubbles.
5. The appropriate confocal microscope instrument parameters for image acquisition and photobleaching must be determined.
6. Acquire a series of five to ten prebleach images to measure the total steady-state fluorescence intensity before photobleaching.
7. Selectively photobleach the region of interest using the established conditions.
8. Acquire one image of the whole cell and photobleach the region of interest again. Repeat the process until the fluorescence intensity of the structure of interest is similar to background.
9. FLIP data processing:
   a. Determine the raw fluorescence intensity measurements in the regions of interest in the cell.
   b. Subtract background.
   c. Correct for laser fluctuations and photobleaching during acquisition.
   d. Normalize data to compare different experiments.

6.4.4 *Photo-activatible GFP to track chromatin loci in living* Drosophila *embryos [23, 24]*

Application:
Recommended for tracking chromatin loci in living *Drosophila* embryos. In this technique, a histone 2A variant photo-activatible GFP (paGFP) protein, which can be activated by light, can be expressed at multiple loci. Excitation of paGFP with specific wavelengths leads to photoconversion of the fluorophore, resulting in a 100-fold increase in the fluorescence. The variant paGFP can be activated or photoconverted using one- or two-photon microscopy and simultaneously tracked. This technique allows one to quantitate the mobility and turnover rates of fluorescently marked chromatin loci.

Formula:
Tyrode's Solution:

1 M NaCl	137 pt
1 M KCl	2.7 pt
1 M $MgCl_2$	1 pt
1 M $CaCl_2$	1.8 pt
1 M Na_2HPO_4	0.2 pt
1 M $NaHCO_3$	12 pt
1 M D-glucose	5.5 pt
dH_2O	839.9 pt

Procedure:
1. Obtain *Drosophila* embryos expressing a paGFP marker using standard genetic strategies.
2. Collect embryos for 1–4 hr on apple juice agar plates.
3. Dechorionate the embryos as described in Chapter 5 of this manual.
4. Place the embryos in a drop of Tyrode's solution in a two-well chamber slide and place a 1–2 mm slice of apple juice agar on top to prevent desiccation.
5. Image embryos using one-photon or two-photon fluorescent microscopy.
6. The appropriate photoactivation conditions must be established and will vary according to the photoconvertible fluorophores used.

7. Acquire a series of one to five pre-activation images. Use 488-nm excitation for paGFP.
8. Selectively photoactivate the region of interest, using the established conditions.
9. Acquire a series of post-activation images using the excitation settings used for the pre-activation images.

Note:
1. Embryos remain viable for >4 hr.

6.4.5 *Live redox imaging using genetically encoded probes in* Drosophila *[25]*

Application:
This technique makes use of genetically encoded reduction/oxidation (redox) probes in *Drosophila* to map glutathione redox potential and H_2O_2 in the cytosol and mitochondria. Here, a redox-sensitive GFP (roGFP) probe containing an engineered dithiol/disulfide switch on its surface functions as a redox sensor. The redox equilibrium of engineered cysteines on the roGFP probes is associated with changes in fluorescence that can be measured using fluorescence microscopy. Tissue-specific expression of the redox probes using the GAL4/UAS system allows detection of redox differences and changes in a given tissue of interest. This protocol allows redox state detection in subcellular compartments of either live larvae or in dissected tissues from fixed adult specimens in which the redox state has been conserved.

Formula:
Glycerol mounting medium in PBS:

Glycerol	80%
N-propyl gallate	0.4%

Procedure:
I. Live Imaging of Third Instars
1. Remove third instars from the food vial, clean briefly with PBS, and gently fix with adhesive tape on a cover glass.
2. Image larvae immediately, using a confocal microscope as described in Chapter 7 of this manual.
3. For calibration of the biosensor response, dissect larvae and salivary glands in PBS and incubate with 10 mM DTT (5 min) or 1 mM DA (2 min) at room temperature, respectively.

II. Imaging of Freshly Dissected Tissue
1. Dissect larval organs, e.g., the wing disc or salivary glands, in PBS.
2. Transfer specimens immediately to a slide and image using a confocal microscope as described in Chapter 7 of this manual.
3. To fully reduce and oxidize the tissue, incubate the dissected organs in 10 mM DTT (5 min) or 1 mM DA (2 min), respectively.

III. Conservation of the Biosensor Redox State with the alkylating agent N-Ethylmaleimide (NEM)
1. To conserve the redox state of roGFP-based biosensors in dissected and fixed tissue, perform all dissections in the presence of 20 mM NEM (n-ethyl maleimide).
2. Briefly anaesthetize adult flies with N_2 and decapitate.
3. Cut and open the abdomen immediately in the presence of 20 mM NEM.
4. Incubate samples with NEM for 10 min at room temperature.

5. Remove remaining NEM by rinsing once with PBS.
6. Fix samples with 4% PFA for 15 min at room temperature.
7. Remove remaining PFA by washing twice with PBS for 10 min.
8. Equilibrate samples in glycerol mounting medium overnight at 4°C and mount the next day.
9. Store samples horizontally at 4°C.
10. For probe calibration, perform parallel measurements (side by side, keeping the exact same microscope settings) on specimens that are fully reduced or fully oxidized (20 mM DTT or 2 mM DA for 10 min) during dissection and before being blocked with NEM. This allows determination of the *in situ* dynamic range and allows one to infer the degree of oxidation of the probe in the samples.

6.4.6 *Fluorescent fusion protein knockout mediated by Degrade green fluorescent protein (deGradFP) [26]*

Application:
Recommended for the specific depletion of a fluorescently marked protein of interest in *Drosophila*. In degrade GFP (deGradFP), the degradation capacity of the ubiquitin proteasome system is harnessed to bring about the rapid depletion of a protein of interest fused to a fluorescent marker. In this system, expression of a transgenic protein containing the E3 ligase F-box domain of the ubiquitin proteasome system, fused to a single-domain antibody fragment directed against fluorescent proteins, leads to the removal of fluorescent proteins. By combining deGradFP with the GAL4/UAS system, a given protein of interest can be depleted in a specific tissue. This technique thus allows the specific and efficient removal of a fluorescently marked protein of interest that can be easily monitored by fluorescence microscopy.

Procedure:
1. Obtain *Drosophila* embryos expressing combinations of GFP, EYFP, RFP1 and/or mCherry deGradFP genes fused to a gene encoding the protein to deplete.
2. Collect embryos and dechorionate in 4% bleach.
3. Mount specimens in 400-5 mineral oil between a glass coverslip and a gas-permeable plastic foil.
4. Image specimens using fluorescent microscopy at 25°C. Target protein degradation can be monitored by testing for the presence or absence of signal from the fluorescently marked fusion protein.

References

1. Sweeney S.T., Hidalgo A., de Belle J.S. & Keshishian H. X-Gal staining of the central nervous system in adult *Drosophila*. *Cold Spring Harbor Protoc.* **2**, 239–241 (2012).
2. Wolff T. Beta Galactosidase activity staining of frozen adult *Drosophila* retinas. *Cold Spring Harbor Protoc.* **5**, (2010).
3. Brand A.H. & Perrimon N. Targeted gene expression as a means of altering cell fates and generating dominant phenotypes. *Development* **118**, 401–415 (1993).
4. Phelps C.B. & Brand A.H. Ectopic gene expression in *Drosophila* using GAL4 system. *Methods* **14**, 367–379 (1998).
5. Chalfie M., Tu Y., Euskirchen G., Ward W.W. & Prasher D.C. Green fluorescent protein as a marker for gene expression. *Science* **263**, 802–805 (1994).
6. Maroniche G.A. *et al.* Development of a novel set of Gateway-compatible vectors for live imaging in insect cells. *Insect Mol. Biol* **20**, 675–685 (2011).

7. Lippincott-Schwartz J. & Patterson G.H. Development and use of fluorescent protein markers in living cells. *Science* **300**, 87–91 (2003).
8. Osanai-Futahashi M. *et al.* A visible dominant marker for insect transgenesis. *Nat. Commun.* **3**, 1295 (2012).
9. Kokoza V.A. & Raikhel A.S. Targeted gene expression in the transgenic *Aedes aegypti* using the binary Gal4-UAS system. *Ins. Biochem. and Mol. Biol.* **41**, 637–644 (2011).
10. Evans I.R., Zanet J., Wood W. & Stramer B.M. Live imaging of *Drosophila melanogaster* embryonic hemocyte migrations. *J Vis. Exp* **12**, (2010).
11. Kuznicki M.L. & Gunawardena S. *In vivo* visualization of synaptic vesicles within *Drosophila* larval segmental axons. *J. Vis. Exp.* **15**, (2010).
12. Xu T. & Rubin G.M. Analysis of genetic mosaics in developing and adult *Drosophila* tissues. *Development* **117**, 1223–1237 (1993).
13. Theodosiou N.A. & Xu T. Use of FLP/FRT system to study *Drosophila* development. *Methods* **14**, 355–365 (1998).
14. Evans C.J. *et al.* G-TRACE: rapid Gal4-based cell lineage analysis in *Drosophila*. *Nat. Meth.* **6**, 603–605 (2009).
15. Tare M., Puli O.R., Moran M.T., Kango-Singh M. & Singh A. Domain specific genetic mosaic system in the *Drosophila* eye. *Genesis* **51**, 68–74 (2013).
16. Nakazawa N., Taniguchi K., Okumura T., Maeda R. & Matsuno K. A novel Cre/loxP system for mosaic gene expression in the *Drosophila* embryo. *Dev. Dyn.* **5**, 965–974 (2012).
17. Siegal M.L. & Hartl D.L. Transgene coplacement and high efficiency site-specific recombination with the Cre/loxP system in *Drosophila*. *Genetics* **144**, 715–726 (1996).
18. Hadjieconomou D. *et al.* Flybow: genetic multicolor cell labeling for neural circuit analysis in *Drosophila melanogaster*. *Nat. Meth.* **8**, 260–266 (2011).
19. Hampel S. *et al.* *Drosophila* Brainbow: a recombinase-based fluorescence labeling technique to subdivide neural expression patterns. *Nat. Meth.* **8**, 253–259 (2011).
20. Tran V., Lim C., Xie J. & Chen X. Asymmetric division of *Drosophila* male germline stem cell shows asymmetric histone distribution. *Science* **338**, 679–682 (2012).
21. Mavrakis M., Rikhy R., Lilly M. & Lippincott-Schwartz J. Fluorescence imaging techniques for studying *Drosophila* embryo development. *Curr. Protoc. Cell Biol.* (2008).
22. Althoff F., Karess R.E. & Lehner C.F. Spindle checkpoint-independent inhibition of mitotic chromosome segregation by *Drosophila* Mps1. *Mol. Biol. Cell* **23**, 2275–2291 (2012).
23. Patterson G.H. & Lippincott-Schwartz J. A photoactivatable GFP for selective photolabeling of proteins and cells. *Science* **297**, 1873–1877 (2002).
24. Post J.N., Lidke K.A., Rieger B. & Arndt-Jovin D.J. One- and two-photon photoactivation of a paGFP-fusion protein in live *Drosophila* embryos. *Febs Letters* **579**, 325–330 (2005).
25. Albrecht S.C., Barata A.G., Grosshans J., Teleman A.A. & Dick T.P. *In Vivo* mapping of hydrogen peroxide and oxidized glutathione reveals chemical and regional specificity of redox homeostasis. *Cell Metab.* **14**, 819–829 (2011).
26. Caussinus E., Kanca O. & Affolter M. Fluorescent fusion protein knockout mediated by anti-GFP nanobody. *Nat. Struct. Mol. Biol.* **19**, 117–121 (2011).

7 Fluorescence

7.1 Introduction

The availability of fluorescence microscopy and imaging has revolutionized *in vivo/in situ* studies, elevated the utility of genetic markers, and improved sub-cellular resolution, all while simultaneously taking the beauty of histological preparations to new bounds. In Chapter 5, Immunostaining Techniques, fluorogenic staining methods allow for tighter and more specific staining compared to standard chromogenic labeling. The fluorescent labels can be quantified over a dynamic range of intensities not possible by manual visual analysis of chromogenic stains. Fluorescent immunostaining also facilitates co-localization studies for determining if independent proteins or markers are in the same cell and/or subcellular compartments. Many of the methods in this chapter build upon those for fluorescent genetic markers detailed in Chapter 6 (Use of genetic markers in insect histology), providing detailed preparation of samples for *in vivo/in situ* localization of individual fluorescently tagged proteins. Finally, fluorescence allows the "painting" of chromosomes through *in situ* hybridization techniques that identify genes within the chromosomes.

7.1.1 *Preparation of* Drosophila *brains for live imaging [1]*

Larval Brains

Procedure:

1. Prepare egg laying chamber and collect embryos on a fruit agar plate.
2. Wait until embryos start to hatch (~21 hr at 25°C) and remove all hatched larvae from plate.
3. After 2 hr collect all newly hatched larvae and allow them to develop in a vial to the necessary stage.
4. Collect the larvae with a paint brush dampened with phosphate buffered saline with 0.3% Triton X-100 (PBTx).
5. Place larvae and PBTx in a dissecting dish.
6. Hold the larval body with one pair of forceps and the mouth hooks with a second pair. Pull the forceps apart gently to pull the mouth hooks with the brain attached to it.
7. While still holding the mouth hooks, gently remove excess tissue surrounding the brain.
8. Still holding the mouth hooks, transfer the brain to a drop of PBS on a microscope slide and carefully separate the eye discs and mouth hooks from the brain.

Insect Histology: Practical Laboratory Techniques, First Edition.
Pedro Barbosa, Deborah L. Berry and Christina S. Kary.

Companion Website: www.wiley.com/go/barbosa/insecthistology

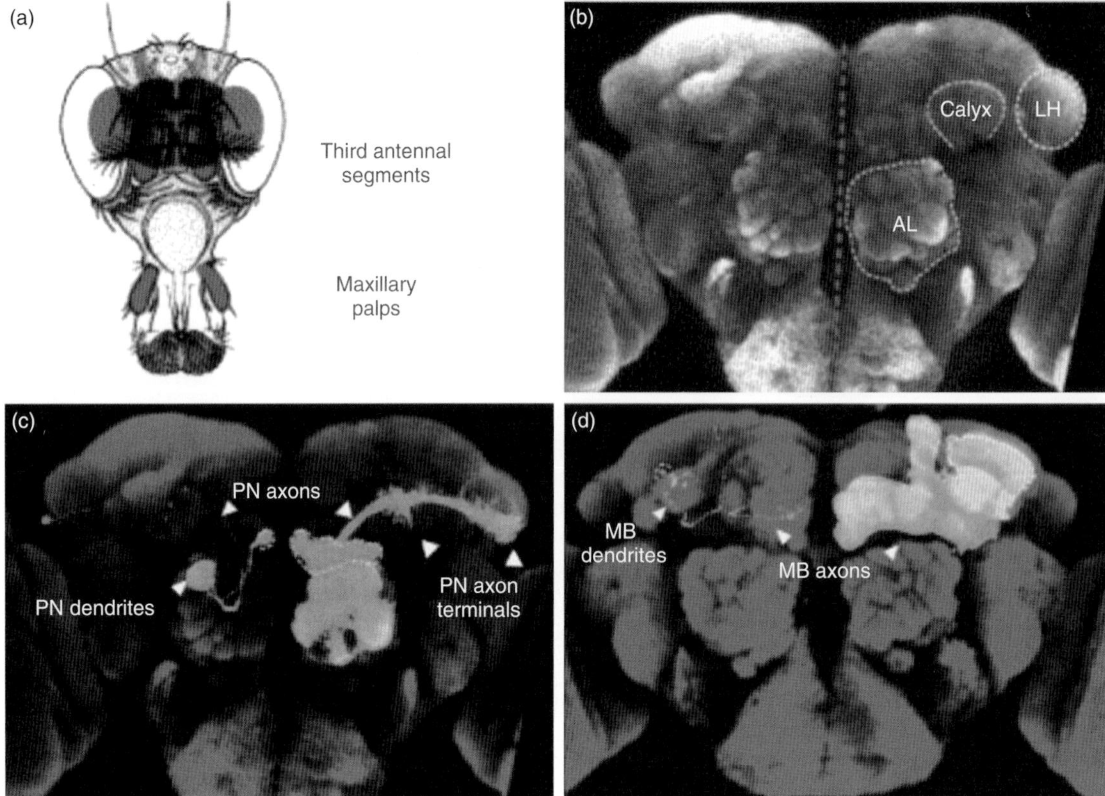

Fig. 7.1 Schematic of the *Drosophila* head (anterior view), with the brain in dark gray. Olfactory appendages are shown in red (a). A brain in the same orientation, stained with the monoclonal antibody nc82, a presynaptic marker recognizing neuropil. Major brain structures are labeled: AL, antennal lobe; calyx, mushroom body calyx; LH, lateral horn. Dotted red line represents midline (b). A brain showing MARCM clones. Shown in green on the left is a projection neuron (PN) single-cell clone; on the right is a PN neuroblast clone. nc82 staining is shown in red. White dotted lines outline cell bodies (c). A brain showing MARCM clones. Shown in green on the left is an MB single-cell clone of the '/' type; on the right is an MB neuroblast clone. nc82 staining is shown in red. White dotted lines outline cell bodies (d). Reprinted with permission from references 8, 9. (Source: Wu & Luo 2006. Reproduced with permission of Nature.) See plate section for the color version of the figure.

9. Orient the brain(s) so that the anterior side of the antennal lobes is facing upward.
10. Arrange two broken coverslips on the microscope to form a bridge around the brains.
11. Cover the bridge with another coverslip and image.

Pupal Brains

1. Collect white pre-pupae with a paint brush dampened with phosphate buffered saline (PBS).
2. Place 0 hr pupae in a small Petri dish in a humidified chamber at 25°C to age the pupae to the appropriate developmental stage.
3. Place Petri dish on ice to stop development.
4. Transfer a pupa to PBS in a dissecting dish.
5. Hold the puparium, dorsal slide up, with forceps. Squeeze a second pair of forceps closed and use the sharp tips to puncture a hole into the abdominal end of the pupa.

6. Open the forceps to tear the puparium apart and pull the pupa out.
7. Use forceps to gently remove abdominal tissue, up to the level of the thorax.
8. Gently hold the pupa at the very end of the head case and use the side of the second pair of forceps to gently push the brain from the head and out through the thorax.
9. Gently remove excess fat particles.
10. Transfer the brain to a drop of PBS on a microscope slide.
11. Orient the brain(s) so that the anterior side of the antennal lobes is facing upward.
12. Arrange two broken coverslips on the microscope to form a bridge around the brains.
13. Cover the bridge with another coverslip and image.

Adult Brains

1. Anesthetize adult flies and place into a dish on ice.
2. Transfer one fly in PBTx, in a dissection dish.
3. Hold the fly belly up with forceps and insert one side of a second pair of forceps into the cavity just below the eye to obtain a grip of the eye. Be careful to avoid internal structures, notably the brain.
4. Gently pull the head off the fly and discard the body.
5. Use the free forceps to obtain a grip of the other eye from the underside, and gently pull the forceps apart to open the head cuticle.
6. Remove the brain from the head cuticle and carefully remove the surrounding trachea.
7. Transfer the brain to a drop of PBS on a microscope slide.
8. Orient the brain(s) so that the anterior side of the antennal lobes is facing upward.
9. Arrange two broken coverslips on the microscope to form a bridge around the brains.
10. Cover the bridge with another coverslip and image.

7.1.2 *Preparation of* Drosophila *ommatidium for live imaging [2]*

Procedure:

1. Anesthetize adult flies, select an individual fly and place it in a small dish of phosphate buffered saline.
2. Under a dissecting scope, using two sets of forceps, pinch the proboscis with one and the neck with the other set and pull apart to detach the head from the body.
3. While holding the head by the proboscis, cut the fly head in half along the left/right midline.
4. Use a pipette tip to pick up the retina and transfer to a dish of Schneider medium (see Appendix).
5. Peel away the head cuticle and remove the brain with forceps. Keep the retina sandwiched between the lamina and the cornea.
6. Transfer the retina onto a drop of fresh Schneider medium on a glass slide.
7. Grasp the most dorsal or ventral end of the retina with forceps. Using microscissors, cut the retina in half along the dorsal/ventral midline to expose the ommatidia in the middle of the eye.
8. Hold the retina so that the cornea is facing down and the lamina are facing up. Using a fine tungsten needle, detach the ommatidia from the lamina and the cornea.
9. Single ommatidium and groups of ommatidia will collect on the bottom of the well or slide. Remove any larger debris that may have collected.
10. Ommatidia expressing fluorescent markers can be imaged directly.

7.1.3 *Preparation of live testis squashes in* Drosophila *[3]*

Formula:

Testis Buffer:

1M EDTA	1 pt
1M KCL	183 pt
1M NaCl	47 pt
1M phenylmethylsulfonyl fluoride (PMSF)	1 pt
1M Tris-HCl	10 pt

Add distilled water to 1 L, adjust pH to 6.8.

Procedure:

1. Dissect testes in testis buffer for *Drosophila* larvae, pupae, or young adults.
2. Transfer testes into a 2 μl drop of testis buffer on coverslip. Tear open the adult testes using either very thin forceps or tungsten needles.
3. Place a clean slide over the coverslip without pressing and invert the sandwich.
4. Image using phase contrast optics.

7.1.4 *Preparing Drosophila egg chambers for live imaging [4]*

Procedure:

1. Put two day old females and males (at a 2:1 ratio) into egg laying chamber for 24–60 hr at 25°C.
2. Anesthetize the flies and under a dissecting microscope, use fine nosed forceps to grasp a female at the anterior end of the abdomen and at the extreme posterior of the fly, and pull to remove the posterior tissues.
3. Maintain the forceps at the anterior end and use cleaned forceps to gently squeeze or massage the abdomen from the anterior to the posterior end to expel the ovaries (two large opaque structures).
4. Place ovaries into a prepared drop of halocarbon oil on a coverslip.
5. Repeat steps 2–4 until there are 4–6 ovaries in the oil. Each ovary should contain 6–10 egg chambers at different stages.
6. In the oil, separate the two ovaries by removing the connective tissue at the posterior (older stages) end with closed forceps.
7. Isolation of ovarioles
   a. Orient an individual ovary with the oldest stages to the left and youngest stage to the right. The younger staged egg chambers are identifiable by their transparency.
   b. Hold the posterior end of the ovary with forceps and use a dissecting probe or tungsten needle to tease apart the individual ovarioles.
   c. Once the ovariole has been separated from the ovary, drag it to the center of the oil and orient as appropriate for the experiment. The ovarioles in oil will settle and adhere to the glass.
   d. Repeat steps 6–9 until there are 15–25 individual ovarioles. Carefully remove any excess ovarian tissue out of the oil.
8. Isolation of individual late-stage egg chambers
   a. Grasp the posterior end of the ovary with forceps and insert dissecting needle into the posterior end of the ovary near the forceps.
   b. Pull the dissecting needle through the ovary until it exits at the early stage egg chambers. When done correctly, the needle will remove the connective tissue holding the ovarioles and late stage egg chambers in place.
   c. Repeat steps 8a and 8b until the ovarioles are spread out on the coverslip.

9. Oocytes expressing fluorescent genetic markers can be visualized directly under a fluorescent microscope.
10. Oocytes can also be injected with transgene RNA tagged with fluorescent markers for in vivo studies.

7.1.5 *Preparation of* Drosophila *embryos for fluorescence microscopy [5]*

Application:
Recommended as an alternative to dechorionation in halocarbon oil to prevent anoxia in the embryos allowing for live imaging.

Procedure:
1. Prepare egg laying chambers and collect eggs on a fresh juice agar plate for 1 hr at 25°C. Discard this "pre-lay" plate, replace with a fresh juice agar plate and incubate at 25°C to obtain embryos at the desire stage of development.
2. Add distilled water to the juice agar plate and gently brush the embryos into suspension. Pour the water containing the embryos into a prepared egg basket. Wash embryos in the basket with distilled water to remove all remnants of yeast paste.
3. Place the egg basket into freshly prepared 50% bleach and agitate gently so that the bleach solution disperses the embryos. Incubate for 1 min with periodic gentle agitation.
4. Wash embryos thoroughly with distilled water to remove all bleach. Use a squirt bottle of distilled water to dislodge any embryos stuck to the side of the egg basket.
5. Blot the mesh containing the embryos on filter paper. Working quickly under a dissecting microscope, use a fine paintbrush to pick the dechorionated embryos from the mesh and mount them on a chambered coverglass on their dorsolateral/ventrolateral sides. Disperse embryos so that they are not touching or too crowded, to avoid anoxia and defects in development.
6. Slowly add phosphate buffered saline from the side of the chamber, avoiding air bubbles, until there is sufficient saline to cover the embryos and compensate for evaporation during imaging. Remove any air bubbles that form with a paintbrush.
7. Image embryos using appropriate settings for the genetic background of the embryos.

Note:
1. If dorsal appendages are still observed under the dissecting scope prior to mounting, place embryos back in bleach solution for 5 to 10 sec and rinse thoroughly.
2. Overexposure of embryos to bleach will damage the embryos so treat embryos the shortest period possible to sufficiently remove the chorion.

7.1.6 *Hanging drop protocol for* Drosophila *embryos [6]*

Application:
Recommended as a preparation method with improved viability of embryos.

Formula:
Hydrocarbon Oil:
Preparation:
1. Mix halocarbon oil (Halocarbon Products Corp.®) viscosity series 700 and series 5 at a ratio of 1:1.

Procedure:

1. Prepare a dechorionation slide by attaching a piece of double-sided tape to a microscope slide and removing the backing from the tape.
2. Prepare a coverslip with a drop of the mixed halocarbon oil.
3. Collect appropriately staged embryos from juice-agar plates using either fine forceps or a fine paint brush.
4. Gently lower the embryos to the surface of the tape on the dechorionation slide.
5. Gently nudge or stroke the embryos with the side of the forceps to break open the outer waxy chorion without rupturing the inner vitelline membrane.
6. Once the outer chorion has been ruptured, tease the embryo from the chorion; the dechorionated embryo tends to adhere to the surface of the forceps. Be sure to avoid touching the dechorionated embryo to the surface of the tape.
7. As soon as an embryo is dechorionated, quickly transfer it to the drop of halocarbon oil by gently placing the forceps into the oil, which will dislodge the embryo.
8. Clean any oil from the forceps and continue dechorionation. Transfer embryos until an appropriate number has been reached.
9. Once the embryos are collected, use forceps or a paint brush to push the embryos to the bottom of the oil drop and arrange them in the desired position.
10. Invert the coverslip over the well of a live imaging chamber and tape the sides of coverslip in place. The embryos should float up to the top of the chamber for live imaging.
11. Poke a few holes in the tape for ventilation and image the embryos as appropriate.

7.1.7 *Preparation of* Drosophila *embryos for live imaging – embryo fixation without methanol and hand devitellinization [7]*

Application:

Recommended for visualization of nuclei and F actin along cell membranes for morphological analysis.

Formula:

Fixative:

Heptane	1 pt
37% Formalin	1 pt

Procedure:

1. Prepare an egg laying chamber with a fruit juice agar plate and allow files to lay eggs.
2. Cover the plate with 100% bleach and brush the embryos with a fine paintbrush to detach them from the plate.
3. Pour the bleach with embryos into an egg basket and place the basket on an empty Petri dish. Fill the dish with enough bleach to cover the embryos and leave for 2 min.
4. Rinse embryos extensively in tap water and transfer the dechorionated embryos to a glass tube.
5. Add 1 ml of fixative and fix for 45 min at room temperature.
6. Transfer the fixed embryos to a small strip of 3MM paper and wait for the heptane to evaporate (~15 sec).
7. Put the 3MM paper strip with the embryos facing down gently onto a strip of double-stick tape in a Petri dish. Press very gently in the center and sides so that the embryos stick firmly to the tape.
8. Remove the paper and cover the embryos with 200 μl of phosphate buffered saline with 0.05% BSA and 0.2% Triton X-100, 0.02% sodium azide (PBTA).

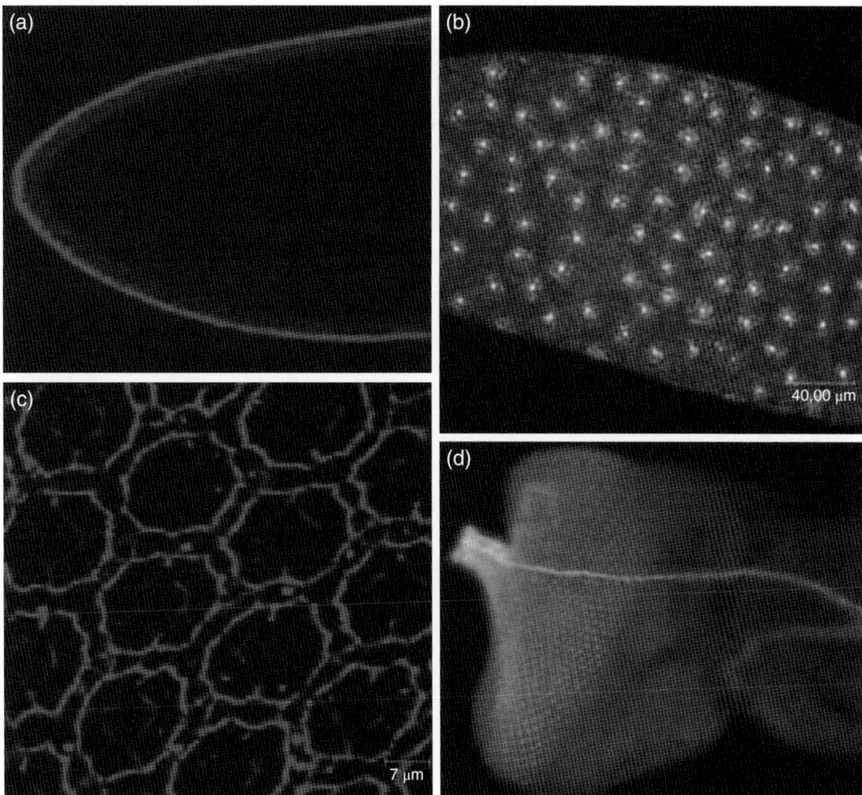

Fig. 7.2 Examples of fluorescence preparations of *Drosophila* whole mounts using the protocols described in this chapter. All confocal images were obtained with a LeicaTCS4D confocal microscope. (a) Confocal optical section of a *D. melanogaster* embryo whole mount at blastoderm stage, double stained with phalloidin–rhodamine (red) and DAPI (blue) to allow simultaneous visualization of nuclei and cortical actin around cell membranes. Anterior is to the *left*. (b). Confocal image of a syncytial stage *D. melanogaster* embryo whole mount with nuclei synchronously undergoing mitoses. The preparation was triple stained with phalloidin–rhodamine (red to visualize F-actin), DAPI (blue; DNA) and antigamma- tubulin antibody (yellow; centrosome, visualized with goat anti mouse Alexa Fluor® 488 secondary antibody. (c) Confocal image of a *Drosophila* pupal retina, approximately 25% after puparium formation, immunostained with a mouse monoclonal antibody against the cell adhesion glycoprotein Roughest (Rst; *see* refs. [15] and [16]) and visualized with a Cy3 Goat anti-mouse secondary antibody. Note: The strong immunoreactivity (*red*) at the borders between primary pigment cells and interommatidial cells (a single layer of pigment cells that surround each ommatidium in the adult eye of *Drosophila*) , but not between the interommatidial cells themselves (For more details *see* refs. [17] and [18]). (d) Photomicrograph taken with a Zeiss Axiophot fluorescence microscope of a *Drosophila* eye/antennal imaginal disc whole mount stained with an antibody specific for sensory neurons (Mab22C10; *see* ref. [19]) and visualized with a FITC anti-mouse secondary antibody. Posterior is to the *left*. The ommatidial clusters containing the differentiating photoreceptors, behind the morphogenetic furrow, and the larval visual nerve, traversing the disc from the optic stalk are clearly stained (For details *see* ref. [20].). (Source: Ramos *et al.* 2010. Reproduced with permission of Springer.) See plate section for the color version of the figure.

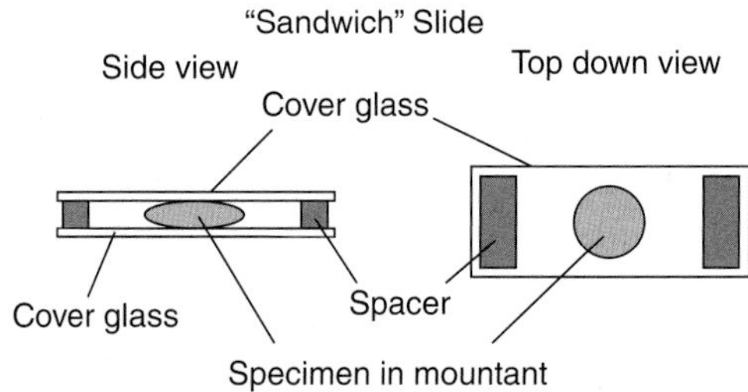

Fig. 7.3 Schematic diagram of "sandwich" slide mounting technique. Note: The rectangular top and bottom cover glasses are 50 mm × 22 mm, #1.5 cover glass, nominal thickness 170 μm, (Fisher® brand catalog # 12-544B, Fisher Scientific®, Pittsburgh, PA) and the spacers are made from 22 mm × 22 mm, #1.5 cover glasses, nominal thickness 170 μm, Fisher® brand catalog # 12-541B. (Source: Schawaroch & Li 2013. Reproduced with permission of John Wiley & Sons.)

9. Remove the vitelline membrane by hand, individually, under a dissecting microscope using an odontological needle, by making a small hole in the vitelline membrane at one end of the embryo.
10. The devitellinzed embryos will float in the PTBA solution. Use a pipette to transfer the embryos to a 0.5 ml microfuge tube (embryos can be stored up to one week in PBTA at 4°C).
11. Wash the embryos two times with 400 μl PBTA.
12. Add a 1:500 dilution of phalloidin in PBTA and incubate for 20 min at room temperature in the dark.
13. Wash the embryos five times with 400 μl PBTA for 5 min each.
14. Add a 1:2000 dilution of DAPI in PBTA and incubate for 4 min at room temperature, in the dark.
15. Wash the embryos five times with 400 μl PBTA for 5 min each.
16. Remove as much PBTA as possible and replace with 50% glycerol.
17. To avoid crushing the embryos, use nail polish to mount a small coverslip on each end of a slide to form a bridge. Place the embryos in 50% glycerol in the middle of the slide and cover with a coverslip so that the coverslip rests on top of the other two coverslips.
18. Image embryos using appropriate fluorescent filters.

7.1.8 *Optimal mounting medium for confocal microscope three dimensional imaging [8]*

Application:
Recommended as a mountant for insect genitalia that gives minimal tissue compression and background noise.

Protocol:
1. Dissect adult male *D. melanogaster* to obtain the genital apparatus.
2. Clear the tissue in 10% KOH for 1.5–2.5 hr at 70°C.
3. Rinse in distilled water for 5–10 min.
4. Wash two times in 70% ethanol for 5 min each.
5. Transfer genitalia to glycerin for final dissection to disarticulate the dorsal and ventral halves of the male genitalic apparatus.
6. Prepare a sandwich slide mount by gluing two small coverslips to the ends of a slide.
7. Orient the genitalia in molten 1% agarose in between the two coverslips.
8. Cover with a long coverslip that sits on the two smaller coverslips.
9. Allow mountant to firm and image using a confocal microscope.

Fig. 1.1 Beetles have a hardened cuticle. See chapter text for full caption.

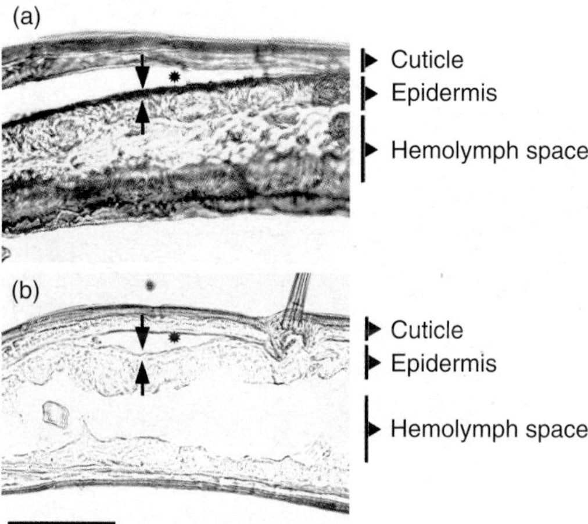

Fig. 1.6 Localization of chitin synthase in the epiproct of Periplaneta americana. See chapter text for full caption.

Insect Histology: Practical Laboratory Techniques, First Edition.
Pedro Barbosa, Deborah L. Berry and Christina S. Kary.

Companion Website: www.wiley.com/go/barbosa/insecthistology

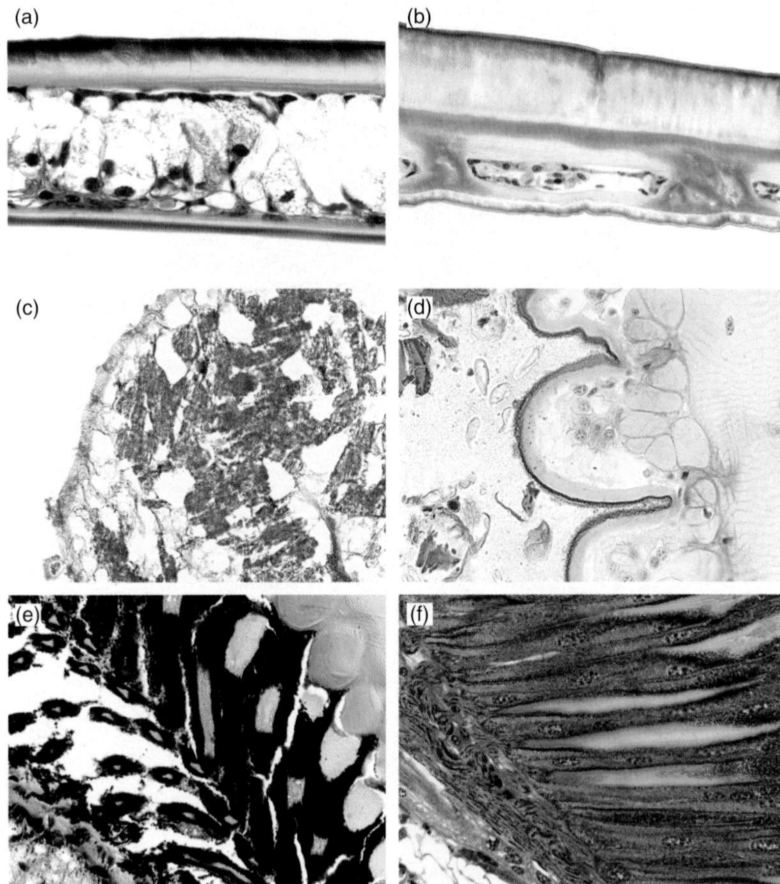

Fig. 2.1 Comparison of fixation techniques for individual tissues from the cockroach Blatella germanica. See chapter text for full caption.

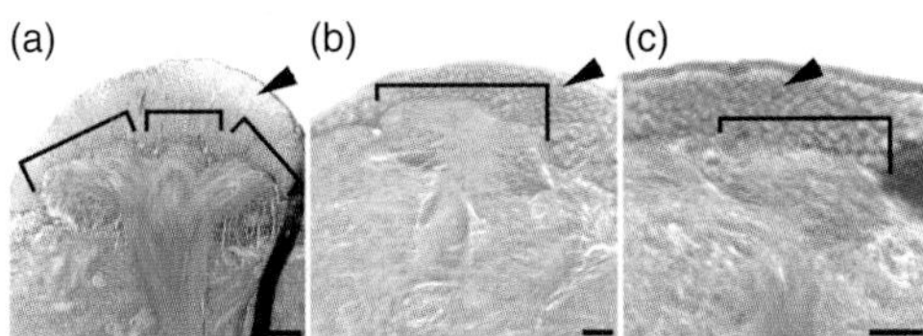

Fig. 2.6 Differences in calyx structure in generalist and specialist predators. (a) Tripartite gyrencephalic calyx (brackets) of the generalist predator Scarites subterraneus (Carabidae). (b) Lissencephalic calyx of the carabid Scaphinotus elevatus, a specialist predator of snails and slugs. (c) Minute lissencephalic calyx of the ladybird beetle Harmonia axyridis (Coccinellidae), a specialist predator of aphids and scale insects. See chapter text for full caption.

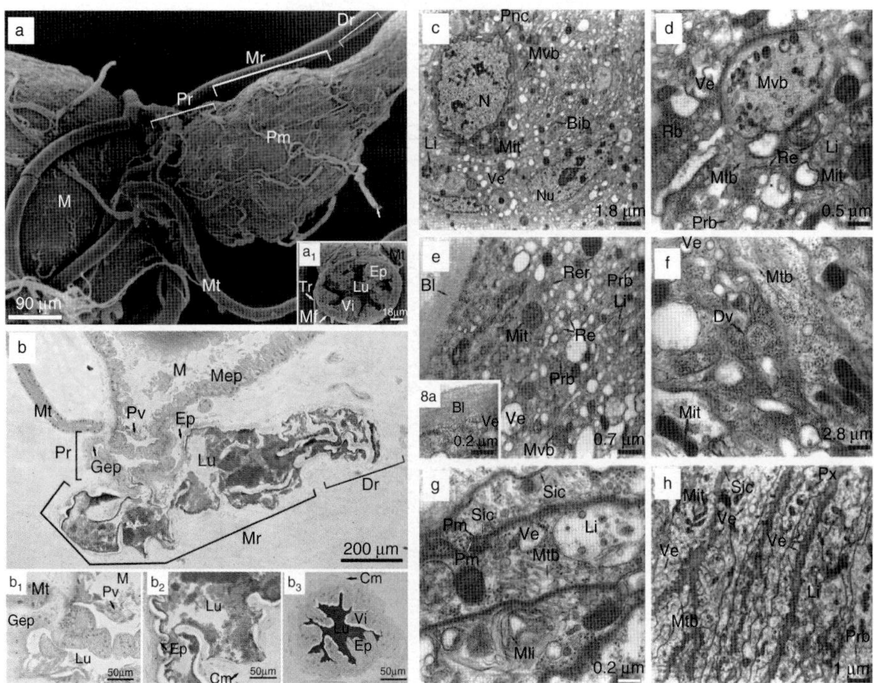

Fig. 2.9 The ileum was fixed in 4% paraformaldehyde for 2 h and then buffered in sodium phosphate pH 7.4 for 24h. See protocol 2.6.5 Karnovsky's solution. See chapter text for full caption.

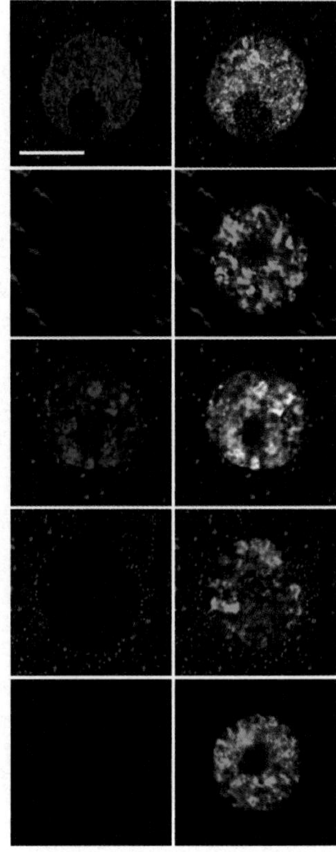

Fig. 2.10 Importin-β11 and Importin-α2 are expressed in Drosophila muscle cell nuclei. See protocol 2.6.7 Mosca and Schwarz. See chapter text for full caption.

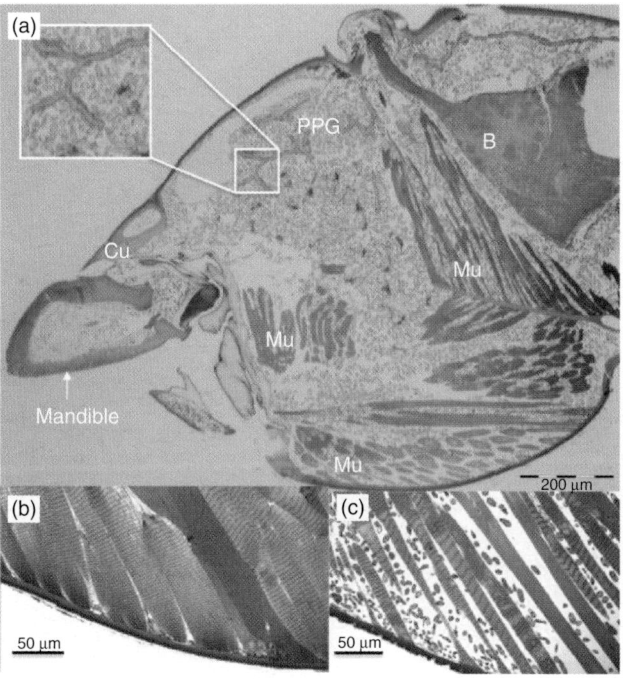

Fig. 2.11 Simultaneous Fixation of Tissue Types - Heads of manipulated ants colonized by the fungus Ophiocordyceps unilateralis s.l,. See chapter text for full caption.

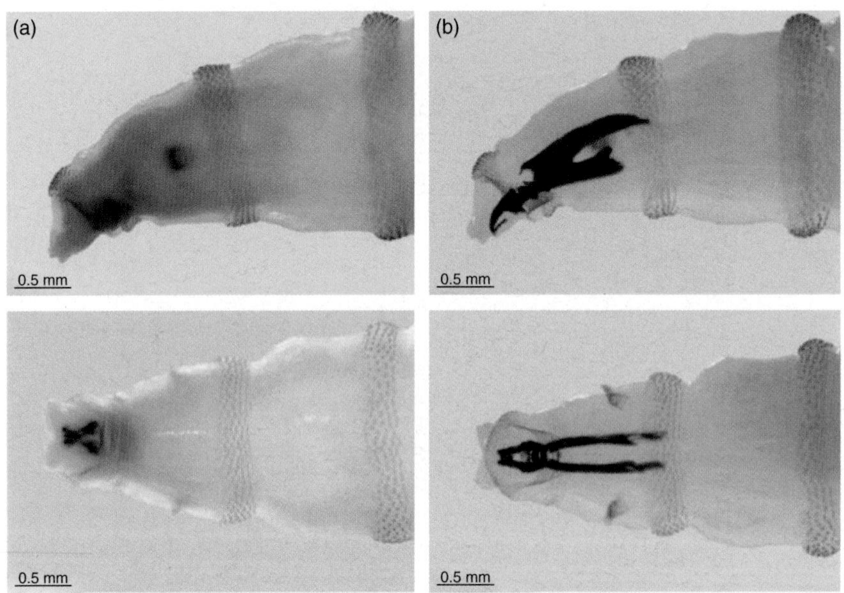

Fig. 3.2 Before (a) and after (b) clearing in methyl salicylate. Lateral (top) and ventral (bottom) views of a third instar Calliphora vomitoria maggot. See chapter text for full caption.

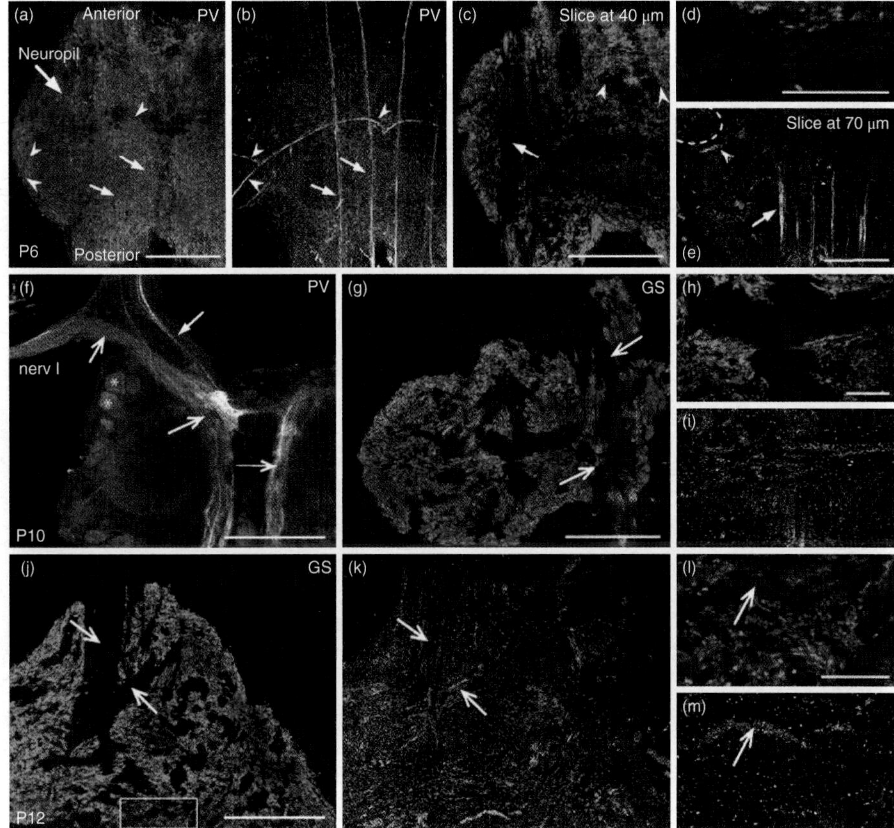

Fig. 3.4 TM-MFasII expression in the mesothoracic ganglion in pupal stages P6, P10, and P12. See protocol 3.2.11 Methylsalicylate. See chapter text for full caption.

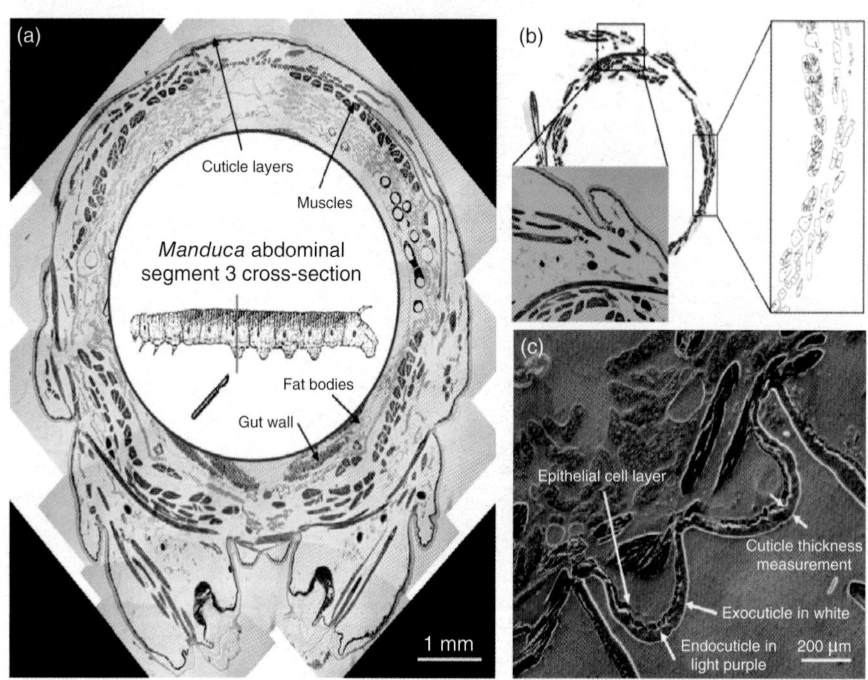

Fig. 4.2 Histology of a Manduca sexta caterpillar. Hematoxylin and eosin (H&E) stained most of the soft tissues inside the body. See chapter text for full caption.

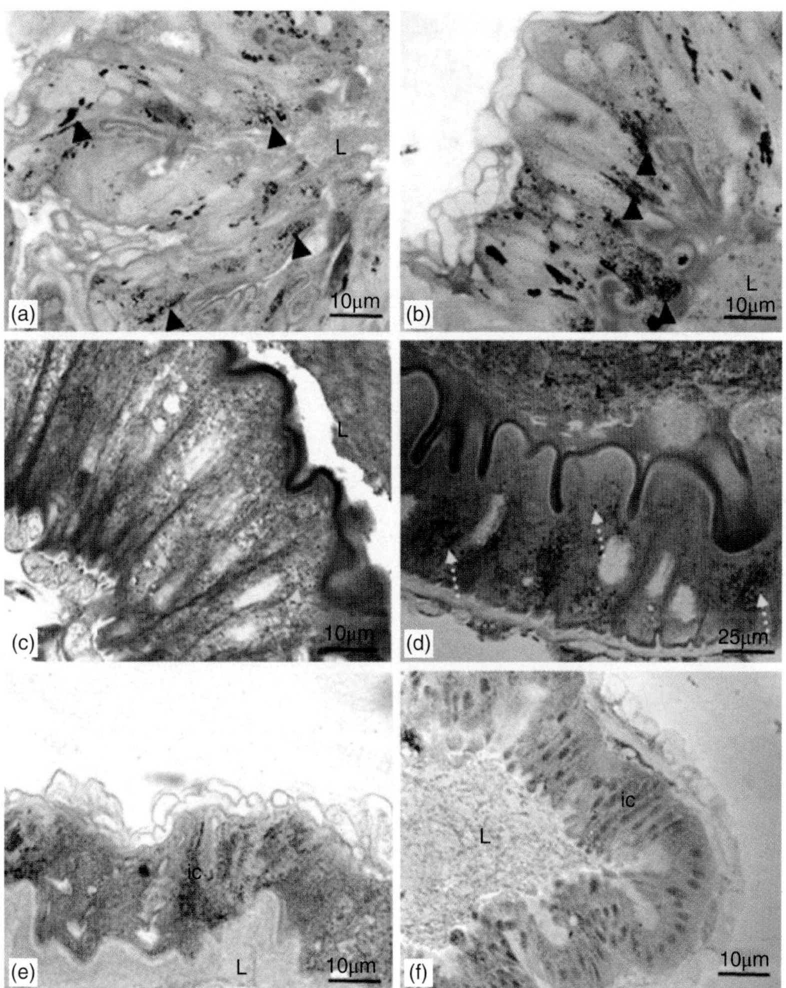

Fig. 4.3 Light microscopy of the median region of the midgut of P. nigrispinus.

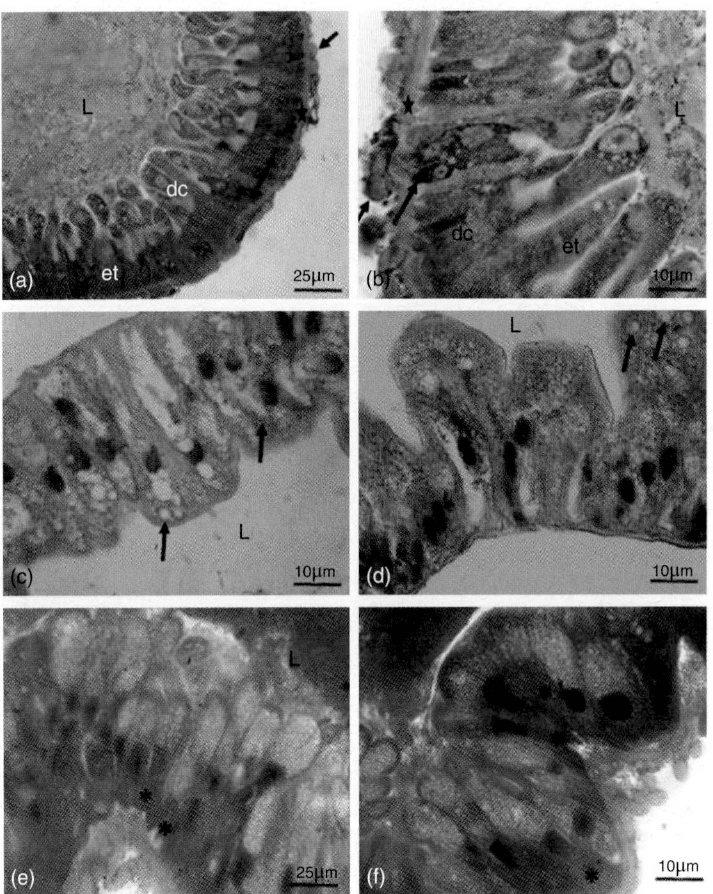

Fig. 4.4 Light microscopy of the median region of the midgut of P. nigrispinus. See chapter text for full caption.

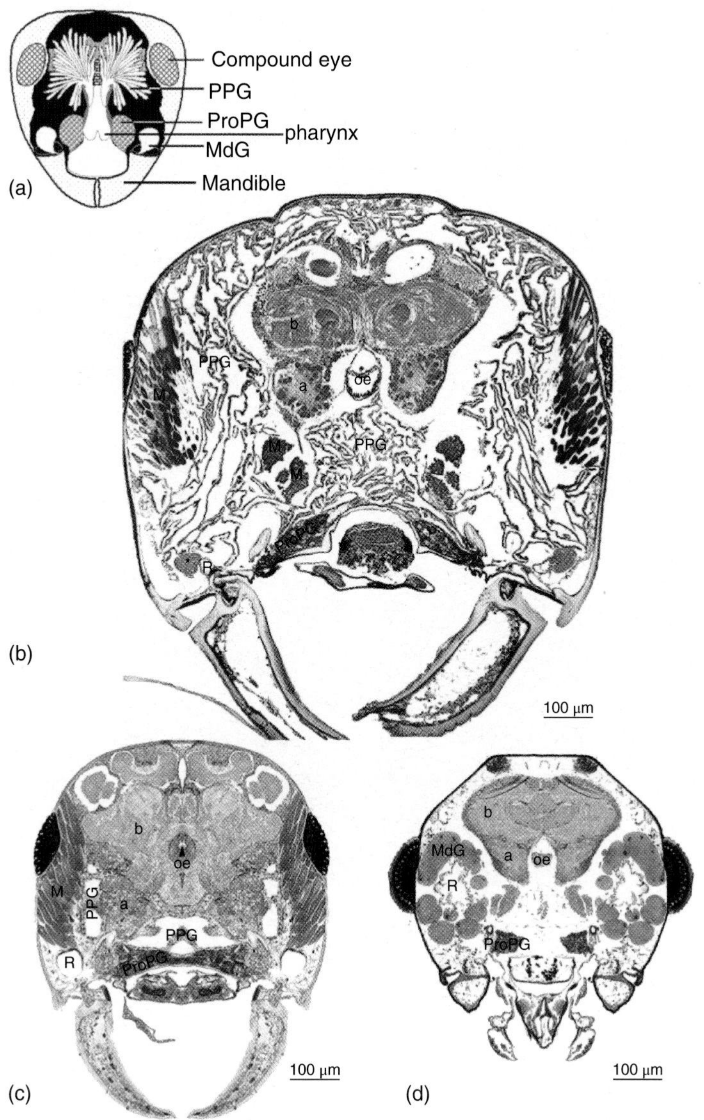

Fig. 4.5 Histological aspects of the head glands of Lasius niger. See chapter text for full caption.

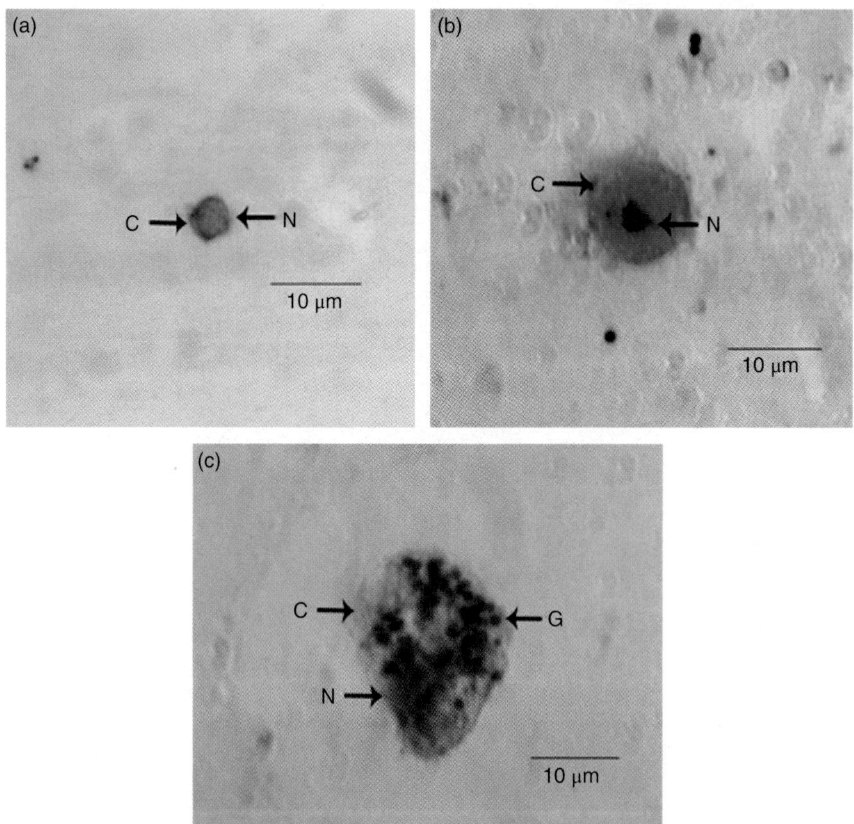

Fig. 4.6 Light microscope images of a representative HEMA 3TM (which is similar to Giemsa stain) fixed, stained prohemocyte (a), oenocytoid (b), and granulocyte (c) from an Ae. aegypti adult female. See chapter text for full caption.

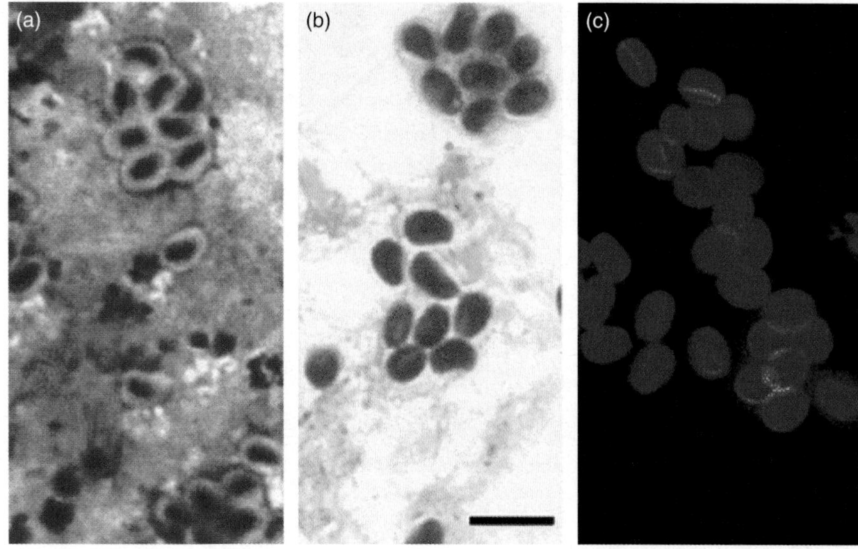

Fig. 4.7 Comparison of Different Stains. Light microscopy images of Thelohania solenopsae spores in smears of tissue from Solenopsis invicta. See chapter text for full caption.

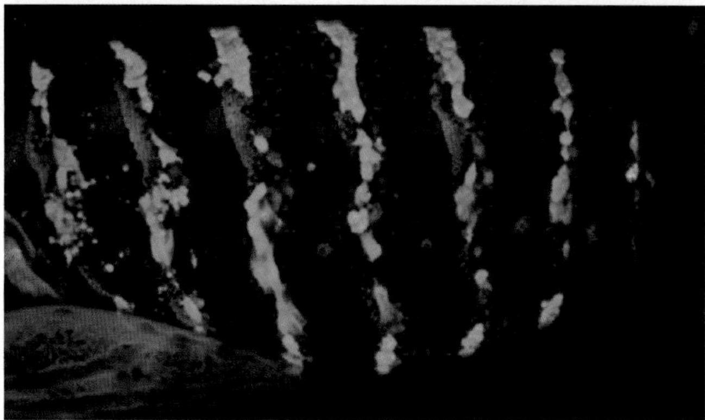

Fig. 5.4 Segmentation gene products in Drosophila pupae. See protocol 5.2.10 Immunohistochemistry of Drosophila pupal abdomen. See chapter text for full caption.

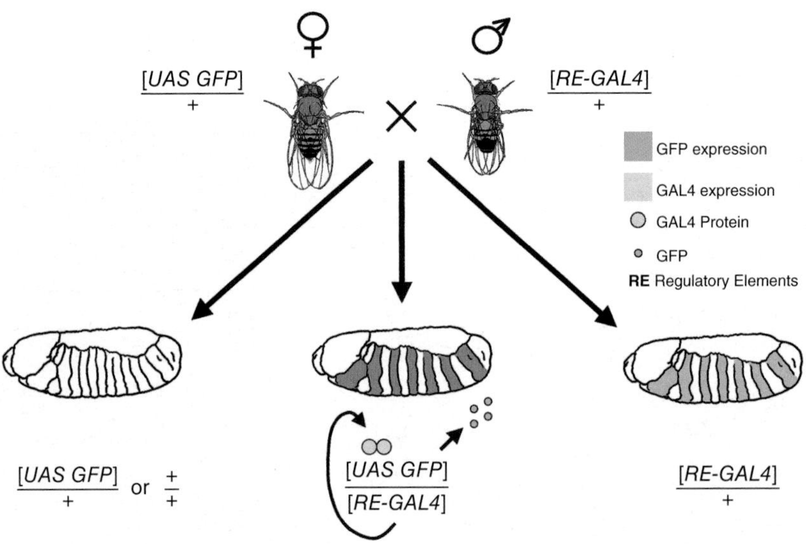

Fig. 6.1 The bipartite UAS/GAL4 system in Drosophila. See chapter text for full caption.

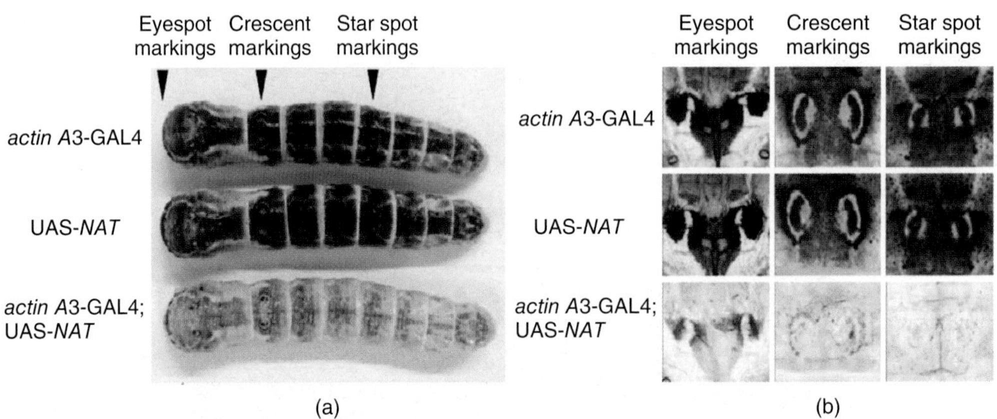

Fig. 6.2 The fifth-instar silkworm larvae shown here have the dominant striped (pS) genetic background. See protocol 6.2.1 A visible dominant marker for insect transgenesis. See chapter text for full caption.

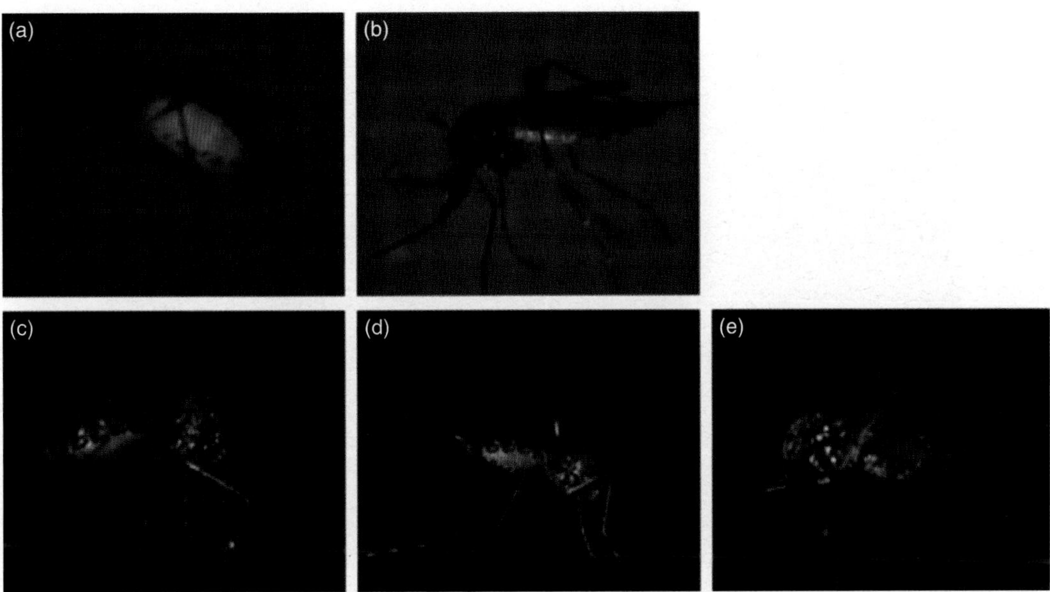

Fig. 6.3 Tissue- and stage-specific EGFP reporter expression in the Vg-Gal4/UAS-EGFP hybrid female mosquitoes after blood meal activation. See protocol 6.2.2 Targeted gene expression in the transgenic Aedes aegypti using the Gal4-UAS system. See chapter text for full caption.

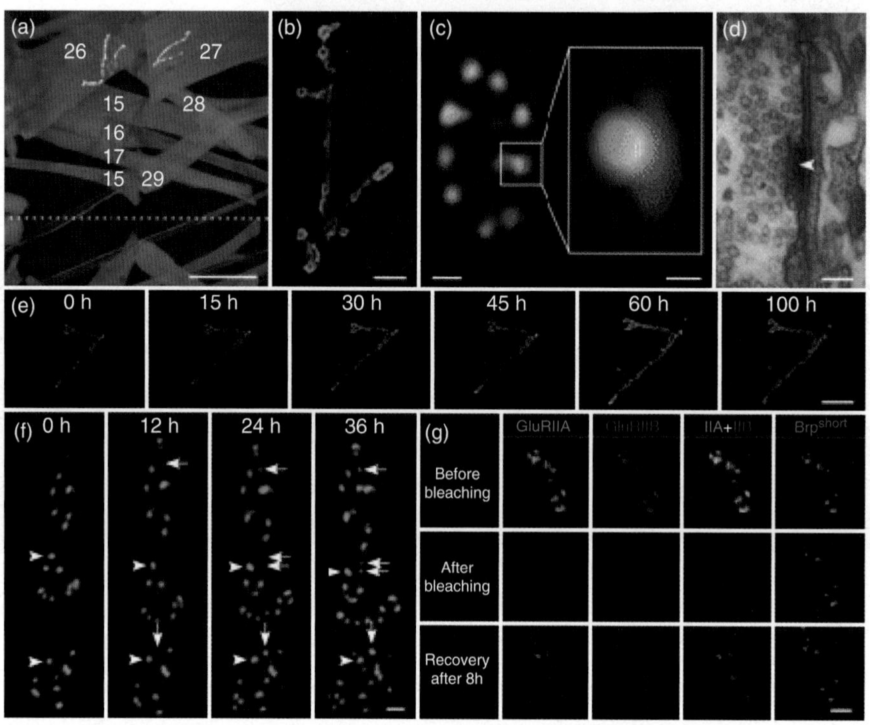

Fig. 6.6 Intravital imaging of neuromuscular junctions (NMJs). See protocol 6.2.4 In vivo visualization of synaptic vesicles within Drosophila larval segmental axons. See chapter text for full caption.

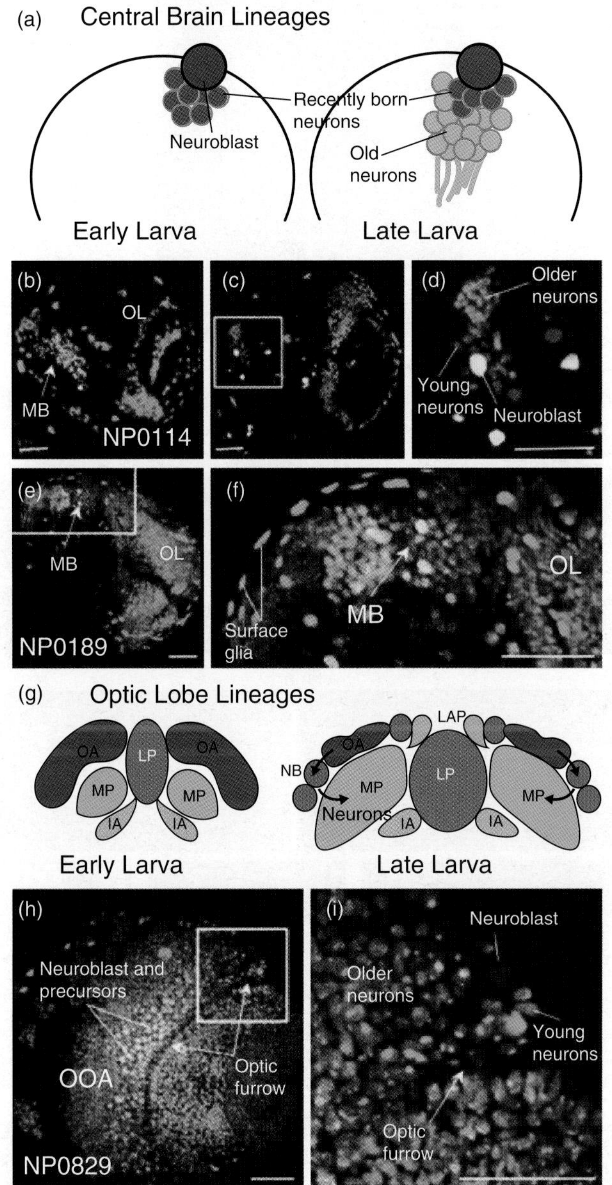

Fig. 6.7 G-TRACE reveals novel cell lineage markers. See chapter text for full caption.

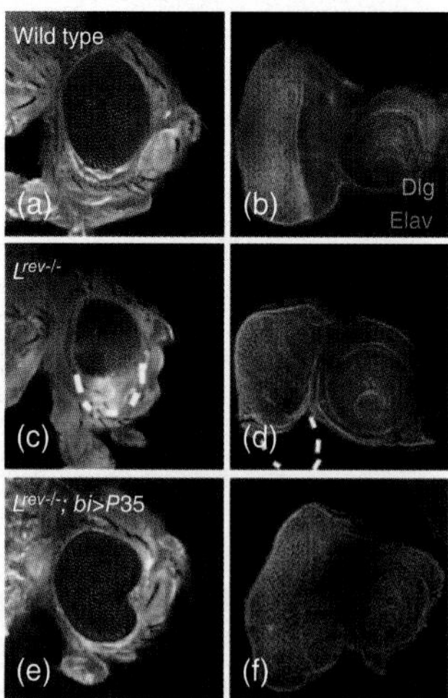

Fig. 6.8 Phenotypes of loss-of-function clones on dorsal and ventral eye margins in Drosophila. See protocol 6.3.1 Domain specific genetic mosaic system in the Drosophila eye. See chapter text for full caption.

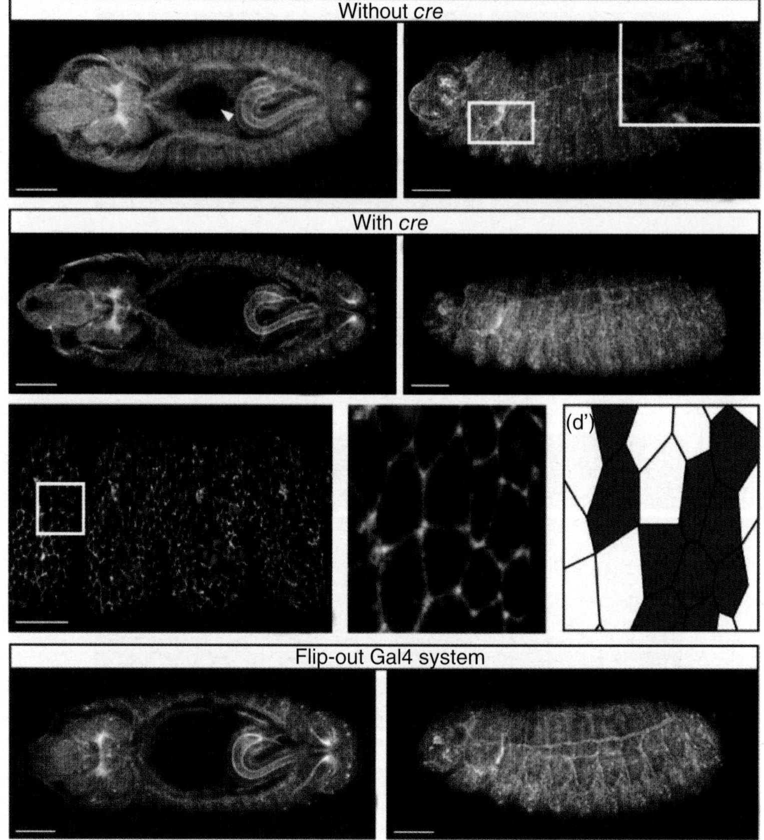

Fig. 6.9 Gal4 mosaic expression is driven by Aloxg-Gal4 in the presence of Cre. See protocol 6.3.2 Mosaic gene expression in the Drosophila embryo using the Cre/loxP system. See chapter text for full caption.

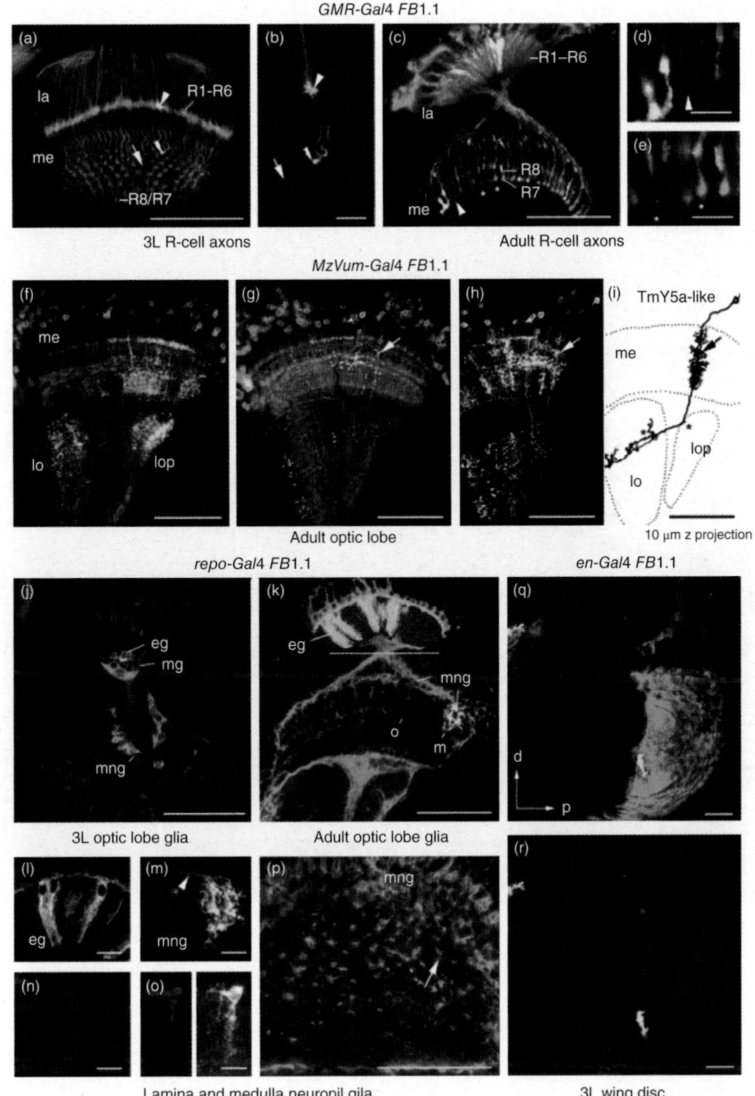

Fig. 6.10 See protocol 6.3.3 Flybow: Genetic multicolor cell labeling for neural circuit analysis in Drosophila. See chapter text for full caption.

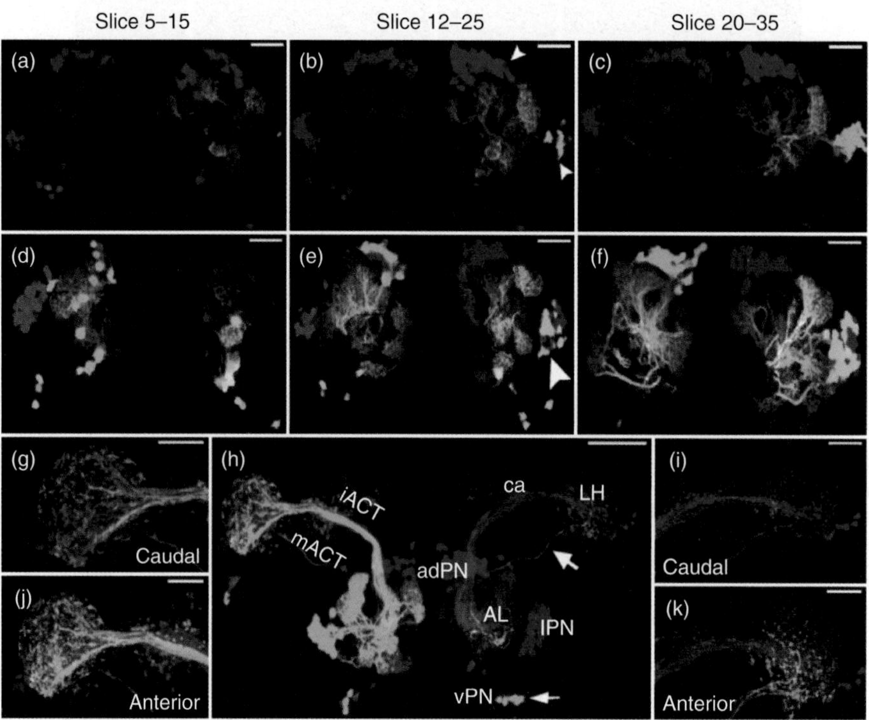

Fig. 6.11 See protocol 6.3.4 Brainbow: Fluorescence labeling to subdivide neural expression patterns in Drosophila. See chapter text for full caption.

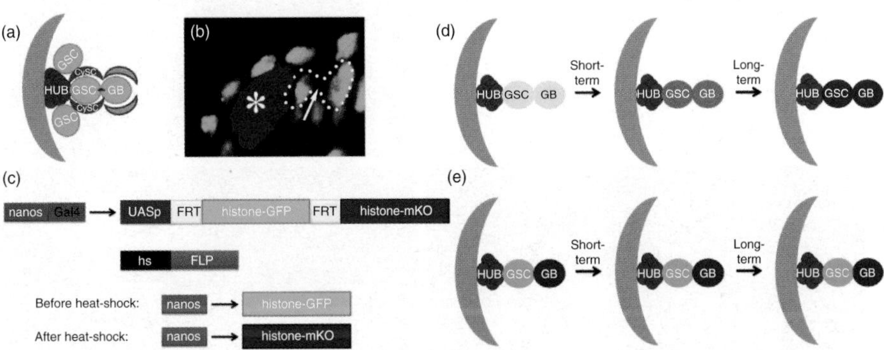

Fig. 6.12 See protocol 6.3.5. Histone labeling in Drosophila male germline stem cells. See chapter text for full caption.

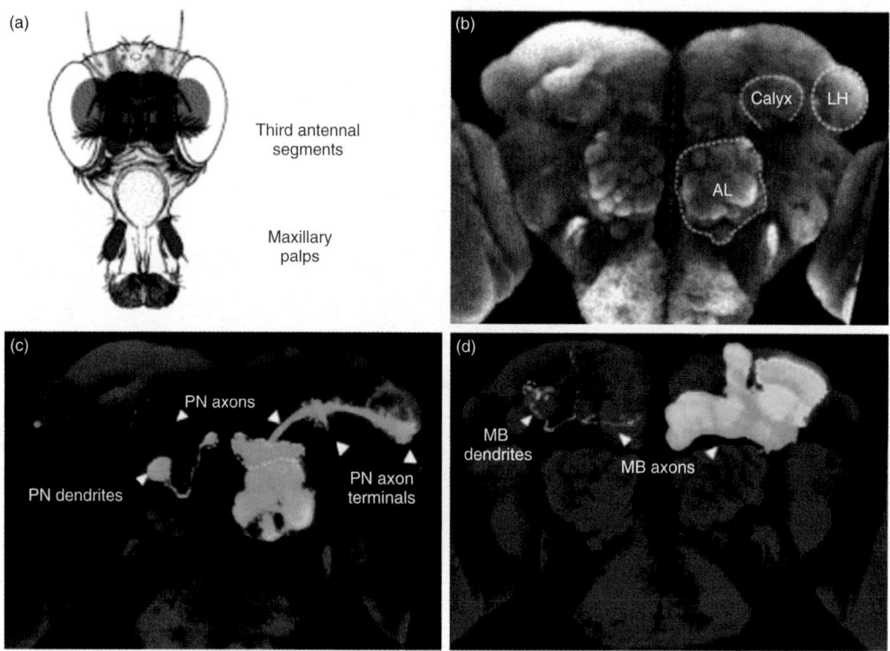

Fig. 7.1 Schematic of the Drosophila head (anterior view), with the brain in dark gray. Olfactory appendages are shown in red. See protocol 7.1.1 Preparation of Drosophila brains for live imaging. See chapter text for full caption.

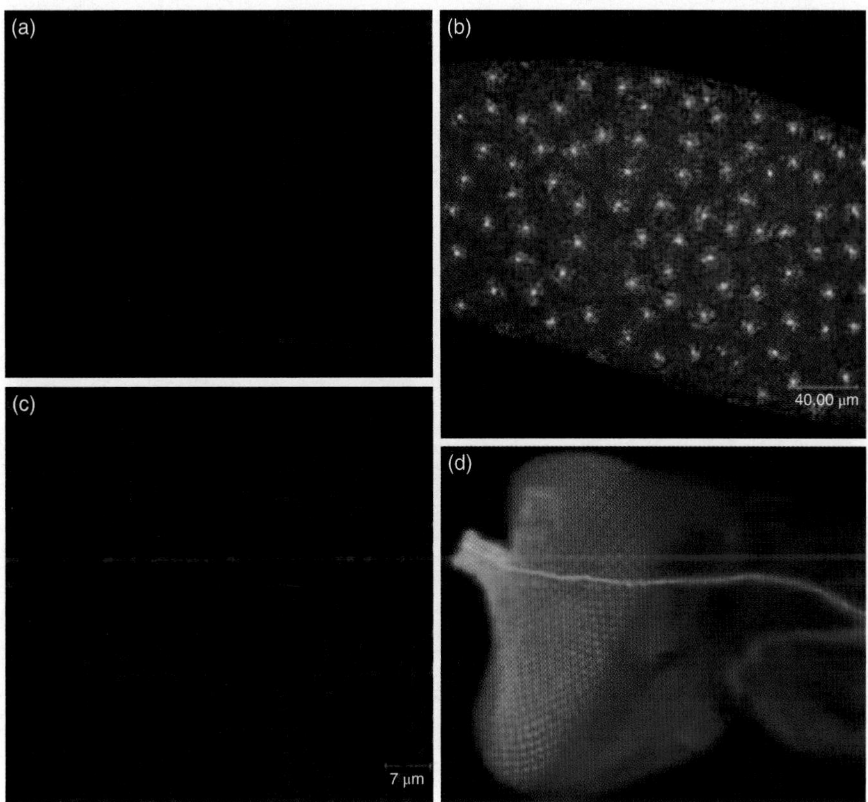

Fig. 7.2 Examples of fluorescence preparations of Drosophila whole mounts using the protocols described in Chapter 7. See protocol 7.1.7 Preparation of Drosophila embryos for live imaging âŁ" embryo fixation without methanol and hand devitellinization. See chapter text for full caption.

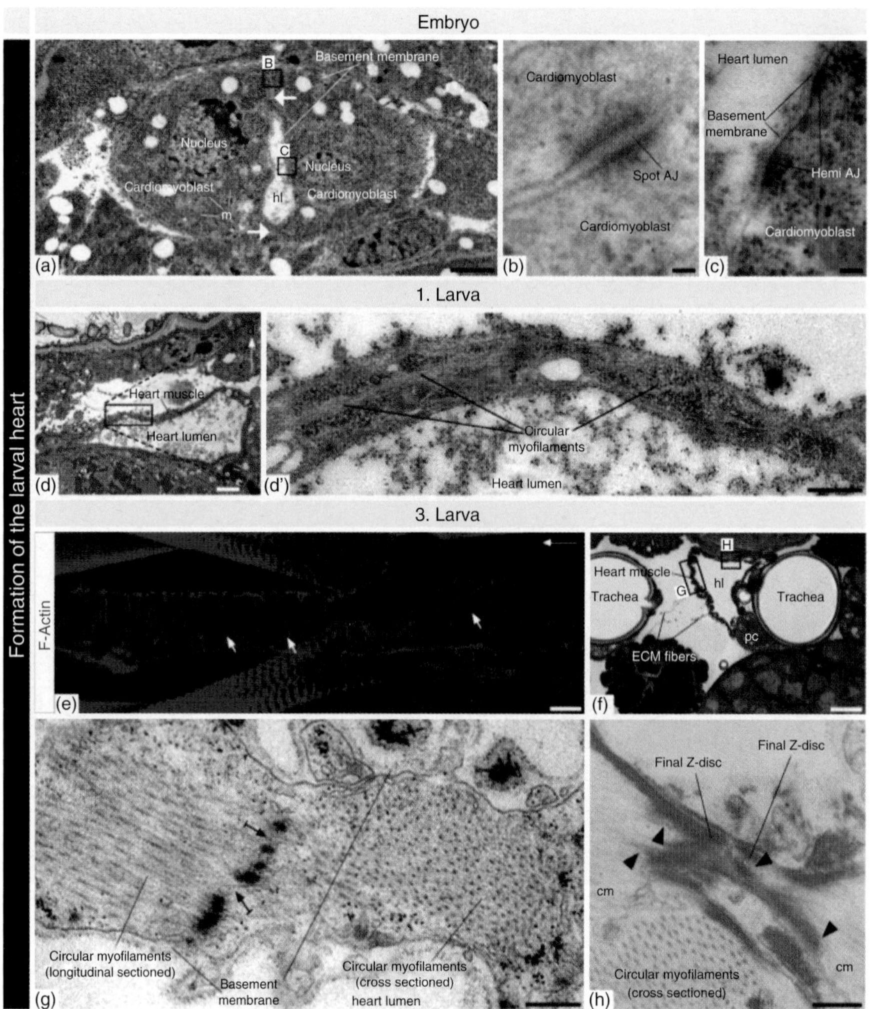

Fig. 7.4 Differentiation process of cardiomyoblasts into larval cardiomyocytes. See protocol 7.1.18 Phalloidin staining of arthropod hearts. See chapter text for full caption.

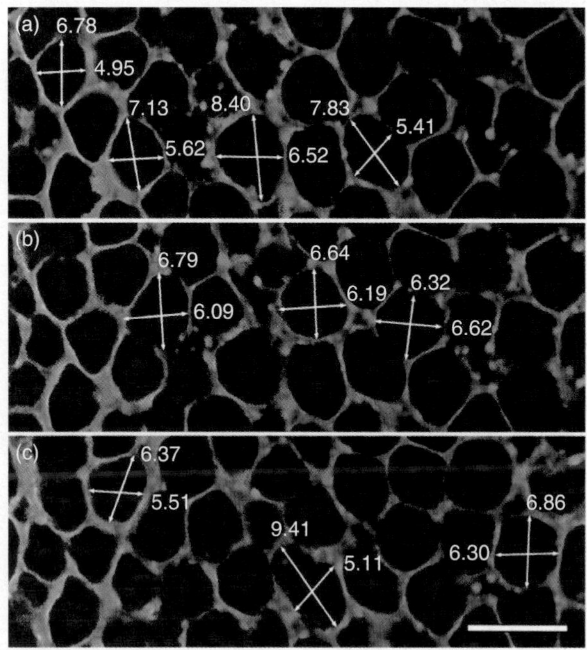

Fig. 7.5 Serial confocal images of Kenyon cell bodies. See protocol 7.1.21 Fluorescent neuropil staining. See chapter text for full caption.

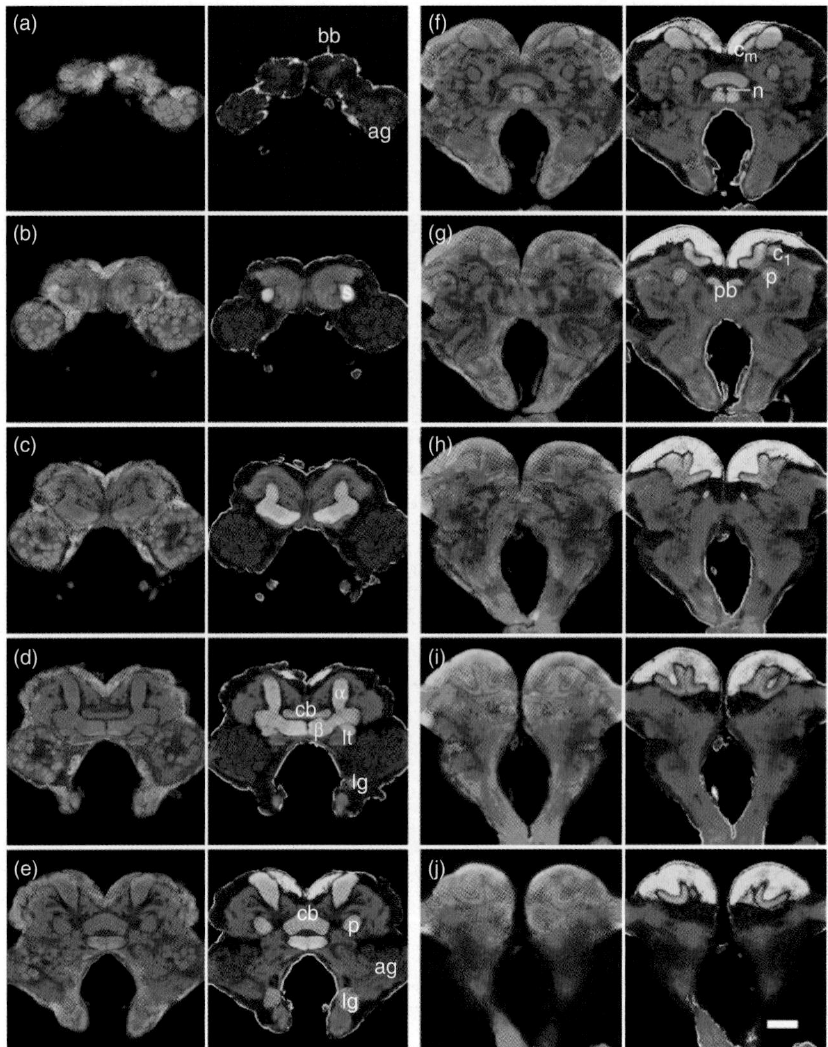

Fig. 7.6 Representative confocal stack of a central brain from an adult female Diploptera punctata. See protocol 7.1.21 Fluorescent neuropil staining. See chapter text for full caption.

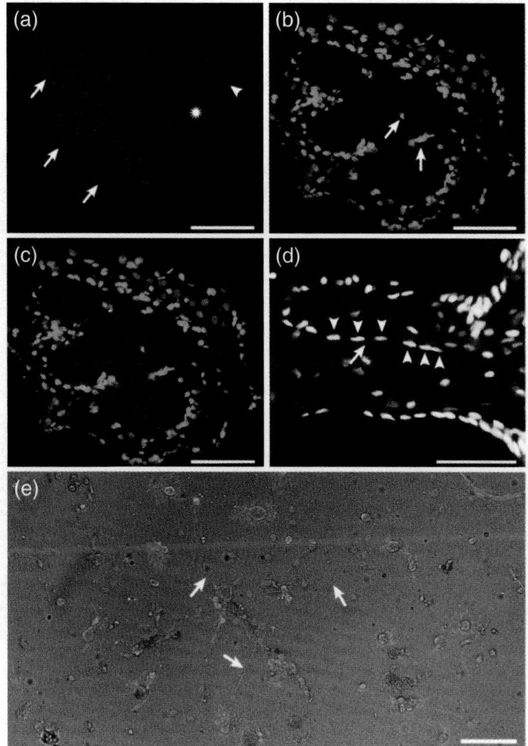

Fig. 7.7 (a) Confocal laser-scanning image of HRP-staining in a section through the antennal lobe of the cockroach Leucophaea maderae. See protocol 7.1.22 HRP neronal stain. See chapter text for full caption.

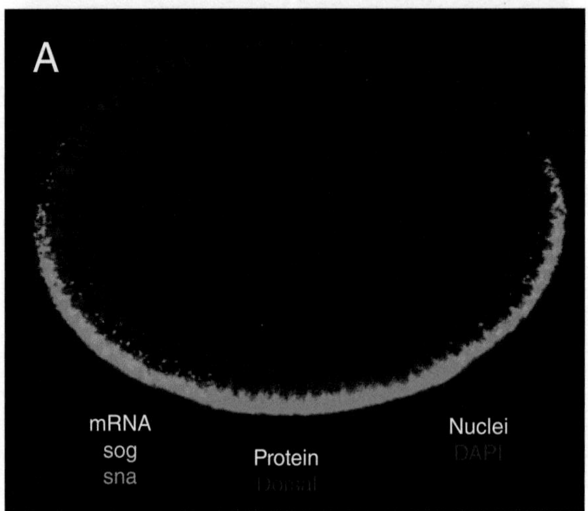

Fig. 8.2 Upright optical slice of an intact Drosophila embryo in blastoderm stage. Double staining for mRNA and protein. See protocol 8.1.31 Upright imaging of embryos. See chapter text for full caption.

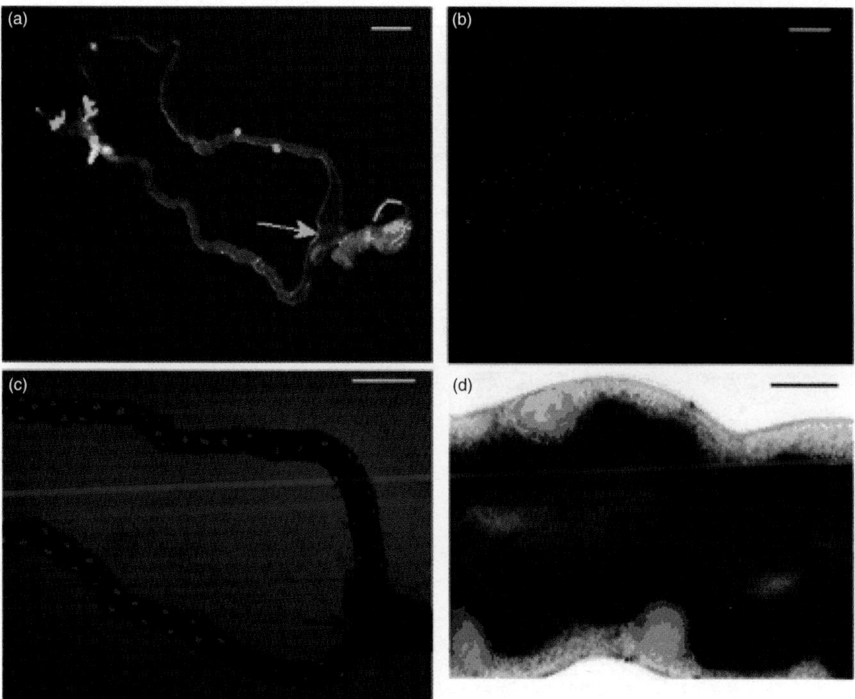

Fig. 10.2 Alkaline phosphatase domain in an intact adult D. melanogaster tubule pair. See protocol 10.1.5 Alkaline phosphate technic. See chapter text for full caption.

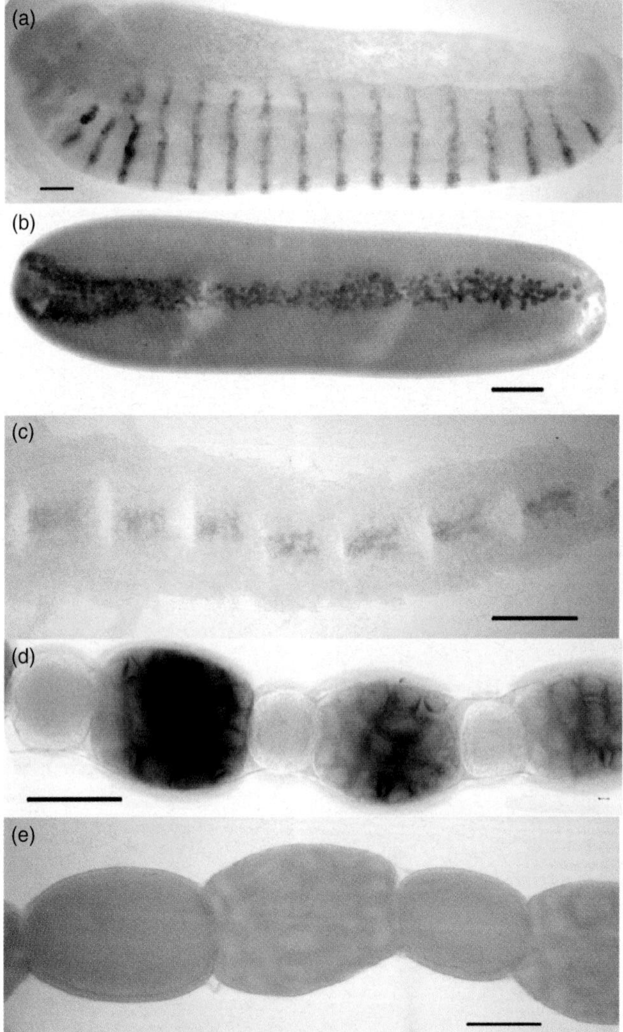

Fig. 10.4 Honeybee embryos (a, b and c) and ovaries (d and e) hybridized with a selection of DIG labelled RNA probes as described. See protocol 10.1.22 Osborne and Deardon in situ hybridization. See chapter text for full caption.

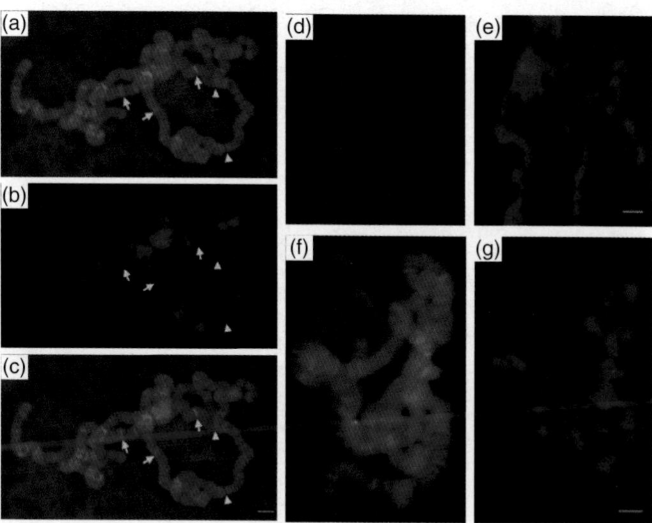

Fig. 12.4 Visualization of psoralen signals on polytene chromosomes. See protocol 12.1.41 Matsumoto technic. See chapter text for full caption.

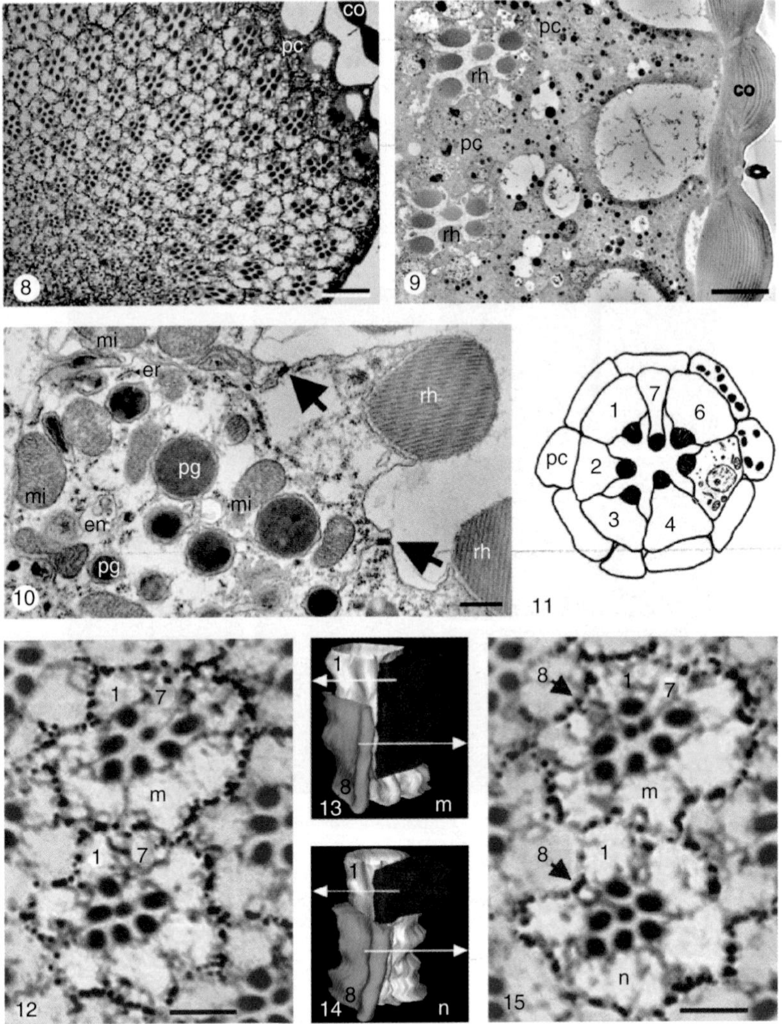

Fig. 13.1 LM-overview on 0.5-lm thick semi-thin cross section through p14-depleted Drosophila facet eye showing the cornea (co), pigment cells (pc), and the hexagonal unit eyes, the ommatidia. See protocol 13.1.29 Vademecum for standardized fixation, embedding and sectioning of the Drosophila facet eye. See chapter text for full caption.

7.1.9 *Manual dissection and fixation of* Drosophila *egg chambers for whole mount FISH [9]*

Formula:

MRM (5X):

1M Sodium Acetate	275 pt
1M Potassium Acetate	40 pt
1M Sucrose	100 pt
1M Glucose	10 pt
1M $MgCl_2$	2.2 pt
1M $CaCl_2$	0.05 pt
1M HEPES (pH 7.4)	100 pt

Procedure:

1. Anesthetize *Drosophila* females in CO_2 or ether and transfer to a drop of 1X MRM.
2. Rupture the abdomen with forceps, and transfer the intact ovaries to a fresh drop of MRM.
3. Fix for 4 min in freshly prepared 5% formaldehyde in cacodylate buffer pre-warmed to 37°C.
4. While in fixative, separate ovarioles by teasing them apart with forceps.
5. Transfer ovaries to a drop of 2X SSC with 0.1% Tween and complete separation of ovaries into individual ovarioles.
6. Transfer to a 0.5 ml microfuge tube and wash two times in 2X SSC with 0.1% Tween.
7. Repeat steps 1–5 until sufficient egg chambers have been collected and stain using standard FISH protocols.

7.1.10 *FISH on mitotic chromosomes [10]*

Formula:

Hybridization mixture for FISH:

Formamide	5 pt
50% Dextran sulfate	2 pt
20X SSC	1 pt
dH_2O	2 pt

Denaturation solution for FISH:

Formamide	35 pt
20X SSC	5 pt
dH_2O	10 pt

Protocol:

1. Prepare mitotic chromosomes on slides using one of the methods described in Chapter 13 (Chromosomes).
2. Label DNA probes with biotin-11-dUTP or digoxigenin-11-dUTP and prepare 10 µl of labeled probe per slide.
   a. Mix labeled DNA (40–80 ng per slide) with sonicated salmon sperm DNA (3 µg per slide).
   b. Add 0.1 volume 3M sodium acetate (pH 4.5) and 2 volumes of cold absolute ethanol.
   c. Incubate at -80°C for 15 min.
   d. Centrifuge at 13,600 g for 15 min to pellet DNA.
   e. Aspirate ethanol and dry the pellet.
   f. Re-suspend labeled DNA pellet in 10 µl hybridization mixture per slide.
3. Denature the probe mix for 8 min at 80°C, and place on ice.

4. Prepare a Coplin jar of denaturation solution for FISH and warm to 70°C.
5. Dehydrate mitotic chromosome squashes by immersion in 70%, 90% and absolute ethanol for 3 min each at room temperature. Air-dry the slides.
6. Place slides into 70°C denaturation solution for 2 min.
7. Transfer slides to -20°C 70% ethanol for 3 min.
8. Dehydrate for 3 min each through 4°C 90% and absolute ethanol. Air-dry the slides.
9. Apply 10 μl of the prepared probe mixture to each denatured slide.
10. Cover each slide with a 22x22-mm coverslip and seal the coverslip with rubber cement.
11. Hybridize in a humid chamber overnight at 37°C.
12. Remove the rubber cement and gently remove the coverslip.
13. Wash slides three times for 5 min each in 2X SSC/50% formamide at 42°C.
14. Wash slides three times for 5 min each in 0.1X SSC at 60°C.
15. Remove excess liquid around the specimen and block in 100 μl blocking solution (4X SSC with 3% BSA and 0.1% Tween-20), under a coverslip, for 30 min at 37°C.
16. Remove coverslip and blot excess liquid from around specimen.
17. For probes labeled with biotin:
    a. Expose to 80 μl of 3.3 μg/ml fluorescein-conjugated avidin in 4X SCC with 1% BSA and 0.1% Tween-20 under a coverslip for 30 min at 37°C in the dark.
    b. Remove coverslip and wash three times 5 min each in SSC/Tween-20 at 42°C.
18. For probes labeled with digoxigenin:
    a. Expose to 80 μl of 2 μg/ml rhodamine-conjugated sheep IgG, Fab fragments in 4X SSC with 1% BSA and 0.1% Tween-20, under a coverslip, for 30 min at 37°C, in the dark.
    b. Remove coverslip and wash three times for 5 min each in SSC/Tween at 42°C.
19. Counterstain with 0.2 μg/ml DAPI (in 2X SSC) at room temperature for 5 min.
20. Wash in 2X SSC at room temperature.
21. Mount slides in Vectashield® H-1000 and seal coverslip edges with rubber cement.

7.1.11 *Whole mount FISH for* Drosophila *tissues [11]*

Formula:

Hybridization Solution (1.1X):

Dextran Sulfate	1 pt
20X SSC	1.5 pt
Formamide	5 pt
dH_2O	Up to 9 ml

Glycerol/NPG

Preparation:

1. Dissolve 4% NPG (*n*-propyl gallate) (w/v) in high quality glycerol. Agitate for several hr. This solution is stable at room temperature indefinitely.
2. Adjust the pH by adding 70 μl of 2M Tris base (no HCL) and 30 μl of water to 900 μl of glycerol/NPG.
3. Mix thoroughly and use the pH adjusted solution within 1 to 2 days.

Procedure:

1. Prepare *Drosophila* whole mount tissues using formaldehyde fixation as described in Chapter 9.

2. Wash the slide(s) with fixed tissue, in a Coplin jar, with 2X SSC with 0.1% Tween (2X SSCT) three times for 10 min each.
3. (Optional) Treat the samples with 10100 μg/ml of RNase A.
4. Incubate in 2X SSCT/25% formamide for 10 min.
5. Incubate in 2X SSCT/50% formamide for 10 min.
6. Transfer to fresh 2X SSCT/50% formamide (prehybridization solution) and place staining jar into a 37°C water bath. Allow the temperature to equilibrate, then incubate at 37°C for 30 min.
7. To prepare probe solution, add 10–200 ng of labeled probe to 1.1X hybridization solution, adjust final volume of the probe mixture to 15 μl with dH_2O.
8. Remove the slides from the prehybridization solution and drain the excess liquid on paper towels. Carefully remove as much liquid as possible with an aspirator or Kimwipe.
9. Pipette probe solution onto a 22x22-mm coverslip. Touch the slide with the samples to the drop and invert the slide.
10. Denature the probe and chromosomal DNA by placing the slide into a humidified denaturation chamber. Cover with humidifying lid and incubate for 2 min at 94°C.
11. Transfer the slides to a humidified incubation chamber at 37°C and incubate for several hr to overnight.
12. Transfer the slides to a Coplin jar with 2X SSCT/50% formamide, at the annealing temperature of the probe.
13. Transfer the slides to fresh 2X SSCT/50% formamide, at the annealing temperature of the probe, and incubate for at least 1 hr.
14. Transfer the slides to fresh 2X SSCT/50% formamide at room temperature and incubate for 10 min.
15. Wash three times with 2X SSCT.
16. Block the sample in 0.5% protein; block solution for 30 min.
17. Add diluted fluorescently conjugated primary antibody or avidin (depending on label in probe) and incubate for 2 hr in the dark.
18. Wash three times in 2x SSCT for 10 min each.
19. Counterstain with DAPI to label nuclei.
20. Pipette 12–15 μl glycerol/NPG onto 22x22 mm coverslip.
21. Rinse the slide briefly in 50 mM Tris-Cl (pH 7.5) to remove salt.
22. Drain the slide and use Kimwipes or an aspirator to remove as much buffer as possible.
23. Touch the sample to the drop of mounting medium and invert the slide. Seal with clear nail polish and image using a fluorescent microscope.

7.1.12 *mRNA* in situ *hybridization [12]*

Application:
Recommended for the detection of mRNAs in *Drosophila* brain.

Formula:
Fixing Solution:

37% Formaldehyde	1.1 pt
1M PIPES	1 pt
10% Triton X-100	0.3 pt
0.1M EGTA	0.2 pt
1M $MgSO_4$	0.01 pt
dH_2O	7.4 pt

Hybridization Buffer:

Formamide	25 pt
20X SSC	10 pt
10% Tween-20	0.5 pt
50 mg/ml Heparin	0.05 pt
RNAse free dH_2O	14.5 pt

MABT:

1M Maleic Acid	100 pt
1M NaCl	150 pt
Tween-20	0.1%
Adjust pH to 7.5	

Procedure:

1. Dissect larvae in Schneider's medium and remove brains.
2. All incubation steps should include gentle rocking.
3. Place brains in microfuge tube and add 500 µl of fixing solution. Incubate for 23 min at room temperature.
4. Wash the brains in 500 µl phosphate buffered saline with 0.3% Triton-X (PBTx), two times quickly, then two times for 20 min each at room temperature.
5. Incubate for 1 hr in 3% hydrogen peroxide in PBTx at room temperature.
6. Wash the brains in 500 µl PBTx, two times quickly, then two times for 20 min each, at room temperature.
7. Wash brains in 400 µl of 1:1 mixture of PBTx and hybridization buffer for 30 min at room temperature.
8. Wash the brains in hybridization buffer three times for 10 min, at room temperature.
9. Incubate the brains in 400 µl hybridization buffer for 2 hr, at 55°C.
10. Denature 400 ng of a digoxigenin-labeled riboprobe in 400 µl hybridization buffer for 5 min, at 90°C. Cool on ice for 5 min then keep at 55°C.
11. Incubate the brains in 150–200 µl of probe/hybridization buffer mixture for at least 16 hr at 55°C.
12. Remove the probe/hybridization buffer mixture.
13. Wash the brains two times in a pre-warmed 1:1 mixture of hybridization buffer and 2X SSC for 30 min at 55°C.
14. Wash two times in pre-warmed 2X SSC for 30 min at 55°C.
15. Wash two times in pre-warmed 0.2X SSC for 30 min at 55°C.
16. Quickly wash the brains two times in MABT for 20 min at room temperature.
17. Incubate in 10% heat-inactivated horse serum in MABT for 1 hr, at room temperature.
18. Incubate in 1:1000 dilution anti-DIG-POD in MABT with 10% heat-inactivated horse serum, overnight at 4°C.
19. Wash the brains in MABT, six times 20 min each at room temperature.
20. Incubate brains in 10 mM imidazole in PBTx for 30 min at room temperature.
21. Incubate in a 1:1000 dilution of FITC-tyramide solution in PBTx with 10 mM imidazole for 30 min at room temperature in the dark. All subsequent steps should be performed in the dark.
22. Develop the signal by adding hydrogen peroxide to a final concentration of 0.002%–0.01%. Incubate for 45 min at room temperature.

23. Wash the brains in 500 µl PBTx, two times quickly, then two times for 30–60 min each, at room temperature.
24. Mount brains and image on a fluorescent microscope.

7.1.13 *mRNA whole mount* in situ *hybridization in GFP-marked mosaic clones [13]*

Application:
Recommended for the detection of mRNA in genetic mosaic clones.

Formula:
Fixative:

Formaldehyde	0.4%
1M Pipes	100 pt
0.5 M EGTA	4 pt
1M $MgSO_4$	1 pt

Hybridization Buffer:

Formamide	50%
SSC	5X
Tween-20	0.1%
Heparin	50 µg/ml

Protocol:
1. Dissect out imaginal discs (IDs) bearing clones that are positively or negatively marked by GFP (or other epitope tags) and fix in fixative for 15–30 min.
2. Rinse IDs, four times for 5 min each, in phosphate buffered saline with 0.1% Tween-20 (PBTw).
3. Incubate IDs in 1:1 PBTw:hybridization buffer, then in hybridization buffer for 10 min.
4. Boil 100 µg/ml sonicated salmon sperm DNA in hybridization buffer for 10 min, then store on ice.
5. Block IDs in 100 µl denatured DNA/hybridization buffer mixture at 60°C for 1 hr.
6. Heat-denature 100 µl gene-specific DIG-labeled riboprobe diluted to the desirable working condition for 5 min, cool on ice and add to IDs. Hybridize overnight at 60°C in a shaking water bath.
7. Remove the probe and wash the IDs six times in pre-warmed PBTw at 60°C for 30 min each.
8. Wash IDs two times in PBS with 0.1% Triton-X 100 (PBTx) at room temperature for 15 min each.
9. Add 100 µl of a 1:20 dilution of anti-Dig Rhodamine antibody (Roche®) in PBTx with 5% normal goat serum (NGS) for 2 hr at room temperature.
10. Wash IDs in PBTx, at room temperature, gently rocking four times for 5 min each, then three times for 30 min each.
11. Incubate IDs in 100 µl of a 1:20 dilution of rabbit anti-GFP N-terminal (Sigma®) in PBTx with 5% NGS overnight at 4°C.
12. Wash IDs in PBTx, at room temperature, gently rocking, four times for 5 min each then three times for 30 min each.
13. Incubate IDs in 100 µl of a 1:200 dilution of FITC goat anti-rabbit antibody (Jackson Immunoresearch®) in PBTx with 5% NGS for 2 hr at room temperature.
14. Wash IDs in PT at room temperature, gently rocking four times for 5 min each then three times for 30 min each.

15. Rinse IDs in PBS for 5 min, then re-suspend in 300 µl of 70% glycerol.
16. Clear for several hr in 70% glycerol at room temperature, then transfer to fresh 70% glycerol with 2.5% DABCO for at least 1 hr.
17. Image discs using appropriate microscope settings.

Note:
1. This protocol can be modified to detect proteins other than GFP just by changing the primary antibodies for co-localization of mRNA and protein (see Broihier[14]).

7.1.14 *Lysotracker staining of ommatidia [2]*

Application:
Recommended for the detection of acidic vacuoles.

Procedure:
1. Follow dissection of ommatidia for live imaging in Chapter 7.
2. Dilute Lysotracker® (Molecular Probes®) 1:10 in Schneider medium then add 0.5 µl of the 1:10 dilution to a 50 µl drop of Schneider medium (see Appendix) containing dissected ommatidia (final dilution of Lysotracker® is 1:1000). Mix and wait for 10 sec.
3. Remove excess liquid from the slide leaving ommatidia behind.
4. Add a drop of Vectashield® to cover the ommatidia, apply coverslip and seal with nail polish.
5. Visualize Lysotracker® staining on a fluorescent microscope at the appropriate wavelength for the lysotracker probe utilized.

7.1.15 *Chromatin staining of* Drosophila *testes – Hoescht staining [15]*

Formula:
Hoescht Buffer:

1M NaCl	150 pt
1M KCl	30 pt
1M Na_2HPO_4	10 pt

Preparation:
1. Add distilled water to 1 L, adjust pH to 7.0

Procedure:
1. Prepare *Drosophila* testes slides using one of the methodologies above.
2. Stain with 0.5 µg/ml Hoeschst 33258 dissolved in Hoescht Buffer for 10 min.
3. Rinse for 1 min in distilled water.
4. Mount in Hoescht Buffer.
5. Seal with rubber cement.

7.1.16 *Chromatin staining of* Drosophila *testes – DAPI staining [15]*

Procedure:
1. Prepare *Drosophila* testes slides using one of the methodologies above.
2. Stain with 0.2 µg/ml DAPI dissolved in phosphate buffered saline (PBS) for 10 min.
3. Rinse for 1 min in distilled water.
4. Mount in PBS.
5. Seal with rubber cement.

7.1.17 *Chromatin staining of* Drosophila *testes – Propidium iodide or TOTO-3 iodide staining [15]*

Formula:

NPG Mounting Medium:

90% Glycerol
2.5% *n*-propyl gallate

Procedure:

1. Prepare *Drosophila* testes slides using one of the methodologies above.
2. Mount in NPG mounting medium containing 1 µg/ml propidium iodide or 1/200 TOTO-3 iodide.
3. Seal with rubber cement.

7.1.18 *Phalloidin staining of arthropod hearts [16]*

Protocol:

1. Dissect hearts from larva or adult arthropods in phosphate buffered saline (PBS).
2. Fix hearts in 4% paraformaldehyde for 1 hr.
3. Wash in PBS several times.
4. Stain in 100 µg/5ml Phalloidin-TRITC/ethanol diluted 1:50 in PBS for 1 hr.
5. Embed hearts in Fluoromount-G and image using appropriate fluorescent settings.

7.1.19 *Calcium green*

Application:

Recommended for detecting changes in calcium signaling in the honeybee.

Formula:

Calcium Green Dye

Preparation:

1. Dissolve 50 µg Calcium Green AM dye (Molecular Probes®) in 50 µl Pluronic F-127 (20% in dimethylsulfoxide DMSO, Molecular Probes®).
2. Vortex for 1 min.
3. Dilute in 800–950 µl saline solution.
4. Vortex for 1 min, sonicate for 5 min.

Protocol:

1. Being careful to avoid injuring the brain, cut open the head of the bee to expose the brain.
2. Apply 50 µl freshly prepared dye to the open brain cavity.
3. Place the bee in dark moist environment at 13°C for 1 hr.
4. Following staining, place the bee in a dark box at room temperature for 15 min.
5. Rinse repeatedly with fresh saline.
6. Image the brain using appropriate fluorescent filters.

Note:

1. If no staining occurs, dilute the prepared dye further. Although counterintuitive, dilution can improve the staining.

7.1.20 *Endogenous alkaline phosphatase [17]*

Application:

Recommended for the detection of endogenous alkaline phosphatase.

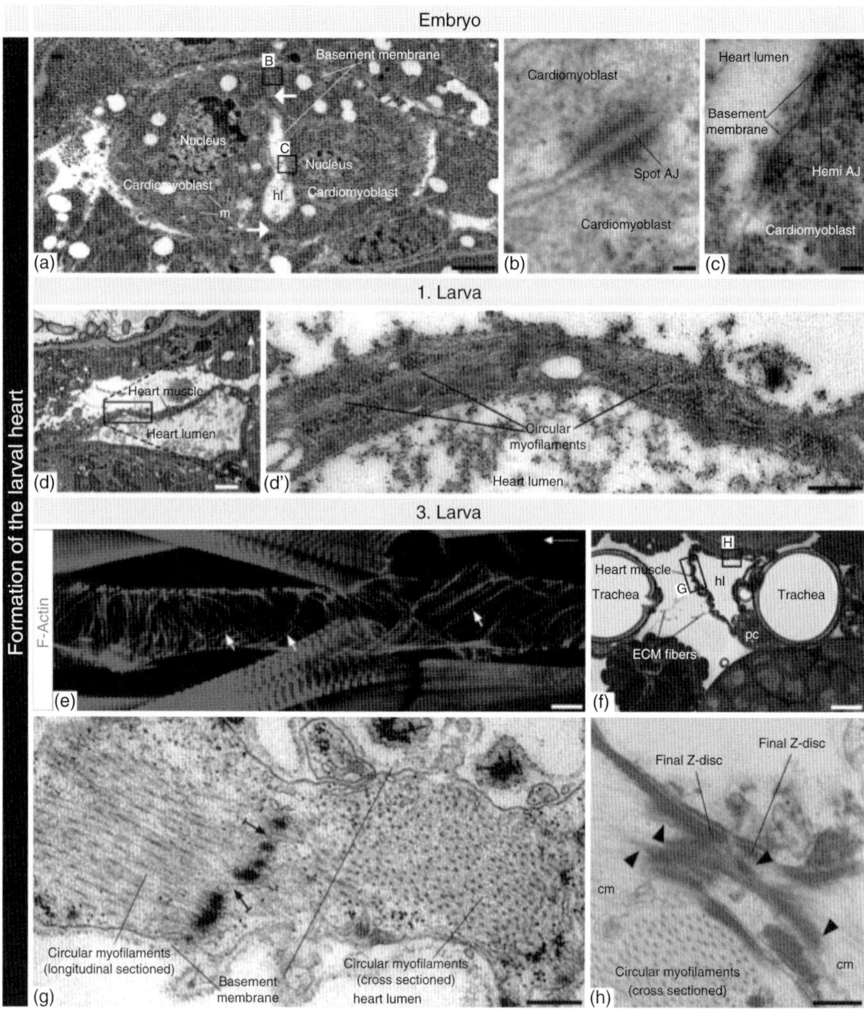

Fig. 7.4 Differentiation process of cardiomyoblasts into larval cardiomyocytes. (a–c) Cross ultra-thin sections through undifferentiated embryonic cardiac cells (cardiomyoblasts, cb) of the heart proper. The cbs are mononucleated cells and enclose the heart lumen. The cbs are dorsally and ventrally connected to each other by spot-adherens junctions (white arrows in a and detail b). Toward the luminal and abluminal cell side a layer of extracellular matrix (basement membrane) surrounds the cbs. Hemi-adherens junctions connect cbs toward the extracellular matrix (detail c). (d–d′) Cross ultra-thin sections through cardiomyocytes (cm) of the heart of 1st instar larva. The cms are thin and yet undifferentiated. Note that only a few circularly arranged myofilaments are present. (e) Phalloidin staining of F-Actin in cms demonstrates the myofilament array in the larval heart portion. Note the irregularities in the arrangement of myofilaments, several myofilaments are not orientated in a clear circular manner (white arrows). (f) Semi-thin section through the larval heart. The heart is located next to the tracheae. Cms are fully differentiated and generate contractive activity. (g–h) Cross ultra-thin sections through cms shown in f. (g) Irregularities in the myofilament array are visible. Circularly and irregularly arranged myofilaments alternate randomly within the cms. The cms show discontinuities in Z-disc formation (indicated by black arrows) and surrounding basement membranes toward luminal and abluminal side. (h) Larval cms are closely connected by characteristic junctional complexes. Black arrowheads indicate the firmly adhering myofibrils to the membranes at the junctional complex. Abbreviations: a=anterior, AJ=adherens junction, d=dorsal, cm=cardiomyocyte, hl=heart lumen, m=mitochondria, myo=myofilaments, pc=pericardial cell. Scale bars: a=2 μm, b–c=30 nm, d=2 μm, d′=0.5 μm, e–f=20 μm, g–h=400 nm. (Source: Lehmacher *et al.* 2006. Reproduced with permission of Elsevier.) See plate section for the color version of the figure.

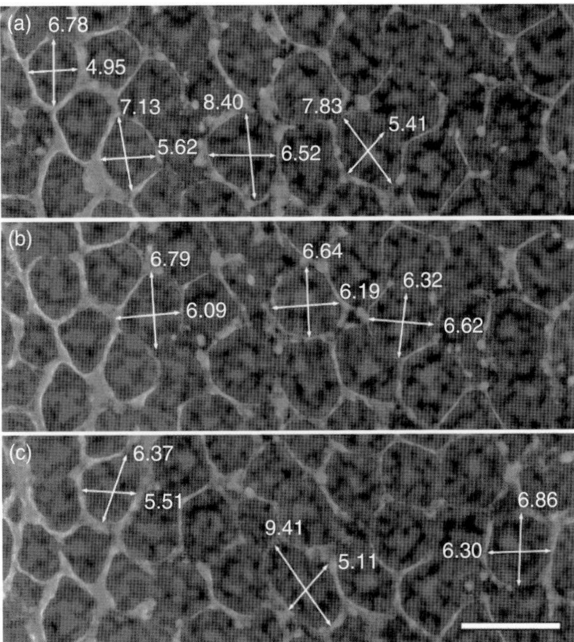

Fig. 7.5 Serial confocal images of Kenyon cell bodies. Nuclei (red color) were stained with propidium iodide. Cytoplasmic membranes (green color) were stained with NBD C6-ceramide. Confocal images were recorded under a 340 C-Apochromat water-immersion objective at 2,048 3 2,048 pixels. a–c show three representative images from seven optical sections at 0.5-mm intervals. The longest axes and shortest axes of the largest cross section of selected cell bodies were measured. Scale bar : 5-10 mm. (Source: Chiang *et al.* 2013. Reproduced with permission of John Wiley & Sons.) See plate section for the color version of the figure.

Protocol:

1. Collect whiteflies (adults and nymphs) and aphids in 70% ethanol or acetone.
2. Fix overnight in ethanolic formaldehyde (3:1 ratio ethanol:formaldehyde).
3. Dehydrate through a graded series of ethanol and clear.
4. Infiltrate and embed in paraffin.
5. Cut 10 μm sections and mount onto glass slides.
6. De-paraffinize and hydrate through xylenes and a graded ethanol series into water.
7. Permeabilize the sections with 0.2% Tween-20 in phosphate buffered saline (PBS) for 10 min at room temperature (this step is optional).
8. Make a 1:20 dilution of ELF 97 (Molecular Probes®) phosphatase substrate in detection buffer (Molecular Probes®). Filter diluted ELF 97 through a 0.2 μm pore size filter just before use.
9. Apply diluted substrate to tissue sections and observe the reaction under a fluorescent microscope using appropriate filters. The reaction time is quick, often 30–90 seconds and should be observed to prevent over-staining.
10. Stop the reaction with wash buffer (PBS with 25 mM EDTA and 5 mM levamisol, pH 8.0).
11. Wash the sections with three changes of wash buffer over 10–15 min with gentle agitation.
12. Remove as much wash buffer as possible and mount with ELF 97 mounting medium provided by manufacturer.
13. Image samples using appropriate fluorescent filters.

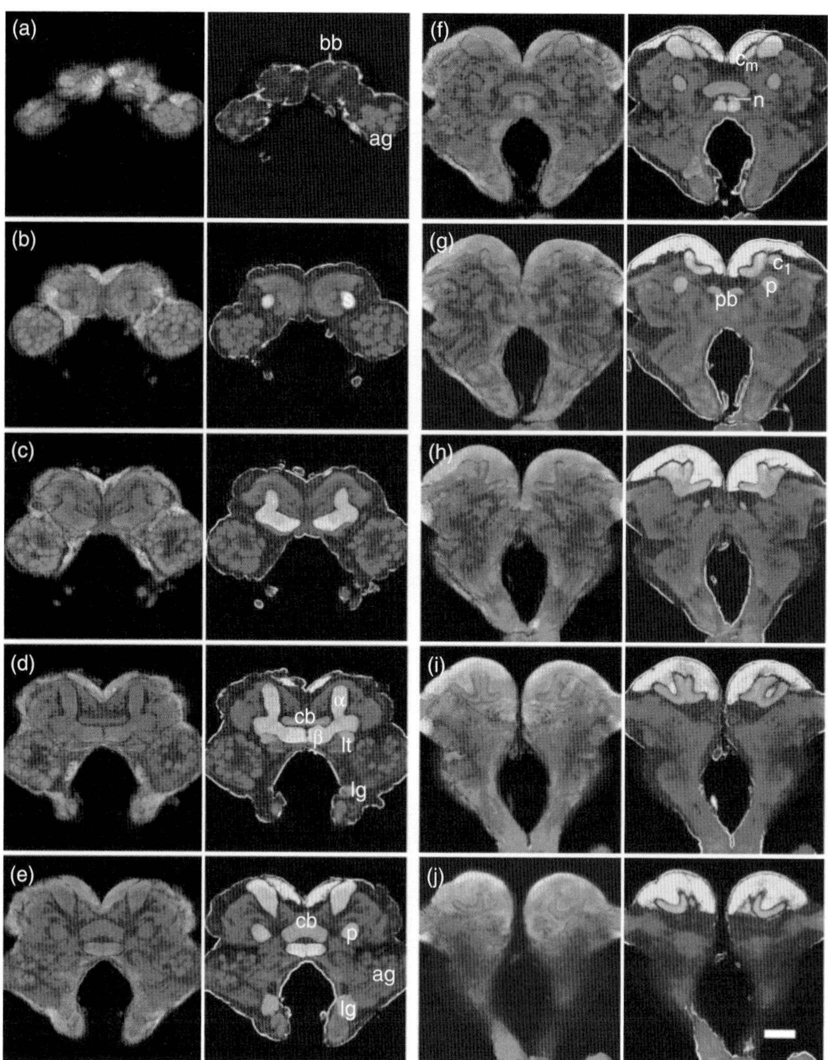

Fig. 7.6 Representative confocal stack of a central brain from an adult female *Diploptera punctata*. Neuropils (green color) were stained with the membrane probe (NBD C6-ceramide), and nuclei (red color) were stained with the DNA probe (propidium iodide); a technique that provided sharp contrast of all internal structures in the brain. Ten images (a–j) out of a total 168 images are separated serially by 50 mm (sequentially, 3.125 mm apart): a is the anterior-most image, (j) is the posterior-most image, and all other images lie in between. Left photomicrographs show unaltered confocal images (primary data set; 1,024 3 1,024 dots per inch), and right photomicrographs show images from a subset of the primary data set (512 3 512 dots per inch). (Source: Chiang *et al.* 2001. Reproduced with permission of John Wiley & Sons.) See plate section for the color version of the figure.

Note:

1. Protocol is described for the whitefly, but is applicable to other insect species (e.g., Li *et al.* [18]).

7.1.21 *Fluorescent neuropil staining [19]*

Application:
Recommended for mapping of brain neuropils in the Pacific Beetle Cockroach *Diploptera punctata*.

Formula:
Saline:

1M NaCl	158.6 pt
1M KCl	17.6 pt
1M $NaHCO_3$	3.86 pt
Na_2HPO_4	8.5 pt
NaH_2PO_4	1.37 pt

1M Glucose	15 pt
dH_2O	Up to 1000 pt
Adjust to pH 7.4	

Procedure:

1. Dissect out brain of newly emerged adult *D. punctate* in saline.
2. Enzymatically de-sheath the brain by agitating it in 0.1% collagenase/dispase in 1X phosphate buffered saline (PBS) for 30 min.
3. Digest further in 0.3% trypsin in 1X PBS for 2 hr at 27°C, then in 0.15% trypsin in 0.5X PBS for 30 min.
4. Pre-heat a microwave safe container of tap water to 50–55°C in a microwave.
5. Place brains in 100 µl 4% paraformaldehyde in PBS and submerge tubes in warmed water. Microwave for 5 min then sit for 5 min.
6. Repeat microwave for 5 min and a rest period 5 min in 4% paraformaldehyde with 1% Triton X-100, then in 8% paraformaldehyde with 2% Triton X-100. The last rest period should be followed by agitation in the same solution for 2 hr at room temperature.
7. Rinse brains two times in PBS with 2% Triton X-100 for 5 min each in microwave.
8. Incubate brains in PBS with 50 µg/ml RNase for 2 hr at 37°C.
9. Rinse in PBS three times for 10 min each.
10. Incubate in PBS with 20 µg/ml propidium iodide overnight.
11. Rinse in PBS and stain with 0.435 mM NBD C_6-ceramide (Molecular Probes®) in dimethyl sulfoxide.
12. Rinse in PBS three times for 10 min each.
13. Clear brains overnight and mount in FocusClear™ (Pacgen®).
14. To prevent compression of brains under the coverslip, mount using spacer rings.

7.1.22 *HRP neuronal stain [20]*

Application:

Recommended for the identification of neuronal cells in the insect central nervous system.

Protocol:

1. Dissect brain of insect of interest under freshly prepared 4% paraformaldehyde in phosphate buffered saline (PBS).
2. Once dissection is complete, fix overnight in 4% paraformaldehyde in PBS at room temperature.
3. Rinse brains in PBS, embed in gelatin/albumin and fix the gelatin/albumin blocks in 10% formalin overnight at 8°C.
4. Cut sections at 50 µm on a vibratome and collect sections into 24-well multi-well plates.
5. Rinse in PBS with 0.1% Triton (PBTx).
6. Block in PBS with 0.5% Triton X-100 and 5% normal swine serum overnight at room temperature.
7. Add a 1:10,000 dilution of anti-HRP serum to the blocking solution and incubate for 48 hr at room temperature.
8. Rinse in PBTx.
9. Incubate sections with a 1:1000 dilution of a Cy3 conjugated goat anti-rabbit (ImmunoResearch Laboratories, Inc.®) in PBS with 0.5% Triton X-100 and 1% normal swine serum overnight at room temperature.

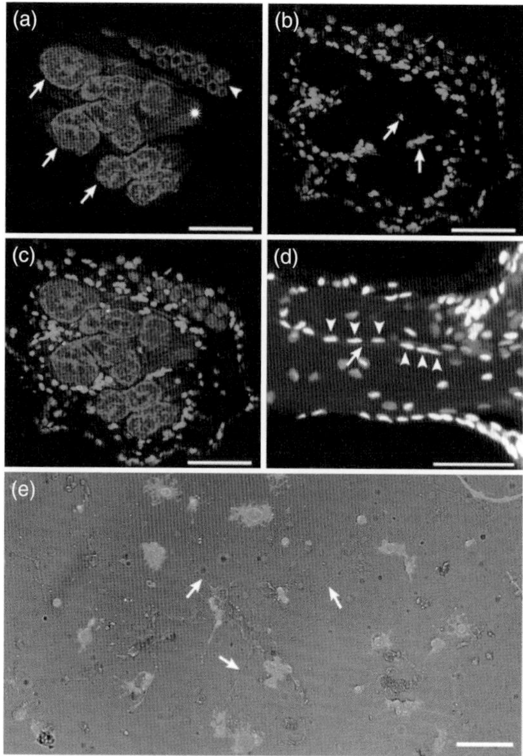

Fig. 7.7 (a) Confocal laser-scanning image of HRP-staining in a section through the antennal lobe of the cockroach *Leucophaea maderae*. Several glomeruli (arrows), axons of olfactory receptors that invade the antennal lobe through the antennal nerve (asterisk), and neuronal cell bodies of the lateral cluster (arrowhead) are stained. (b) Identical section as in (a) showing DAPI labeled nuclei. Several of these nuclei (arrows) can already be assigned to glial cells due to their position in the center of the antennal lobe, where no neurons reside. (c) Fluorescent double-labeling as a result of superimposing (a) and (b). The image demonstrates that the antennal lobe contains by far more non-neuronal cells than neurons. (d) DAPI-staining in a thoracic connective of the locust *Locusta migratoria*. Nuclei of tracheal cells (arrowheads) can be distinguished from nuclei of glial cells by their elongated appearance and close proximity to the wall of a tracheal tube (arrow) that exhibits weak auto fluorescence. (e) Primary cell culture derived from the metathoracic ganglion of *L. migratoria*. The image demonstrates that in this culture the number of DAPI-stained glial nuclei (blue) exceeds the number of neurons (red). Glial cells are further characterized by having less perinuclear cytoplasm. Also their processes are straighter and less branched than those of the neurons (arrows). All scale bars: 100 μm. (Source: Loesel *et al.* 2006. Reproduced with permission of Elsevier.) See plate section for the color version of the figure.

10. Remove the secondary antibody and incubate sections with a 1:1000 dilution of DAPI in PBS for 12 min.
11. Rinse sections in PBTx and mount on chrome alum/gelatin-coated slides under glass coverslips using Elvanol®.

7.1.23 *Rhodamine dextran labeling of neruons [21]*

Application:
Recommended for visualization of the olfactory projection in the Honeybee *Apis mellifera*.

Formula:

Pupal Honeybee Physiological Solution:

1M NaCl	130 pt
1M KCl	6 pt
1M $MgCl_2$	2 pt
1M $CaCl_2$	7 pt
1M HEPES	10 pt
1M Glucose	25 pt
1M Sucrose	160 pt
dH_2O	To 1000 pt
pH to 6.7	

Procedure:

1. Anesthetize honeybee pupae by cooling on ice and remove the heads of the pupae from the thorax in physiological saline.
2. Open the head capsule between the antennae, compound eyes and ocelli and remove the glands. Remove the tracheae surrounding the antennal lobes (leave other tracheae intact).
3. Coat the tips of glass microeletrodes with rhodamine dextran (Molecular Probes®) and insert into the calycal neuropil to label processes of the mushroom body intrinsic and extrinsic neurons, including axons and axon terminals of the olfactory projection neurons.
4. Allow the dye to diffuse for 1–4 hr.
5. Fix the heads in 4% paraformaldehyde in phosphate buffered saline (PBS) for 1–2 hr at 20°C or overnight at 4°C.
6. Dissect the brains free, dehydrate in a graded series of ethanol and clear in methyl salicylate.
7. Mount brains in Permount and examine under fluorescent microscope.

7.1.24 *Lucifer yellow labeling of individual neurons [22]*

Protocol:

1. Anesthetize adult locusts by cooling, immobilize the heads in a wax-rosin mixture and remove the legs.
2. Cut a small window into the head capsule between the two compound eyes. Remove the right antenna, the median ocellus, the fat and some of the tracheal sacs.
3. Support and slightly lift the midbrain with a steel platform and de-sheath a small area of midbrain to allow probe penetration.
4. Fill an electrode with 5% Lucifer yellow backed up with 0.1 M LiCl and inject neuron.
5. Dissect brain out of the head capsule and fix for at least 1 hr in 4% paraformaldehyde in phosphate buffered saline (PBS).
6. Dehydrate through a graded ethanol series, clear in methyl salicylate and examine using a fluorescent microscope.

7.1.25 *Neurobiotin labeling of individual neurons [22]*

Protocol:

1. Anesthetize adult locusts by cooling, immobilize the heads in a wax-rosin mixture and remove the legs.
2. Cut a small window into the head capsule between the two compound eyes. Remove the right antenna, the median ocellus, the fat and some of the tracheal sacs.

3. Support and slightly lift the midbrain with a steel platform and de-sheath a small area of midbrain to allow probe injection.
4. Fill an electrode with 4% Neurobiotin backed up with 1 M KCl and inject neuron.
5. Dissect brain out of the head capsule and fix for at least 1 hr in 4% paraformaldehyde, 0.25% glutaraldehyde and 0.2% saturated picric acid in phosphate buffered saline (PBS).
6. Rinse brains in PBS with 0.1% Triton X-100 (PBTx) and embed in gelatin-albumin.
7. Cut sections at 30 μm with a Vibratome.
8. Incubate free-floating sections with a 1:200 dilution of horseradish peroxidase conjugated streptavidin in PBTx for 18 hr.
9. Rinse in PBTx and develop with DAB chromagen.
10. Dehydrate sections through a graded series of ethanol, clear in methyl salicylate and mount on slides for imaging.

7.1.26 Neuronal labeling [23]

Procedure:
1. Anaesthetize bees by cooling on ice and mount in metal holders for labeling of suboesophageal neurons.
2. Label neurones with three different dyes: rhodamine dextran MW2000, Alexa 568 (both from Molecular Probes®) and Lucifer yellow (Sigma®).
3. Fill the tips of glass microelectrodes with Lucifer yellow, Alexa 568 or rhodamine dextran dissolved in distilled water (5% each). Depending on the dye to be used, back fill the shafts of the microelectrodes with 0.1 M LiCl Lucifer yellow, 0.5 M KCl Alexa 568, or 0.2 M K-acetate rhodamine dextran, respectively.
4. Inject the dyes into the cells using 300 ms pulses of 5 nA hyperpolarizing current (Lucifer yellow) or depolarizing current (Alexa 568 and rhodamine dextran) at 2 Hz for 10–60 min. Electrode resistances should be 100–400 MΩ.
5. Allow the dyes to diffuse for at least 1 hr.
6. Remove the heads from the thorax and fix in 4% paraformaldehyde in phosphate-buffered saline (PBS, pH 7.2), for 2 hr at 20°C or overnight at 4°C.
7. Dissect the brains free, dehydrate through a graded series of ethanol, clear in methyl salicylate and image the whole mount brains on a confocal microscope.

References

1. Wu J.S. & Luo L. A protocol for dissecting *Drosophila melanogaster* brains for live imaging or immunostaining. *Nat. Protoc.* **1**, 2110–2115 (2006).
2. Volpi V., Mackay D. & Fanto M. Single *Drosophila* ommatidium dissection and imaging. *J Vis. Exp* 19, (2011).
3. Bonaccorsi S., Giansanti MG, Cenci G & Gatti M. Preparation of live testis squashes in *Drosophila. Cold Spring Harbor Protoc.* **3**, prot5577 (2011).
4. Weil T.T., Parton R.M. & Davis I. Preparing individual *Drosophila egg* chambers for live imaging. *J Vis. Exp* 60, (2012).
5. Mavrakis M., Rikhy R., Lilly M. & Lippincott-Schwartz J. Fluorescence imaging techniques for studying *Drosophila* embryo development. *Curr. Protoc. Cell Biol.* (2008).
6. Reed B.H., McMillan S.C. & Chaudhary R. The preparation of *Drosophila* embryos for live-imaging using the hanging drop protocol. *J Vis. Exp* **13**, (2009).
7. Ramos R.G., Machado L.C. & Moda L.M. Fluorescent visualization of macromolecules in *Drosophila* whole mounts. *Methods Mol Biol* **588**, 165–179 (2010).
8. Schawaroch V. & Li S.C. Testing mounting media to eliminate background noise in Confocal microscope 3-D images of insect genitalia. *Scanning* **29**, 177–184 (2007).

9. Dernburg A.F. Manual dissection and fixation of *Drosophila* egg chambers for whole-mount FISH. *Cold Spring Harbor Protoc.* **12**, 1531–1533 (2011).
10. Pimpinelli S., Bonaccorsi S., Fanti L. & Gatti M. Fluorescent *in situ* hybridization (FISH) of mitotic chromosomes from *Drosophila* larval brain. *Cold Spring Harbor Protoc.* **3**, prot5391 (2010).
11. Dernburg A.F. Hybridization to tissues on slides or coverslips for whole-mount FISH in *Drosophila*. *Cold Spring Harbor Protoc.* **5**, (2012).
12. Daul A.L., Komori H. & Lee C.Y. Multicolor fluorescence RNA *in situ* hybridization of *Drosophila* brain tissue. *Cold Spring Harbor Protoc.* **7**, (2010).
13. VanZomeren-Dohm A., Flannery E. & Duman-Schell M. Whole-mount *in situ* hybridization detection of mRNA in GFP-marked *Drosophila* imaginal disc mosaic clones. *Fly* **2**, 323–325 (2008).
14. Broihier H. Whole-mount fluorescence in situ hybridization and antibody staining of *Drosophila* embryos. *Cold Spring Harbor Protoc.* **8**, (2012).
15. Bonaccorsi S., Giansanti M.G., Cenci G. & Gatti M. Chromatin staining of *Drosophila* testes. *Cold Spring Harbor Protoc.* **8**, (2012).
16. Lehmacher C., Abeln B. & Paululat A. The ultrastructure of *Drosophila* heart cells. *Arthropod Struct. & Dev.* **41**, 459–474 (2012).
17. Funk C.J. Alkaline phosphatase activity in whitefly salivary glands and saliva. *Arch. Ins. Biochem. and Phys.* **46**, 165–174 (2001).
18. Li L., Ge J., Wu H., Xu Q-H. & Yao S.Q. Organelle-specific detection of phosphatase activities with two-photon fluorogenic probes in cells and tissues. *JACS* **134**, 12157–12167 (2011).
19. Chiang AS *et al*. Three-dimensional mapping of brain neuropils in the cockroach, *Diploptera punctata*. *J Comp Neurol* **440**, 1–11 (2001).
20. Loesel R., Weigel S. & Braunig P. A simple fluorescent double staining method for distinguishing neuronal from non-neuronal cells in the insect central nervous system. *J. Neurosci. Methods* **155**, 202–206 (2006).
21. Schroter U. & Malun D. Formation of antennal lobe and mushroom body neuropils during metamorphosis in the Honeybee, *Apis mellifera J Comp Neurol* **422**, 229–245 (2000).
22. Vitzthum H., Muller M. & Homberg U. Neurons of the central complex of the locust *Schistocerca gregaria* are sensitive to polarized light. *J. Neurosci.* **22**, 1114–1125 (2002).
23. Schroter U., Malun D. & Menzel R. Innervation pattern of suboesophageal ventral unpaired median neurones in the honeybee brain. *Cell Tissue Res.* **327**, 647–667 (2007).

8 Mounting

8.1 Introduction

Various mountants have been used in the procedures for histological preparations of insect tissues. Many of these are the same as those recommended for other animal tissue. Nevertheless, particular mountants and procedures for mounting specimens have been suggested specifically for insect tissues. The most widely known and widely used is of course balsam.

8.1.1 *Balsam [1]*

Characteristics:

1. Specimens must be dehydrated (cloudy or milky appearance indicates inadequate dehydration).
2. Dehydration can be accomplished by increasing concentrations of ethyl alcohol.
3. Specimens can be transferred from 99% alcohol or absolute alcohol to a clearing agent.
4. Dehydration may be substituted by transfer to cellosolve.
5. It is reported to shrink delicate insect structures such as antennae and palpi.
6. Mounts of small fragile insects may be inadequate for taxonomic purposes.
7. Use of this mountant may be time consuming.

8.1.2 *Mounting of adult* Drosophila *structures in Canada balsam [2]*

Procedure:

1. Anesthetize and wash adult flies in 70% ethanol.
2. Remove ethanol and replace with 10% KOH. Cover the flies with a glass slide.
3. Incubate at 70°C for ~45 min until the thorax becomes transparent.
4. Bring specimen back to room temperature and remove KOH.
5. Cover specimen with distilled water and transfer to 70°C for 45 min.
6. Bring specimen back to room temperature and remove water.
7. Add 70% ethanol and dissect the specimen.
    a. For head and abdomen, gently pull the head or abdomen off the thorax.
    b. For thorax, remove the wings and legs by holding thorax with one pair of forceps and pulling at the base of the appendages with another pair. Cut the remains of the thorax in two with dissecting scissors, either along the dorsal and ventral midlines or laterally, depending on the regions of greatest interest.
8. Dehydrate the specimens by washing two times, 10 min each with 100% ethanol.
9. Remove ethanol and cover specimens with clove oil for 15 min at room temperature.

Insect Histology: Practical Laboratory Techniques, First Edition.
Pedro Barbosa, Deborah L. Berry and Christina S. Kary.

Companion Website: www.wiley.com/go/barbosa/insecthistology

10. Transfer specimens from clove oil to a drop of Canada Balsam on a clean slide. Arrange as desired.
11. Leave specimen until the surface becomes sticky.
12. Wet a clean coverslip with xylene. Drop the coverslip onto the Canada Balsam at an angle.
13. Dry the slide at 50°C until hard (~3–5 days).
14. Seal the coverslip with nail polish.

8.1.3 *Wirth's Phenol-balsam technic [3]*

Application:
1. Recommended for small insects that have been preserved in alcohol (Procedure I).
2. Recommended for specimens preserved dry, placed in alcohol after dessication or those which have been restored with KOH (Procedure II).
3. Recommended for heavily sclerotized specimens.

Formula:
Solution A:
Saturated solution of melted phenol with absolute ethyl alcohol (maintaining a layer of crystals at bottom).

Solution B:
Equal parts of melted phenol and Canada Balsam.

Solution C:
Potassium Hydroxide (10%).

Procedure I:
1. Transfer to phenol: alcohol (70%) mixture (if specimen is very fragile) for 5 min.
2. Transfer to phenol (liquefied) for 1 to 24 hr.
3. Transfer to drop of solution A (on slide).
4. Place coverslip on specimen.
5. Dry in oven at 135°F for 1 to 2 weeks.
6. Replace evaporated mixture with pure balsam or solution B until no further evaporation occurs.

Procedure II:
1. Transfer to preheated (200°F) 10% potassium hydroxide solution for 5 to 10 min.
2. Transfer to 70–75% alcohol for 30 min.
3. Transfer to a phenol: alcohol (70%) mixture for 10 min.
4. Transfer to solution A for no more than 3 to 4 hr.
5. Arrange specimen in a drop of phenol (on a slide).
6. Replace evaporating phenol with solution B.

8.1.4 *Heming's modified balsam technic [4]*

Application:
Recommended for the taxonomic study of small insects such as thrips, beetles, Hymenoptera, Diptera, etc.

Characteristics:
1. Appendages (antennae, legs, wings, etc.) and genitalia are usually extended.

2. It addresses three problems of regular balsam procedures, a) time expenditure, b) shrinkage and crumbling in clearing and dehydration process, c) the need to spread appendages.
3. Disadvantages of this procedure when used for thrips were, a) terminal portions of reproductive system of the female are everted making comparison between specimens difficult, b) female ovipositors usually extended at right angles to the body.

Procedure:
1. Collect specimens into ethanol (10%) with 0.1% Triton-X emulsifier added.
2. Transfer specimens through successive changes of 35, 50, and 70% ethanol for ½ hr each (Proceed to step 6).
3. If specimen is dark and sclerotized treat in cold 10% KOH or NaOH for ½ hr to several days.
4. Transfer to acetic acid (15%) for 15 min.
5. Transfer to distilled water for 15 min.
6. Transfer to 98% or absolute alcohol for ½ hr.
7. Transfer to cedarwood oil and absolute alcohol layered preparation for about 1 hr.
8. Transfer to pure cedarwood oil.
9. Mount and dry.

Note:
1. The 10% ethanol mixture distends specimen (thrips) and extends the wings in about 2 days to 1 week.
2. High temperature (37°C) aids distention and placing grass seeds in the solution also aids the process.
3. The preparation in step 7 is made by pouring alcohol over cedarwood oil: since they are immiscible, two distinct layers are formed.
4. Specimen will rest at bottom of alcohol layer (top of oil), as it is infiltrated it sinks in the oil.

8.1.5 *Damar balsam [5]*

Characteristics:
Preserves, dyes and will show blues better than Canada Balsam.

Preparation:
1. Melt clear lumps of crude gum over a Bunsen flame.
2. Make a dilute solution with benzol with constant stirring.
3. Filter through glass wool in a funnel.
4. Filter again through two layers of filter paper in a ribbed funnel.
5. Filter again through soft filter paper.
6. Concentrate, by evaporation, in a warm place.

8.1.6 *Fox's creosote technic [6]*

Application:
1. Recommended for general use in the preparation of insects like thrips, *Collembola*, fleas and immature *Diptera*.
2. Recommended for preparation of insect organs such as honeybee stings, mouthparts, etc.
3. Recommended for the preparation of specimens for taxonomic study.

Characteristics:

1. Dispenses with the use of potassium hydroxide.
2. Dispenses with dehydration and clearing as separate processes.
3. Clears soft parts and intestinal contents and prepares specimen for mounting.
4. Doesn't cause the loosening or loss of setae as KOH sometimes does.
5. Causes slight irritation to skin and may have an unpleasant odor.

Procedure:

1. Transfer specimen from any degree alcohol or even water to creosote for 24 hr.
2. If specimen is very delicate (e.g., Collembola integument) transfer to 1:1 mixture of creosote and absolute alcohol for 24 hr.
3. Mount in balsam.

8.1.7 *Piccolyte® [1]*

Characteristics:

1. Synthetic resin that is soluble in xylene.
2. Very clear and dries slower than balsam.
3. Often substituted for balsam.

8.1.8 *Euparal (diaphane) [1]*

Application:

Recommended for whole mounts of thrips, larval mosquitoes, aphids, etc.

Characteristics:

1. Dehydration must be achieved before transfer of these mountants.
2. Generally, transfers occur from 95% ethyl alcohol.

8.1.9 *Hoyer's [1]*

Application:

Recommended for insect eggs and small aquatic larvae or nymphs.

Characteristics:

1. It is a water based mountant.
2. Specimens become clear and near transparent in about 24 to 48 hr.

Formula:

Distilled Water	50 pt
Gum Arabic (tear form)	30 pt
Chloral Hydrate	200 pt
Glycerine	20 pt

Preparation:

1. Dissolve gum Arabic in water with chloral hydrate for 24 to 48 hr.
2. Place glycerine into mixture.
3. Filter.

Procedure:

1. Drain water off live or recently killed specimen.
2. If alive, kill with 2–3 drops of Hoyer's for 1 to 2 min.
3. Orient specimen and add mountant.

Note:
1. Slide preparations should be sealed (ringed) with appropriate materials.

8.1.10 *Mounting of adult* Drosophila *structures in Hoyer's medium [7]*

Formula:
Hoyer's Medium

Preparation:
1. Prepare under a fume hood. Add 15 g gum Arabic to 25 ml of dH_2O in a glass beaker.
2. Heat to 60°C and stir overnight on a magnetic stirrer.
3. Successively add 100 g chloral hydrate.
4. Once chloral hydrate has dissolved, add 10 g glycerol.
5. Centrifuge for 30 min at 10,000 g and filter the solution through glass wool.
6. Store the solution at room temperature in a tightly sealed flask.
7. Shortly before use, centrifuge Hoyer's medium for at least 15 min to pellet undissolved particles.

Procedure:
1. Anesthetize adult fly, place in 70% ethanol and dissect necessary structure.
2. Put dissected structure into dH_2O until it rehydrates.
3. Place structures in Hoyer's medium and cover with coverslip.
4. Incubate slides at 50°C until the soft tissues have cleared (several hours or longer).
5. Seal edges of coverslip with nail polish.

8.1.11 *CMC-10 [8]*

Characteristics:
1. Available commercially and used like Hoyer's.
2. Transparency not as great as Hoyer's, specimens remain opaque with external structures evident.
3. It is a water based mountant.

Procedure:
1. Live specimens or specimens killed in boiling water may be placed in medium.
2. If alive, transfer (to kill) to CMC-10 for 15 to 30 min.
3. Mount.

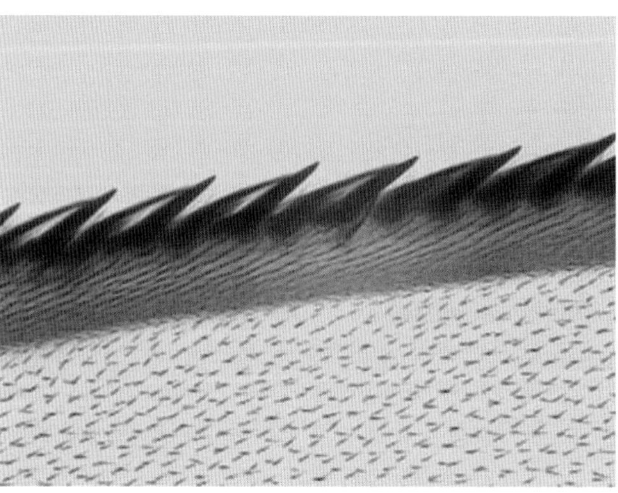

Fig. 8.1 Fly wing collected and preserved in alcohol 70%, and mounted in glycerin. (Source: Walter Dioni.)

8.1.12 *C-M medium*

Characteristics:

1. Cannot be directly transferred from glycerine, bases, or strong acids.
2. Can be transferred from KOH if specimen is washed in acid-alcohol.
3. Specimens may be transferred from xylene, toluene, water, alcohol, cellusolve, lactic acid, lactophenol, etc.
4. Does not turn yellow.
5. No crystallization occurs.

Formula:

Methocellulose	5 pt
Carowax	2 pt
Diethylene Glycol	1 pt
Ethyl Alcohol (95)	25 pt
Lactic Acid	100 pt
Distilled Water	75 pt

Preparation:

1. Mix methocellulose and alcohol.
2. Add mixture to remainder of formula.
3. Filter through glass wool.
4. Place in oven at 40–45°C, until it reaches desired consistency, or about 3 to 5 days.
5. Medium can be thinned with ethyl alcohol (95%) on heat.

Procedure:

1. If specimen is small and delicate it can be transferred directly to medium with clearing.
2. If insect is larger, transfer to cellosolve then into the medium.

8.1.13 *Glycerine [1]*

Characteristics:

1. Used for temporary mounts.
2. Also used as component for other mountants such as glycerine jelly, Farrant's medium, Faur's medium, etc.

8.1.14 *Glycerine jelly [5]*

Application:

Recommended as an aqueous mountant used for mounting aphids and other minute insects.

Characteristics:

It is a temporary mountant that must be ringed with gold size, asphaltum or a varnish.

Formula:

Water	42 pt
Gelatin	6 pt
Glycerine	50 pt
Phenol crystals	2 pt

Preparation:

1. Soak the gelatin in water for 30 min.
2. Dissolve gelatin with heat.

3. Add 5 pt of egg white and heat (not above 75°C) for 30 min.
4. Filter through moist, hot funnel or glass wool (pressed into funnel that is maintained hot) to remove egg but not jelly.
5. Add glycerine and phenol.
6. Reheat and stir for 10 to 15 min until homogenous.

8.1.15 *Mounting in methyl salicylate [9]*

Application:
Recommended for improved clearing of structures for imaging.

Procedure:
1. Collect larvae to be cleared.
2. Douse larvae with boiling water for 30 sec to prevent shrinkage and browning.
3. Transfer larvae to 70% ethanol for 2 hr.
4. Transfer larvae to 96% ethanol for 2 hr.
5. Transfer larvae to methyl salicylate for 3 hr.
6. Mount and image larvae.

8.1.16 *Faure's gum chloral [10]*

Formula:

Distilled Water	50 pt
Gum Arabic	30–40 pt
Glycerine	20 pt
Chloral Hydrate	50 pt

Preparation:
1. Dissolve the chloral hydrate.
2. Add glycerine and stir thoroughly.
3. Place pieces of gum Arabic in gauze bag, suspend in the fluid until dissolved.

8.1.17 *Gater's [10]*

Formula:

Distilled water	10 pt
Gum Arabic	8 pt
Chloral Hydrate	7 pt
Glucose Syrup	5 pt
Glacial Acetic Acid	3 pt

Note:
1. This is a modification of Faure's medium.
2. Glucose syrup is prepared by dissolving 98 pt of glucose in 100 pt of water.

8.1.18 *Imm's gum chloral [11]*

Application:
Recommended for the preparation of slide mounts of minute, delicate hexapod species such as those in the *Protura*.

Formula:

Chloral Hydrate	160 pt
Gum Arabic	15 pt
Glucose Syrup	10 pt
Acetic Acid	5 pt
Water	20 pt

Procedure:
1. Collected specimens are placed in alcohol (96%).
2. Place in glacial acetic acid for a few minutes or until slightly extended.
3. Place a drop of Imm's on a clean slide.
4. Place specimen into the medium and place coverslip on preparation.
5. Bake slide gently for an hour or two.
6. Let stand for 2 to 3 weeks before ringing.

Note:
1. If specimen has been mounted and still needs further cleaning heat gently.

8.1.19 *Doetchman's gum chloral stain mountant [10, 12]*

Application:
Recommended for ticks, mites, fleas, lice, bedbugs, mosquitoes (larvae and adults), gnats, midges, stigmata of dipterous larvae, aphids, scale insects, ova and insect embryos.

Characteristics:
It kills, fixes, dehydrates, clears, stains, and mounts a specimen with one preparation.

Formula:

Water	35 pt
Chloral Hydrate	20 pt
Gum Arabic	20 pt
Glycerine	20 pt
Glucose Syrup	3 pt
Basic Fuchsin (in 95% alcohol)	0.3 pt/10 pt alcohol

Note:
1. Specimens that are highly sclerotized may require treatment with KOH.
2. Permanent mounts may be prepared by treatment in oven at 37–50°C.
3. About 10 or more drops of mountant should be used.

8.1.20 *Berlese's gum chloral [13]*

Application:
1. Recommended for small insects and mites.
2. Recommended as particularly useful for all Apterygota, Aphididae, Coccids, Anoplura, Thysanoptera, Psocoptera, and Diptera.

Formula:

Water	20 pt
Gum Arabic	15 pt
Chloral Hydrate	160 pt (approx)
Glucose Syrup	10 pt
Glacial Acetic Acid	5 pt

Preparation:
1. Dissolve gum Arabic in water.
2. Add glucose syrup and chloral hydrate to saturation.
3. Filter (glass-wool, cotton wool).

Note:
1. Mounting on warm slide gives better results.
2. If specimen is heavily sclerotized, treatment with KOH may be required.
3. If specimen is treated (KOH) transfer to 10% acetic acid before mounting.

8.1.21 *Buxton's [5]*

Application:
Recommended for use in mounting mosquito legs, mouthparts, etc.

Formula:

Distilled Water	50 pt
Glycerine	20 pt
Gum Arabic	40 pt
Chloral Hydrate	50 pt
Cocaine Hydrochloride	0.5 pt

Preparation:
1. Dissolve gum Arabic in water with chloral hydrate and cocaine.
2. Add glycerine.
3. Filter with collucotton.
4. Add a drop of HCl.

8.1.22 *Swan's [14]*

Application:
Recommended as a mountant for the preparation of temporary slides.

Formula:

Distilled Water	20 pt
Chloral Hydrate	60 pt
Gum Arabic	15 pt
Glucose Syrup	10 pt
Glacial Acetic Acid	5 pt

8.1.23 *Ewing's modified Berlese's [12]*

Characteristics:
1. Better refractive index than most mountants.
2. It is suggested for permanent mounts.
3. Reduces potential for crystallization.

Formula:

Water	35 pt
Chloral Hydrate	30 pt
Gum Arabic	20 pt
Glycerine	12 pt
Glucose Syrup	3 pt

Note:
1. For permanent mounts, oven treatment at temperatures below 33°C is required.
2. Mountant is not recommended for very heavily pigmented specimens.

8.1.24 *Polyvinyl-lactophenol [10]*

Application:

1. Recommended for insect larvae.
2. Recommended for mosquito genitalia.

Formula:

Polyvinyl Alcohol	56 pt
Phenol	22 pt
Lactic Acid	22 pt

8.1.25 *Ribeiro's formic acid polyvinyl [18]*

Application:

Recommended for mosquito larvae and pupae.

Formula:

Solution A:

Formic Acid (98–100%)	4 pt
Distilled Water	5 pt

Solution B: Polyvinyl Alcohol

Preparation:

1. Dissolve solution B in solution A to desired viscosity.
2. Heat in oven at 75–80°C for 1 hr.
3. Store (do not use cork stopper).

Procedure:

1. Specimens may be dropped alive into formic acid-PVA (and will also be cleared).
2. Mount (permanent mounts may be prepared by heating overnight).

8.1.26 *Ribeiro's polyvinyl alcohol [15]*

Application:

Recommended as general mountant for insects.

Characteristics:

1. The mountant also clears and hardens overnight.
2. The mountant has a refractive index of 1.496.
3. Excess water must be removed from specimen.

Formula:

Polyvinyl Alcohol	6 pt
Formic Acid	10 pt
Chloral Hydrate	72 pt
Phenol	12 pt

Preparation:

1. Dissolve PVA in formic acid.
2. Place in an oven or water bath of 60°C.
3. Add chloral hydrate and place in oven until completely dissolved.
4. Add phenol and stir until dissolved.
5. Store in dark stoppered glass bottle.

8.1.27 *Essig's PVA [16]*

Application:
Recommended as a mountant for aphids.

Formula:
Solution A:
PVA Stock

Distilled Water	80 pt
Polyvinyl Alcohol	20 pt (about)

Preparation:
1. Add PVA in small amounts until a thick consistency mixture of about 100 pt is attained.
2. Heat in water bath until milky and homogenous.
3. Cool.
4. Reheat until clear.
5. Cool.
6. Filter through fine copper gauze or filter hot solution through four to five layers of filter paper.

Solution B:

PVA Mountant	
PVA Stock	56 pt
Phenol Crystals	22 pt
Lactic Acid	22 pt
Picric Acid (for staining)	1 grain

Note:
1. Slides must be ringed.

8.1.28 *Salmon's [17]*

Application:
Recommended for small insects and other arthropods.

Characteristics:
1. Specimens may be mounted from alcohol of any strength or from water.
2. It has a low refractive index.
3. Permanent mounts can be prepared.
4. Mountant is also a clearing agent.

Formula:
Solution A:

Phenol Crystals	45 pt
Lactic Acid	45 pt

Solution B:

Polyvinyl Alcohol	2.5 pt
Distilled Water	10 pt

Preparation:
1. Prepare solution A with heat.

2. Add 25 pt of solution A to solution B.
3. Clear in water bath for several hours.

Note:
1. Work using warmed slide will facilitate preparations.

8.1.29 *Steedman's dimethyl hydantoin formaldehyde [18]*

Application:
Recommended for insects and other arthropods.

Formula:

Dimethyl Hydantoin Formaldehyde	70–80 pt
Distilled Water	20–30 pt
Or	
Ethyl Alcohol (70%)	20–30 pt

Preparation:
1. Grind up resin to a powder.
2. Add to water or alcohol for three to five days.
3. Filter and refilter.
4. If solution is aqueous add phenol sufficient to make a 5% solution.
5. Do not add phenol if solution is alcoholic.

Note:
1. Mounting can be facilitated by working with warmed slides.

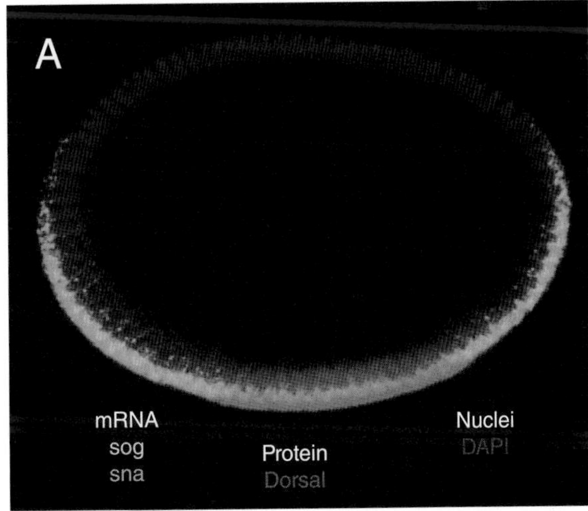

Fig. 8.2 Upright optical slice of an intact *Drosophila* embryo in blastoderm stage. Double staining for mRNA and protein. Target mRNAs for in situ probes (sog (yellow) and sna (green)) and protein stained by primary antibody (anti-dorsal) are indicated in the figure. DAPI was used to label the nuclei. Note: The expression of sog and sna are located apically in the cytoplasm, whereas the expression of Dorsal is nuclear. Strong signals from sog nascent transcripts that are located in nuclei can also be seen at this magnification. (Source: Belu *et al.* 2010. Reproduced with permission of the authors.) See plate section for the color version of the figure.

8.1.30 *Agarose [19]*

Application:
Similar consistency to glycerine mounting, but with less background fluorescence.

Procedure:
1. Warm 1% agarose just to melting point.
2. Place drop of agarose in slide chamber.
3. Add specimen to agarose.
4. Cover with coverslip and allow to harden.

8.1.31 *Upright imaging of embryos [20]*

Application:
Technique for mounting of embryos for coronal viewing.

Procedure:
1. Fix and stain embryos.
2. Place a thin layer of glycerin jelly on a slide.
3. Align embryos on the glycerin jelly layer.
4. Apply a second thin layer of glycerin jelly over embryos and allow the sandwich to solidify.
5. Cut strips of the jelly embedded embryos and flip upright for imaging.

References

1. Peterson A. *Entomological Techniques. How to work with Insects*. Edwards Brothers Inc., Ann Arbor, Michigan (1964).
2. Stern D.L. & Sucena E. Preparation and mounting of adult *Drosophila* structures in Canada balsam. *Cold Spring Harbor Protoc.* 3, 373–375 (2012).
3. Wirth W.W. & Marston N. A method for mounting small insects on microscope slides in Canada balsam. *Ann. Entomol. Soc. Amer.* **61**, 783–784 (2011).
4. Heming B.S. A modified technique for mounting Thysanoptera in Canada balsam. *Entomol. News* **80**, 323–328 (1969).
5. Kennedy C.H. *Methods for the Study of the Internal Anatomy of Insects*. Department of Entomology, Ohio State University (Mineographed) (1932).
6. Fox I. The use of creosote in mounting fleas and other arthropods on slides. *Science* **96**, 478 (1942).
7. Stern D.L. & Sucena E. Rapid mounting of adult *Drosophila* structures in Hoyer's medium. *Cold Spring Harbor Protoc.* **1**, 107–109 (2012).
8. Ribeiro H. A solidifiable formic acid-PVA solution for transporting, preserving and mounting mosquito larvae and pupae. *Stain Technol.* **42**, 159–160 (1962).
9. Niederegger S., Wartenberg N., Spiess R. & Mall G. Simple clearing technique as species determination tool in blowfly larvae. *Forensic Sci. Int.* **206**, e96–e98 (2011).
10. Lee A.B. *Microtomist's Vade-Mecum*. The Blakiston Company, Philadelphia (1950).
11. Womersley H. Notes on the mounting of Protura. *Entomol. Mon. Mag.* **63**, 153–154 (1927).
12. Doetschman W.H. Some suggestions in microtechnique particularly useful in microentomology and parasitology. *Trans. Amer. Microsc. Soc.* **63**, 175–178 (1944).
13. Swan D.C. Berlese's fluid: Remarks upon its preparation and use as a mounting medium. *Bull. Entomol. Res.* **27**, 389–391 (1936).
14. Lane J. *The Preservation and Mounting of Insects of Medical Importance*. World Health Organization Vector Control (1965).
15. Ribeiro H. Un nouveau milieu de montage pour microscopie avec l'alcool vinylique polymerise. *Ann. De Parasitol* **4**, 677–681 (1962).

16. Essig E.O. Mounting aphids and other small insects on microscopic slides. *Pan Pacific Entomol.* **24**, 9–22 (1948).
17. Bellinger P.F. Salmon's fluid, A new mounting medium for slides of small larvae and larval pelts of Lepidoptera. *Lepid. News* **7**, 170–171 (1953).
18. Steedman H.F. Dimethyl hydantoin formaldehyde: A new soluble resin for use as a mounting medium. *Quart. J. Microsc. Sci.* **99**, 451–452 (1958).
19. Schawaroch V. & Li S.C. Testing mounting media to eliminate background noise in Confocal microscope 3-D images of insect genitalia. *Scanning* **29**, 177–184 (2007).
20. Belu M. *et al.* Upright imaging of *Drosophila* embryos. *J. Vis. Exp.* (2010).

9 Preparation of whole mounts

9.1 Introduction

The following are procedures that have been recommended for the preparation of insect whole mounts. Although most are designed for a specific insect group, many also are applicable to insects with similar characteristics.

9.1.1 *Gabaldon's technic [1]*

Application:
Recommended for the preparation of whole mounts of *Anopheline* eggs.

Formula:
Solution A: duNoyer's

Lanolin	20 pt
Rosin	80 pt

Preparation:
1. Boil lanolin until dehydrated.
2. When foam disappears (dehydrated) add rosin.
3. Make homogenous paste.

Procedure:
1. Transfer eggs to formalin (5%).
2. Place coverslip on specimen.
3. Draw off excess fluid with filter paper.
4. Seal with solution A.

9.1.2 *Burton's technic [2]*

Application:
Recommended for the histological preparation of mosquito eggs and larvae.

Procedure:

I. Eggs
   1. Kill in hot water or alcohol (70%).
   2. Dehydrate in alcohol (70%) for 2 hr.
   3. Dehydrate in alcohol (95%) for 2 hr.
   4. Clear if necessary (usually only heavily pigmented eggs or egg rafts).
   5. Mount in Euparal.

Insect Histology: Practical Laboratory Techniques, First Edition.
Pedro Barbosa, Deborah L. Berry and Christina S. Kary.

Companion Website: www.wiley.com/go/barbosa/insecthistology

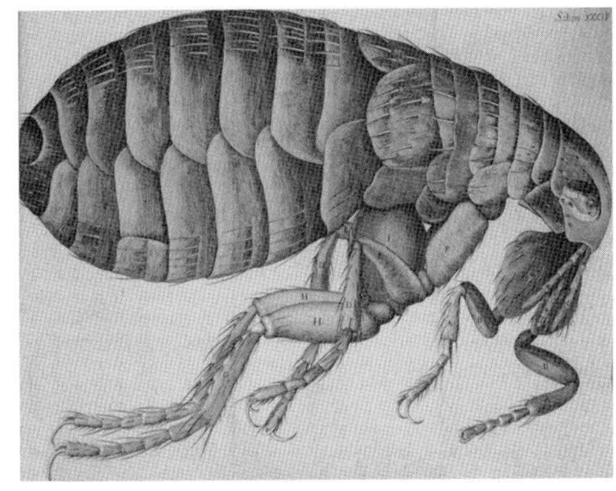

Fig. 9.1 Flea whole mount. (Source: Original micrograph by Robert Hooke published 1665.)

Fig. 9.2 Whole mount of the seven spotted ladybird *Coccinella 7 punctata*. (Source: David Walker, Micscape. Reproduced with permission.)

Fig. 9.3 Whole mount of the Plumed gnat; one of the common names for a fly in the family known as Chironimidae. (Source: David Walker, Micscape. Reproduced with permission.)

II. Larvae and Pupae
1. Kill specimen in hot water or alcohol (70%).
2. Transfer to alcohol (70%) for 2 to 24 hr.
3. Transfer to alcohol (95%) for 2 to 24 hr.
4. Clear in beechwood creosote for at least 10 min.
5. Transfer to xylene for 1 min.
6. Mount in synthetic resin.

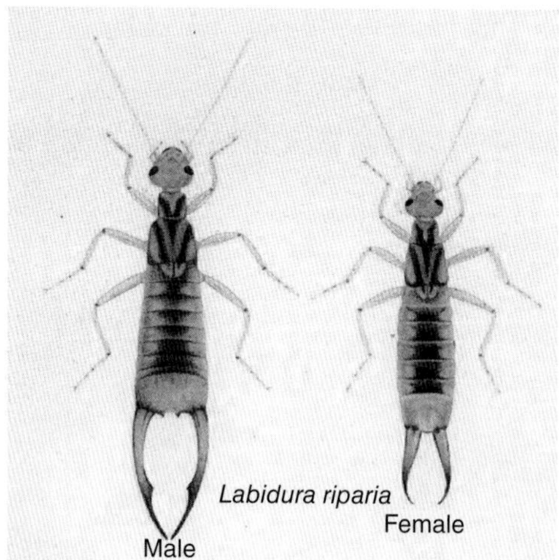

Fig. 9.4 Whole mount of the earwig *Labidura riparia*. (Source: www.earthlife.net/insects/dermapta.html)

9.1.3 *Craig's technic [3–5]*

Application:
Recommended for the accurate examination of the chorion of mosquito eggs in whole mounts.

Formula:
Solution A: Mayer's Chlorine Bleach

Potassium Chlorate	0.1 pt
HCl	0.1 pt
Water	100 pt

Preparation:
1. Mix potassium chlorate and HCl.
2. Let stand until chlorine starts evolving.
3. Add water.

Procedure:
1. Remove gelatinous exchorion by rolling between moist filter paper or in alcohol.
2. Place egg(s) in well slide.
3. Remove cap and tease out larva and vitelline membrane.
4. Transfer to alcohol (after dissection) and wash several times.
5. Drain alcohol, add crystals of potassium chlorate then HCl.
6. Transfer through several changes of absolute alcohol.
7. Mount in euparal.

9.1.4 *Jennings and Addy technic [6]*

Application:
1. Recommended as an egg whole mount procedure for staining and counting.
2. Recommended as particularly useful for distinguishing individual eggs in egg masses of the jack pine budworm.

Procedure:
1. Transfer egg masses to a methylene green (1% aqueous) solution.
2. Differentiate in warm water (60–65°C) for 5 to 10 sec.
3. Cover with water and make counts.

9.1.5 *Gifford and Trahan technic [7]*

Application:

1. Recommended for the display of eggs of insect species that lay eggs within plant tissues.
2. Recommended as particularly useful for use in egg counts of leafhoppers.

Formula:

Solution A:

Phenol	1 pt
Lactic acid	1 pt
Distilled Water	1 pt
Glycerine	2 pt
Acid Fuchsin	to saturation

Procedure:

1. Transfer plant material to boiling water for 3 to 5 min.
2. Transfer through several changes of 95% alcohol for 1 to 3 days.
3. Wash in water.
4. Transfer to solution A (hot or cold) for 15 min (boiling) or 24 hr + (cold).
5. Wash in water.
6. Transfer to 3% KOH (to clear) for 24 hr.
7. Wash in water.
8. Transfer to 2% acetic acid (can be stored for months).

9.1.6 *Peacock's technic [5, 8]*

Application:

Recommended for small insect larvae.

Formula:

Solution A: Polyvinyl Alcohol-Lacto-phenol

Lactic Acid	25 pt
Phenol	22 pt
Polyvinyl Alcohol	56 pt

Preparation:

1. Dissolve phenol crystals in lactic acid.
2. Add polyvinyl alcohol and stir vigorously.
3. Heat in water bath.

Procedure:

1. Clear specimen in potassium hydroxide (10%).
2. Wash in distilled water.
3. Carefully stain with acid fuschin (do not overstain).
4. Wash in distilled water.
5. Mount in solution A.
6. Dry off mountant in a warm oven (gently).
7. Mount in Euparal.

9.1.7 *Lane's technic I [9]*

Application:

Recommended for the preparation of whole mounts of larvae of *Culicidae*.

Formula:

Solution A: MacGregor's

Formalin (40%)	10 pt
Borax (5%)	10 pt
Distilled Water	80 pt
Glycerine	0.25 pt

Solution B: Berlese's Fluid

Distilled water	25 pt
Gum Arabic	15 pt
Chloral Hydrate	100 pt
Glycerine	10 pt
Chloral Hydrate of Cocaine	0.5 pt

Preparation:

1. Mix in water bath (50–80°C).
2. Dissolve gum Arabic in water then add the chloral hydrate.
3. Add remaining ingredients.
4. Filter through glass wool or bolting silk.

Procedure:

1. Kill specimen in hot water and transfer to solution A (preservative).
2. Specimens can be mounted from solution A directly into solution B.

9.1.8 *Foote's technic [10]*

Application:

1. Recommended for the preparation of whole mounts used in the study of chaetotaxy of larval mosquitoes.
2. Recommended as a procedure which prevents subsequent darkening.

Formula:

	Cellosolve	Beechwood Creosote
Solution A:	3 pt	1 pt
Solution B:	3 pt	2 pt
Solution C:	1 pt	1 pt

Procedure:

1. Kill larvae in hot water (180 °F) for 10 sec.
2. Transfer to KOH (10%) until transparent.
3. Wash in distilled water for ½ hr.
4. Drain and transfer to glacial acetic acid for 24 hr.
5. Transfer to cellosolve for 2 hr.
6. Transfer to solution A for 2 hr.
7. Transfer to solution B for 2 hr.
8. Transfer to solution C for 2 hr.
9. Transfer larvae to a drop of balsam.

Note:

1. Specimen should not remain in KOH for more than 24 hr.

9.1.9 *Gater's technic [11]*

Application:
Recommended as a simple and effective method of preparing mounts of mosquito larvae.

Formula:
Solution A: Mounting Medium

Distilled water	10%
Gum Arabic	8%
Chloral Hydrate	74%
Glucose Syrup (98 pt glucose/100 pt water)	5%
Glacial Acetic Acid	3%

Preparation:
1. Dissolve ingredients in order listed in water-bath or oven (50°C).
2. Filter with a Buchner funnel.

Procedure:
1. Place larva on slide.
2. Drain excess water.
3. Place drop of solution A on specimen.
4. Cover with coverslip.
5. For more permanent slides dry preparation and ring.

9.1.10 *Gibbin's technic [12]*

Application:
1. Recommended as a mounting medium for the preparation of mosquito larvae whole mounts.
2. Recommended as a medium that fixes, clears, and preserves.

Formula:
Solution A: Amann's Lactophenol

Carbolic Acid	20 pt
Lactic Acid	20 pt
Glycerine	40 pt
Water	20 pt

Procedure:
1. Construct enclosure or cell on slide (slightly smaller than coverslip).
2. Fill with solution A.
3. Transfer larvae to mountant.
4. Seal with varnish.

Note:
1. Lactophenol can be used to kill larvae.

9.1.11 *Middlekauff's technic [13]*

Application:
Recommended as a rapid procedure for permanent whole mounts of mosquito larvae.

Procedure:
1. Kill larvae in hot water.
2. Transfer to ethyl alcohol (70–75%) for 10 to 15 min.
3. Transfer to alcohol (95%) for 3 to 5 min.
4. Transfer to absolute alcohol for 5 sec.
5. Transfer to creosote for a few minutes until cleared.
6. If specimen is delicate, mix creosote with absolute alcohol (equal parts).
7. Place larvae on a slide and drain excess creosote.
8. Cover larvae with balsam and coverslip.

9.1.12 *Colless' technic I [14]*

Application:
Recommended for the preparation of permanent mounts of mosquito larvae and for other small arthropods.

Formula:
Solution A: Mounting Medium
Canada balsam in phenol

Solution B:

Chloral Hydrate	80 pt
Water	20 pt

Procedure:
1. Kill larvae in hot water or fixative (puncture).
2. Transfer to solution B for 24 hr.
3. Wash in alcohol (90%)
4. Transfer (with pipette) to phenol bath briefly.
5. Mount in solution A.

Note:
1. This method is only appropriate for fresh material, not specimens that have been stored in alcohol.

9.1.13 *Burton's technic (I) [15]*

Application:
Recommended for whole mounts of mosquito larvae, eggs, and pupae.

Procedure:
1. Kill specimen in hot water.
2. Transfer specimen to alcohol (70%) for 2 hr.
3. Transfer specimen to alcohol (95%) for 2 hr.
4. Transfer specimen to beechwood creosote for 10 min to 2 hr.
5. Transfer to xylene.
6. Drain excess xylene.
7. Apply several drops of mounting medium.
8. Cut (with razor blade) specimen, for larvae, at 6th or 7th abdominal segment.
9. Cover with coverslip.

Note:
1. Killing specimen in alcohol (70%) is not suggested since it prevents proper clearing.

2. Dehydration time is generally a function of specimen size.
3. Cutting specimen (at abdomen) enables proper orientation of morphological structures.

9.1.14 *Burton's technic (II) [15]*

Application:
Recommended for whole mount preparations of mosquito larvae, eggs, and pupae.

Formula:
Solution A:

Ethyl Alcohol (95%)	70 pt
Beechwood Creosote (or phenol)	25 pt

Solution B: Lactophenol

Phenol (absolute)	20 pt
Lactic Acid	40 pt
Glycerine	40 pt
Water	20 pt

Solution C: Polyvinyl Alcohol

Polyvinyl Alcohol (stock)	56 pt
Phenol	22 pt
Lactic Acid	22 pt

Procedure:

I. Creosote-alcohol and Phenol-alcohol method
   1. Kill specimen by placing in solution A.
   2. Let stand in solution A for 2 hr.
   3. Transfer specimen to beechwood creosote for 1 to 2 min.
   4. Transfer specimen to clean slide.
   5. Mount in euparal or diaphane.
   6. Cut posterior abdominal portions to view appropriate structures.
   7. Cover with coverslip and dry in oven at 100°F for 30 min.

II. Lactophenol and polyvinyl alcohol method
   1. Kill specimen in solution B.
   2. Transfer to solution C for 1 hr.
   3. Mount in solution C.
   4. Cover with coverslip.

Note:
1. In Method I only, euparal or diaphane can be used.
2. In Method II, specimen can be transferred to solution B either from water or alcohol.
3. Polyvinyl alcohol serves both as a clearing agent and a mounting medium.
4. To prevent bubbles forming in the mountant, apply it with a pipette.
5. Drying of slides in oven is suggested for areas with high humidity.

9.1.15 *Burton's modified technic [16]*

Application:
Recommended for rapid permanent mounts of mosquito larvae.

Procedure:
1. Kill larvae in hot water.
2. Transfer to alcohol (95%) for 2 hr.

3. Transfer to beechwood creosote for 10 min.
4. Transfer to xylene (or toluene) for 30 sec to 1 min.
5. Mount in Permount.
6. Cover with coverslip.

9.1.16 *Rapp and Jones' technic [17]*

Application:
Recommended as a very rapid method for the preparation of mosquito larvae whole mounts.

Procedure:
1. Place specimen in a 3-dram vial.
2. Remove all excess water.
3. Add cellosolve to vial (2/3 full).
4. Transfer to xylol.
5. Mount in synthetic resin.

Note:
1. This method allows for the preparation of a slide mount of field collected specimens.

9.1.17 *Steck and Wharton larval technic [18, 19]*

Procedure:
1. Boil larvae briefly in water.
2. Preserve larvae in 70% ethanol.
3. Treat overnight with 10% KOH for clearing.
4. Mount in Hoyer's medium.

9.1.18 *Puri's technic [20]*

Application:
1. Recommended for the display of lateral tracheal trunks in larval whole mounts.
2. Recommended as particularly useful for the preparation of the sugarcane stem borer (*Argyria stricticraspis*).

Formula:
Solution A:

Turpentine Oil	3 pt
Phenol (heated to fluid form)	5 pt

Procedure:
1. Kill specimen in strong alcohol.
2. Fix in alcohol (70%) for a week.
3. Place in KOH for 24 hr.
4. Wash well in water for 8 to 12 hr.
5. Place in diaphanol.
6. Wash with water.
7. Slowly press specimen between two slides until inner organs are eliminated.
8. Fasten ends of slides with clip and place in standing jar with methyl alcohol for 20 min.
9. Carefully separate slides and place specimen in alcohol.
10. Wash in several changes of alcohol.
11. Remove surface debris with brush.
12. Stain with eosin (in 90% alcohol).

13. Place in solution A for 24 hr.
14. Clear first in turpentine then in cedarwood oil.
15. Mount in balsam.

9.1.19 *Geyer-Duszynska technic [21]*

Application:
1. Recommended for the histological display of pre-cleavage maturation divisions.
2. Recommended for use with several species of *Cecidomyidae*.

Procedure:
1. Coat slide with a thin film of albumen.
2. Place embryo in several rows in drops of water (e.g., 20–30).
3. Let water evaporate.
4. Prick embryos at one end (protoplasm flows out) and after a few seconds cover with Bouin's.
5. Place in moist chamber for 24 hr.
6. Transfer to alcohol (80%) for 24 hr.
7. Transfer to water.
8. Stain with Feulgen method at 56 °C and with 12 min hydrolysis time in 1 N HCl.
9. Transfer through three changes of SO_2 water solution for at least 15 min each.
10. Dehydrate through series, from 50, to 70 and 90% alcohol for 10 min each.
11. Transfer to alcohol (96%) for 30 min.
12. Transfer through three changes of absolute alcohol for 30 min each.
13. Transfer through two changes of xylene for 30 min each.
14. Mount in a thin layer of balsam.

Note:
1. For puncturing embryos it is suggested that thin sharp platinum needles be used or needles flattened and sharpened on porcelain plates.
2. The moist chamber (step 5) can be made by placing wet sand in a Petri dish.
3. The temperature indicated in step 8 should not exceed 58 °C.

9.1.20 *Lane's technic II [9]*

Application:
Recommended for the preparation of whole mounts of muscoid larvae.

Formula:
Solution A: Pampel's Fluid

Glacial Acetic Acid	4 pt
Distilled Water	30 pt
Formalin (40%)	6 pt
Alcohol (95%)	15 pt

Procedure:
1. Kill in hot water.
2. Make lateral incision in larvae.
3. Transfer to boiling KOH or NaOH (10%) for 10 min.
4. Rinse in water two or three times.
5. Transfer through an ascending alcohol series.
6. Transfer to xylol.
7. Mount in Canada balsam.

9.1.21 *Steedman's Ester wax technic [22]*

Application:

1. Recommended for small, soft, relatively impermeable objects.
2. Recommended for blowfly larvae.

Formula:

Solution A: Smith's Formol-dichromate

Potassium dichromate	0.5 pt
Formaldehyde (40%)	10 pt
Glacial acetic acid	2.5 pt
Distilled water	100 pt

Procedure:

1. Fix in solution A for 12 to 24 hr.
2. Transfer through alcohol series to alcohol (70%).
3. Transfer to a mixture of ethanol (70%) and ethylene glycol mono-ethyl ether (equal parts).
4. Transfer to pure ethylene glycol mono-ethyl ether and ester wax (equal parts).
5. Transfer to wax for 5 to 10 min.
6. Embed and section.
7. Place sections on lightly albumenized slide.
8. Flood with 1% Safranin (with 1 drop anilin oil/100 pt).
9. Drain, then flood with distilled water.
10. Drain and let dry at 35–40°C overnight.
11. Dissolve wax in xylene and mount.

Note:

1. Prepare solution A immediately before use.

9.1.22 *Super-skipper technic [23]*

Application:

Provides improved fixation of internal organs to allow for maintenance of insect's shape.

Formula:

Super-skipper Solution:

Kerosene	17 pt
Acetic Acid	11 pt
Ethanol	50 pt
Isobutanol	17 pt
Dioxane	5 pt

Procedure:

1. Fix insect in 2% glutaraldehyde with 4% paraformaldehyde in 0.1M sodium cacodylate buffer pH 7.2 for no more than 30 min.
2. Treat insect with Super-skipper solution for 30–60 sec to inflate integument.
3. Rinse in distilled water.
4. Immerse in 1% OsO_4 and fix for up to 10 days.
5. Rinse repeatedly in distilled water.
6. Dehydrate in graded ethanol series into 100% ethanol.
7. Immerse in 100% acetone.

8. Transfer to a mixture of 1 pt acetone and 1 pt hexamethyldisilazane.
9. Transfer to pure hexamethyldisilazane for 20–30 min.
10. Allow to dry.

9.1.23 *Essig's technic I [24]*

Application:
Recommended for clearing and mounting aphids.

Procedure:
1. Place specimens in alcohol (70%) over a boiling water bath for 10 to 15 min.
2. Transfer to lactic acid (70%) over a boiling water bath for 15 to 20 min (or 10 min for small, light-colored specimens).
3. Transfer to a saturated chloral hydrate solution with a few phenol crystals.
4. Heat over water bath until colorless or disappears.
5. Remove body contents (and embryos).
6. Wash well and dehydrate in alcohol (95%).
7. Clear in clove oil.
8. Mount in balsam or euparal.
9. Or, mount directly in Berlese mixture.
10. Heat lightly and ring.

9.1.24 *Essig's technic II [24]*

Application:
Recommended for clearing and mounting aphids.

Formula:
Solution A:

Lactic Acid	45 pt
Acetic Acid	5 pt
Ethanol	30 pt
Saturated Aqueous Phenol	5 pt
Water	15 pt

Procedure:
1. Transfer specimens to covered stender dish with solution A.
2. Place dish in oven 49 °C for 1 hr (if specimen freshly collected), or 24 to 48 hr (if old alcohol preserved specimens).

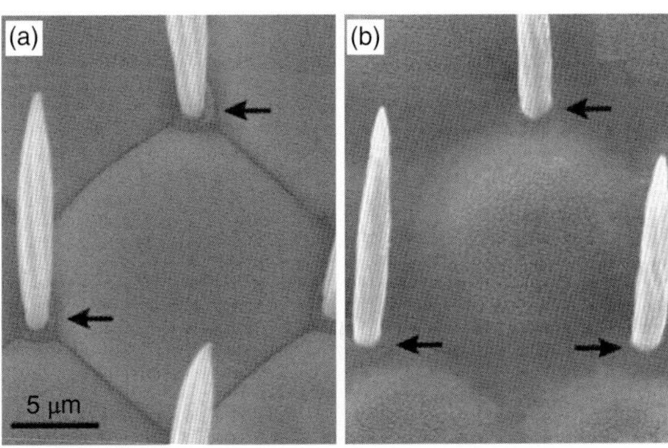

Fig. 9.5 Whole animals were fixed in 2% glutaraldehyde plus 4% paraformaldehyde in 0.1 M sodium cacodylate (pH 7.2 for 3–4 hr) and then 1% OsO4 in sodium cacodylate buffer for 24–96 hr. They were extensively washed in sodium cacodylate buffer and then dehydrated. (Source: Beno *et al.* 2007. Reproduced with permission of John Wiley & Sons.)

3. Remove from solution A and remove body contents.
4. Transfer to alcohol (95%).
5. Place in clove oil and mount.

Note:
1. Specimens preserved in alcohol (80 to 95%) for long periods may require the addition of more lactic acid or water to solution A.
2. Overheating in oven is not harmful so uncleared specimens may be returned after time given in step 2.

9.1.25 *Maltais' technic [25]*

Application:
1. Recommended for whole mounts of winged and wingless aphids.
2. Recommended for whole mounts of other soft-bodied insects.

Formula:
Solution A: Fixative and Preservative

Sugar	10 pt
Formaldehyde	5 pt
Glacial Acetic Acid	2 pt
Water	83 pt
Vatsol OT (wetting agent)	5 drops

Solution B: Carbo-xylol

Phenol crystals	1 pt
Xylol	3 pt

Procedure:
1. Place field collected specimens in solution A.
2. Transfer to ethanol (70%) for 15 min.
3. Transfer to ethanol (95%) for 10 min.
4. Transfer to absolute alcohol for 5 min.
5. Transfer to solution B for at least 12 hr.
6. Transfer to dilute mixture of balsam and solution B.
7. Mount in balsam.

Note:
1. Specimens remain in good condition in solution A for several months.
2. Specimens may be transferred from solution A with a camel's hair brush and transferred through solutions by draining liquids with a pipette.
3. The author notes that Vatsol OT is an American Cyanamid product.

9.1.26 *Mathur's technic [26]*

Application:
Recommended for the histological preparation of aphids as well as other minute insects.

Formula:
Solution A: Fixative

Lactic Acid (50%)	60 pt
Glycerol	10 pt
Glacial Acetic Acid	10 pt

Chloroform	5 pt
Chloral hydrate (50% aqueous-wt. to vol.)	15 pt

Solution B: Mountant

Chloral Hydrate	50 pt
Acacia crystals	30 pt
Phenol	10 pt
Lactic Acid (50%)	10 pt
Glycerol	60 pt
Distilled Water	50 pt

Preparation:
1. Mix well and filter through double silk just before use.

Procedure:
1. Place live specimens into solution A for 10 days.
2. Transfer to lactic acid (50%) for 15 min.
3. Place specimen on slide and orient in a little solution B.
4. Dry at 40°C for 4 days.

9.1.27 *Keifer's technic I [27]*

Application:
Recommended for the preparation of aphid whole mounts.

Formula:
Solution A:

Lactic Acid (85%)	20 pt
Glacial Acetic Acid	4 pt
Phenol (melted)	4 pt
Water	2 pt

Procedure:
1. Transfer to solution A at 60–80°C for several hours.
2. Press out body contents gently.
3. Transfer to phenol-xylene solution.
4. Transfer through xylene to balsam.
5. Or, transfer to fuchsin in isopropyl alcohol after step 2.
6. Proceed through series up to balsam.

Note:
1. Iodine crystals can be added to solution A when working with colorless specimens.

9.1.28 *Richards' technic [28]*

Application:
1. Recommended as a relatively quick method for whole mount preparation of aphids and scale insects.
2. Recommended for the preparation of small flies, thrips, lice, and fleas.

Formula:
Solution A: Lactophenol
Lactic acid saturated with phenol crystals.

Solution B: Glacial Acetic Acid

Glacial Acetic Acid	1 pt
Oil of Wintergreen	1 pt

Solution C:

Water-white Oil of Cedar	1 pt
Terpeneol (Oil of Lilac)	1 pt
Oil of Wintergreen	2 pt
Glacial acetic acid	0.5 pt

Procedure:

1. If specimen is alcohol preserved, decant and replace with conc. KOH.
2. Heat in water bath for about 3 to 5 min.
3. For small flies and thrips use 10% KOH.
4. Transfer specimens to solution A until cleared.
5. Place with concentrated KOH until solution A has been replaced.
6. If staining is desired add KOH pellets to a saturated aqueous solution of acid fuchsin until solution is straw yellow in color.
7. Add 6 to 12 drops of this solution to KOH solution used in step 5.
8. Transfer specimens to solution B for 24 hr.
9. Transfer to solution C for 24 hr.
10. Mount in neutral solution of balsam in xylene.
11. Bake slides at 50°C until set.

Note:

1. Aphids may be placed live at time of collection into lactophenol.
2. In step 2, it is not necessary to clear with KOH but merely to soften.
3. Aphids in KOH are ready when distended.
4. Exchange of solution within specimen in step 5 is usually visible.
5. The degree of staining is correlated to the length of time of exposure.
6. In steps 8 and 9 the process may be hastened by heating slowly for 5 to 10 min although this evaporates the oils.
7. Formation of crystals in this process (note 6) can be obviated by adding acetic acid.
8. Balsam may be diluted to the needed consistency by adding oil of lilac.

9.1.29 *Kingsbury-Johannsen technic [29]*

Application:

Recommended for whole mount preparation of scale insects.

Formula:

Solution of Picro-Creosote

Picric Acid	1 pt
Beechwood Creosote	100 pt

Procedure:

1. Place specimen in caustic potash solution (10%) for 5 min.
2. Transfer to glacial acetic acid solution (70–80%).
3. Transfer to distilled water for 5 to 15 min.

4. Transfer through two or more changes of distilled water.
5. Transfer to alcohol (95%) for 10 min.
6. Place specimen in either alcoholic acid fuchsin solution (1/2%) or in gentian violet for 24 hr.
6a. Or, place in picro-creostote solution for 3 to 4 days after dehydration in alcohol.
7. Clear in beechwood creosote.
8. Mount in xylene balsam.

Note:
1. Some specimens require staining for 3 to 4 days to demonstrate specific areas.

9.1.30 *Keifer's technic II [27]*

Application:
Recommended for the preparation of whole mounts of scale insects and mealybugs.

Procedure:
1. Boil insect in lactic acid (85%)-phenol mixture until penetration occurs.
2. Transfer to NaOH.
3. Boil and gently press out body contents.
4. Transfer to phenol and boil until body parts are gone.
5. Transfer to fairly strong fuchsin (in isopropyl alcohol).
6. Transfer to xylene.
7. Destain with isopropyl, dropwise under the microscope.
8. If still dark, drop specimen in acetic acid to destain.
9. Transfer to xylene.
10. Transfer to mounting medium.

Results:
Chitin is light in color and sclerotin is bright magenta.

9.1.31 *Glance's technic [30]*

Application:
Recommended for whole mounts of *Collembola*.

Formula:
Solution A: Hoyer's Fluid

Distilled Water	50 pt
Gum Arabic	30 pt
Chloral Hydrate	200 pt
Glycerine	20 pt

Preparation:
Filter before use.

Solution B: Gage's Acid Fuchsin

Acid Fuchsin	0.5 pt
HCl (10%)	25 pt
Distilled Water	300 pt

Procedure:

I. In general, for light colored specimens without well developed muscle system.
1. Place in acetic acid (20%) for at least 15 min and as long as 24 hr.
2. Mount in solution A.
3. To dry and clear further place in oven at 45°C.

II. Modification A, for opaque muscular specimens, e.g., larger collembolans in the families Isotomidae and Entomobryidae.
1. Place in lactic acid (50%) until specimen begins to clear (about 1 to several hours).
2. Transfer to acetic acid (20%).
3. Mount in solution A.

III. Modification B, for collembolans in the genera *Achorutes*, *Anurida*, and families Onychiurdae, Sminthuridae, and small Isotomidae.
1. Puncture or behead specimen.
2. Place in hot lactic acid for at least an hour or until body contents are removed.
3. Transfer to acetic acid (20%) for at least ½ hr.
4. Transfer to solution B.
5. Transfer to ethanol (95%) for about 5 min.
6. Transfer to clove oil for a few seconds.
7. Mount in damar.

Note:
1. In modification B the volume of lactic acid should be maintained while heating.
2. Specimens left in solution B overnight will not overstain.
3. Puncturing of specimens should be around the second pair of legs.

9.1.32 *Wilkey's technic [31]*

Application:
Recommended for small insect types including *Collembola*, *Protura*, *Thysanura*, *Psocoptera*, *Mallophaga*, *Embioptera*, *Anoplura*, and *Homoptera*.

Formula:
Solution A: Essig's Aphid Fluid

Lactic Acid (95%)	20 pt
Phenol (sat. aqueous solution)	2 pt
Glacial Acetic Acid	4 pt
Distilled Water	1 pt

Solution B: Stain

Acid Fuchsin (2% aqueous solution)	20 drops
Acid Fuchsin (2%)	20 drops
Lignin pink	20 drops

Procedure:
1. Transfer specimen to about 3 pt solution A in closed container and heat (at 120° to 130°F) until clear for 30 to 60 min.
2. Tease out loosened body contents or repeat step 1 until clear.
3. Add three or four drops of solution B.

4. Heat until specimen absorbs stain, for 15 to 30 min.
5. Transfer to tetrahydrofuran.
6. Or, for delicate specimens transfer to absolute alcohol for 15 to 30 min.
7. Transfer to xylene for 10 to 15 min.
8. Tease out remaining body content and stain.
9. Let stand in tetrahydrofuran for 1 to 5 min.
10. Mount in balsam and dry at 100°F for ½ to 1 hr.

Note:
1. Clearing of larger insects will be aided by puncturing carefully.
2. As little tetrahydrofuran as possible should be carried over when the specimen is being transferred.
3. Other stains also are recommended including chlorazol Black E, aniline blue and acid yellow.

9.1.33 *Osorno-Mesa's technic [32]*

Application:
Recommended for preparing permanent mounts of *Phlebotomus* (a genus of sand fly) and other small insects.

Formula:
Solution A: Potassium Hydroxide – Trisodium Phosphate

KOH	20 pt
Trisodium Phosphate	0.4 pt
Distilled Water	100 pt

Solution B: Mercurochrome – formol

Mercurochrome	2 pt
Formalin (40%)	10 pt
Distilled Water	100 pt

Procedure:
1. Wash specimen in water at 50°C.
2. Transfer to sodium hypochlorite (5 to 10%) for 5 min.
3. Remove wings and transfer to ethanol (90%) for a few minutes and then replace with solution B.
4. Transfer specimen to solution A.
5. Place in oven at 50°C for a day or until clear.
6. Wash several times with water at 50°C.
7. Wash with acidified (add acetic acid) water.
8. Wash several times with warm water.
9. Transfer to solution B for at least 2 days.
10. Wash several times with ethanol (70%) then with ethanol (90 to 100%) for 5 min each.
11. Transfer to clove oil: ethanol (100%) mixture (1:1) for 30 min.
12. Let ethanol evaporate and add more clove oil.
13. Dissect abdomen and place in watch glass with Euparal diluted with ethanol 100% until the abdomen stops moving.
14. Mount and position head, thorax, and abdomen in small drops of diluted Euparal on a cover glass.

15. Wash wings (step 3) with ethanol (90%) and mount on same cover glass.
16. Let dry.
17. Invert cover glass on supports (glass chips on a slide) and add undiluted Euparal.

Note:
1. A slender dish is suggested for additions of solutions to specimen to avoid excess handling.
2. Freshly preserved specimens result in better preparations.

9.1.34 *Fairchild and Hertig's technic [33]*

Application:
1. Recommended for making permanent slide mounts of small and delicate insects without shrinkage and distortion.
2. Recommended as particularly useful for the preparation of sand fly whole mounts.
3. Recommended for work under tropical conditions.

Procedure:
1. Treat with KOH (10–20%) by either boiling, heated in solution or in cold solution for several hours.
2. Rinse in water.
3. Remove loose or excess tissue.
4. Place specimen in concentrated (1:1) acid fuchsin:phenol mixture for 20 min.
5. If overstained, destain in phenol with a little KOH.
6. Rinse off the excess stain with pure phenol.
7. Mount in balsam or copal dissolved in phenol.
8. Support cover slip with coverslip chips.

Note:
1. Treatment with KOH is designed to remove all tissue and leave sclerotized structures.
2. It is recommended that wings should be removed, not treated with KOH and placed in stain.
3. It is suggested that the head, abdomen, and thorax be separated before KOH treatment.
4. Specimens stored in alcohol for prolonged periods do not make good mounts and require longer KOH treatment.

9.1.35 *Sinton's technic [9]*

Application:
1. Recommended for permanent preparation and for identification of small insect adults.
2. Recommended for preparation of whole mounts of sand fly adults.

Procedure:
1. Place specimen in a single drop of alcohol (70%).
2. Add aqueous caustic potash solution (10%) for at least 12 hr.
3. Wash in three or four changes of water for 15 to 30 min each.
4. Remove water and add aqueous carbol fuchsin stain solution for 12 hr.
5. Remove half of stain and replace with absolute alcohol for 30 min.

6. Remove solution and replace with alcohol (70%).
7. Dehydrate with absolute alcohol.
8. Clear in xylol or oil of turpentine.
9. Mount in Canada Balsam.

9.1.36 *Batte's technic [34]*

Application:
Recommended for the preparation of ectoparasites to facilitate their identification.

Formula:
Solution A:

	Alcohol (20%)	97 pt
	Ether	3 pt

Solution B:

	Phenol crystals	5 pt
	Xylol	50 pt

Procedure:
1. Place specimen in potassium hydroxide (10%) at room temperature until color disappears.
2. Transfer to two or three changes of water for 12 hr each.
3. Transfer to 40% and 50% alcohol for 1 hr each.
4. Stain in pyrogallic acid (in 60% alcohol) for 6 to 8 hr (if light tan is desired) or 24 hr (if brown color is desired).
5. Wash specimens in two or three changes of isopropyl alcohol for 30 min each, to destain.
6. Transfer through 80, 95, and 100% ethanol for 1 hr each.
7. Transfer to solution B for 2 hr.
8. Transfer through three changes of xylol.
9. Mount in Diaphane or Euparal with proper support for specimen to avoid distortion.

9.1.37 *Rapp's technic [35]*

Application:
1. Recommended as an easy, fast, and simple method of making permanent slide mounts.
2. Recommended for mounts of *Siphonoptera*, *Mallophaga*, and *Anoplura*.

Procedure:
1. Clear specimen in potassium hydroxide (10%) at 50°C (water bath) for 5 hr.
2. Wash in acetic acid (1%).
3. Transfer to alcohol (70%) for 1 hr.
4. Clear in beechwood creosote for 1 hr.
5. Transfer to xylol.
6. Mount in synthetic resin.

Note:
1. Alcohol (70%) is suggested for storage of specimens.
2. Sodium hydroxide may be substituted in step 1.

9.1.38 *Benton's technic [36]*

Application:
Recommended as a rapid method for the preparation of slide mounted fleas.

Procedure:
1. Place specimen in KOH (10%) for 6 to 12 hr (light-colored specimens) or 24 hr. (dark specimens).
2. Briefly rinse in water.
3. Transfer to cellosolve (ethyl cellosolve) for about 24 hr.
4. Mount in Canada Balsam.

Note:
1. Specimens may be left in cellosolve indefinitely or until ready to mount.

9.1.39 *Jordan's technic [9]*

Application:
Recommended for identification, examination and permanent preparation of fleas.

Procedure:
1. Boil specimen in sodium or potassium hydroxide solution (10%) for 10 min.
2. Place in cold water for 1 hr.
3. Place in distilled water acidified with drop of glacial acetic acid for 30 min.
4. Rinse in water.
5. Dehydrate through graded series of alcohol up to absolute alcohol.
6. Clear in xylol.
7. Mount in Canada Balsam.

Note:
1. An alternative step 1 can be to place specimen in cold solution overnight.

9.1.40 *Colless' technic II [37]*

Application:
Recommended for small arthropods, including small *Diptera*, *Collembola*, fleas, ticks, and mites.

Formula:
Solution A:

Chloral Hydrate	80 pt
Acetic Acid	20 pt

Procedure:
1. Transfer live (or fresh) specimen to solution A for 2 hr until clear.
2. Wash in alcohol (70–90%) for 10 min.
3. Transfer to liquid phenol.
4. Transfer to drop of phenol on slide (about a 0.3 to 0.5 pt drop).
5. Orient specimen.
6. Add about 0.02 pt of powdered balsam to drop.
7. Dry in oven at 110°F overnight.
8. Add a little xylol-balsam and cover with coverslip.

Note:

1. Step one and two are only suggested for live or freshly killed specimens.
2. Step one and two will result in very transparent mounts.
3. For preserved specimens or where transparency is not desired, kill in warm water and start with step 2 (opening made with fine needles will provide better penetration.
4. The amount of balsam required for step 6 is dependent on desired consistency.

References

1. Gabaldon A. A method for mounting *Anopheline* eggs. *J. Parasitol.* **25**, 281 (1939).
2. Burton G.J. Some techniques for mounting mosquito eggs, larvae, pupae and adults on slides. *Mosq. News* **13**, 7–15 (1953).
3. Craig G.B. Jr. Preparation of the chorion of eggs of *aedine* mosquitoes for microscopy. *Mosq. News* **15**, 228–231 (1955).
4. Gray P. *The Microtomist's Formulary and Guide*. The Blakiston Company, Inc., New York (1954).
5. Lee A.B. *Microtomist's Vade-Mecum*. The Blakiston Company, Philadelphia (1950).
6. Jennings D.T. & Addy N.D. A staining technique for Jack-Pine budworm egg masses. *J. Econ. Entomol.* **61**, 1766 (1968).
7. Gifford J.R. & Trahan G.B. Staining technique for eggs of rice water weevils oviposited intracellularly in the tissue of the leaf sheaths of rice. *J. Econ. Entomol.* **62**, 740–741 (1969).
8. Peacock H.A. *Elementary Technique*. Edward Arnold Ltd., London (1966).
9. Lane J. *The Preservation and Mounting of Insects of Medical Importance*. World Health Organization Vector Control (1965).
10. Foote R.H. A method of making whole mounts of mosquito larvae for special study. *J. Parasitol.* **38**, 494–495 (1952).
11. Gater B.A.R. An improved method of mounting mosquito larvae. *Bull. Entomol. Res.* **19**, 367–368 (1929).
12. Gibbins E.G. A simple method of making permanent microscope mounts of mosquito larvae. *Bull. Entomol. Res.* **21**, 429–430 (1930).
13. Middlekauff W.W. A rapid method for making permanent mounts of mosquito larvae. *Science* **99**, 206 (1944).
14. Colless D.H. An improved method for mounting mosquito larvae. *Nature* **166**, 486–487 (1950).
15. Burton G.J. Rapid permanent mounts of mosquito larvae with creosote-alcohol, phenol-alcohol, lactophenol and polyvinyl alcohol. *Mosq. News* **14**, 72–75 (1954).
16. Burton G.J. A revised procedure for making rapid permanent mounts of mosquito larvae. *Mosq. News* **15**, 147 (1955).
17. Rapp W.F. & Jones D.S. A rapid technique for mounting mosquito larvae. *Entomol. Gaz.* **12**, 120 (1961).
18. Steck G.J. & Wharton R.A. Descriptions of immature stages of Eutreta (Diptera, Tephritidae). *J. Kansas Entomol. Soc.* **59**, 296–302 (1986).
19. Frias D. Morphology of immature stages in the neotropical *Nonfrugivorous Tephritinae* fruit fly species *Rachiptera limbata Bigot* (Diptera: Tephritidae) on *Baccharis linearis* (R. et Pav.) (Asteraceae). *Neotropical Entomol.* **37**, 536–545 (2008).
20. Puri V.D. Preparation of whole mounts of larvae of sugarcane moth borers to study their tracheal system. *Curr. Sci.* **23**, 21 (1954).
21. Geyer-Duszynska I. A quick cytological method for mounting embryos of some insects (Diptera). *Zool. Poloniae* **7**, 411–423 (1956).
22. Steedman H.F. *Section Cutting in Microscopy*. Charles C. Thomas Publishing, Springfield, Illinois (1960).

23. Beno M., Liszekova D. & Farkas R. Processing of soft pupae and uneclosed pharate adults of *Drosophila* for scanning electron microscopy. *Microsc. Res Tech.* **70**, 1022–1027 (2007).
24. Essig E.O. Mounting aphids and other small insects on microscopic slides. *Pan Pacific Entomol.* **24**, 9–22 (1948).
25. Maltais J.B. A simple method of mounting aphids on microscope slides. *Canad. Entomol.* **77**, 103–104 (1945).
26. Mathur L.M.L. An improved technique for mounting aphids. *Science and Culture* **36**, 238 (1970).
27. Keifer H.H. Isopropyl alcohol and phenol used in entomological microtechnique. *J. Econ. Entomol.* **39**, 665–666 (1946).
28. Richards W.R. A short method for making balsam mounts of aphids and scale insects. *Canad. Entomol.* **96**, 963–966 (1964).
29. Kingsbury B.F. & Johannsen O.A. *Histological Technique*. John Wiley & Sons, Inc., New York (1927).
30. Glance G. Slide mounting of *Collembola*. *Ann. Entomol. Soc. Amer.* **49**, 132–133 (1956).
31. Wilkey R.F. A simplified technique for clearing, staining, and permanently mounting small arthropods. *Ann. Entomol. Soc. Amer.* **55**, 606 (1962).
32. Osorno E., deOsorno F. & Alarcon A.M. A technique for permanent mounts of *Phlebotomus* applicable to other small insects. *J. Med. Entomol.* **3**, 124–126 (1966).
33. Fairchild G.B. & Hertig M. An improved method for mounting small insects. *Science* **108**, 20–21 (1948).
34. Batte E.G. A technique for mounting ectoparasites. *J. Econ. Entomol.* **41**, 523–524 (1948).
35. Rapp W.F. A technique for mounting Siphonaptera, Mallophaga and Anoplura. *Entomol. Gaz.* **11**, 6 (1960).
36. Benton A.H. A modified technique for preparing whole mounts of Siphonaptera. *J. Parasitol.* **41**, 322–323 (1955).
37. Colless D.H. An improved technique for permanent mounts of small insects and nematodes. *Bull. Entomol. Res.* **49**, 45–47 (1958).

10 Preparation of whole mounts for staining

10.1 Introduction

The following are procedures that have been recommended for the preparation of insects for whole mount staining. Although most are designed for a specific insect group, many are also applicable to insects with similar characteristics.

10.1.1 *Exposed whole mount technic [1]*

Procedure:
1. Freeze fresh tissue or insect in liquid nitrogen.
2. Cut sections sagittally on cryostat to varying depths by counting the number of 50 μm sections.
3. Once area of interest has been reached, immerse tissue in 4% paraformaldehyde at room temperature to dissolve any embedding medium.
4. Transfer to fresh 4% paraformaldehyde.
5. After fixation, proceed with staining methods required.

10.1.2 *Karr and Albert's embyro technic [2]*

Application:
Recommended for embryos for removal of chorion and vitelline membranes to allow for adequate penetration of fixative, nucleic acids, and immunoglobulins.

Formula:
Solution A:

Commercial Bleach	1 pt
Distilled Water	1 pt

Solution B:
PEM Buffer:

1 M Pipes	1 pt
10 mM $MgCl_2$	1 pt
10 mM EGTA	1 pt
Distilled Water	7 pt
pH adjusted to 6.9 with KOH	

Solution C:

0.5 M Taxol	0.01 pt
Heptane	5 pt

Insect Histology: Practical Laboratory Techniques, First Edition.
Pedro Barbosa, Deborah L. Berry and Christina S. Kary.

Companion Website: www.wiley.com/go/barbosa/insecthistology

Solution D:

	20% Formaldehyde	1 pt

Solution E:

	Heptane	1 pt
	90% Methanol with 0.05M Na_3EGTA	1 pt

Procedure:

1. Remove chorion in solution A for 1.5 min.
2. Rinse and remove bleach in NaCl/Triton solution.
3. Add 5 ml solution B, then add 5 ml solution C and shake vigorously for 0.5 min.
4. Add solution D and shake for 10 min.
5. Rinse embryos in phosphate buffered saline, transfer to ice cold solution E, and shake for 10 min.
6. Rinse with methanol, and rehydrate through to PBS (Phosphate-buffered saline).

10.1.3 *Prpic and Damen spider embryo TUNEL technic [3, 4]*

Application:
Recommended for the detection of apoptotic cells in the spider embryo.

Formula:
PEMS Buffer:

	1 M PIPES	1 pt
	10 mM EDTA	1 pt
	10 mM $MgSO_4$	2 pt
	Distilled Water	6 pt

TdT Buffer:

	1M Sodium cacodylate	140 pt
	1 M Cobalt Chloride	1 pt
	1M Tris-HCl pH 7.2	30 pt
	dH_2O	To 1000 pt

Procedure:

1. Dechorionate embryos in 50% bleach and fix overnight in a mixture of equal parts heptane and 5.5% formaldehyde in PEMS buffer with gentle agitation.
2. Transfer embryos to methanol and remove the vitelline membranes manually using forceps.
3. Rehydrate embryos through graded methanols and wash in PBS with 0.5% Tween (PBTw) with 10 mg/ml BSA and 2% sheep serum (PBTw-block) two times for 5 min.
4. Partially digest the embryos with 5 μg/ml proteinase K in PBTw-block for 4 min at room temperature.
5. Rinse PBTw-block three times and post fix in 1 ml PBT-block with 125 μl formaldehyde for 20 min.
6. Wash with PBTw-block three times for 5 min each.
7. Incubate in 0.1% sodium borohydrate in water for 20 min at room temperature.
8. Wash embryos in TdT Buffer once quickly and once for 5 min.
9. Incubate in TdT buffer with 20 μM dig-UTP and 0.3 U/μl TdT enzyme at 37°C for 2 hr.

10. Wash three times 5 min each in TBS with 0.5% Tween (TBST).
11. Incubate embryos in TBST for 20 min at 70°C.
12. Wash embryos in (PBTw) three times for 5 min each.
13. Block in PBTw-block for 1 hr.
14. Expose to a 1:2000 dilution of anti-DIG-AP (Roche®) antibody in PBT-block for 3 hr at room temperature.
15. Wash in PBSTw with several changes over 2 hr, then overnight at 4°C, then once more for 30 min at room temperature.
16. Develop with appropriate alkaline phosphatase substrate.

10.1.4 *Acridine orange staining [5]*

Application:
Recommended for the detection of cell death in embryos.

Formula:
Dilute acridine orange to 5 µg/ml in 0.1M phosphate buffer.

Procedure:
1. Collect embryos and transfer to tubes containing a mixture of 1 pt heptane and 1 pt acridine orange.
2. Shake tube vigorously by hand for 3–5 min.
3. Remove liquid and add fresh heptane.
4. Transfer embryos to glass slides and soak up extra heptane.
5. Mount with halocarbon oil and a coverslip.

10.1.5 *Alkaline phosphatase technic [6]*

Formula:
Solution A:

75 mg/ml 4-nitroblue tetrazolium chloride (NBT)	4.5 pt
50 mg/ml 5-bromo-4-chloride-3-indolyl-phosphate (BCIP)	3.5 pt
DIG kit detection buffer	1000 pt

Procedure:
1. Dissect tissues and fix in 4% paraformaldehyde for 20 min.
2. Wash in PBS + 0.1% Triton (PBSTr).
3. Immerse in solution A until color develops.
4. Rinse in PBS.

10.1.6 *Ethanolic phosphotungstic acid (E-PTA) method [7]*

Application:
Recommended for the detection of basic proteins.

Procedure:
1. Fix tissues in 2.5% glutaraldehyde in 0.1M cacodylate buffer, pH 7.2 for 24 hr at 4°C.
2. Wash in cacodylate buffer.
3. Dehydrate through graded series of alcohols into 100% ethanol.
4. Treat tissue with 2% phosphotungstic acid in absolute ethanol for 2 hr at room temp.
5. Embed in resin.

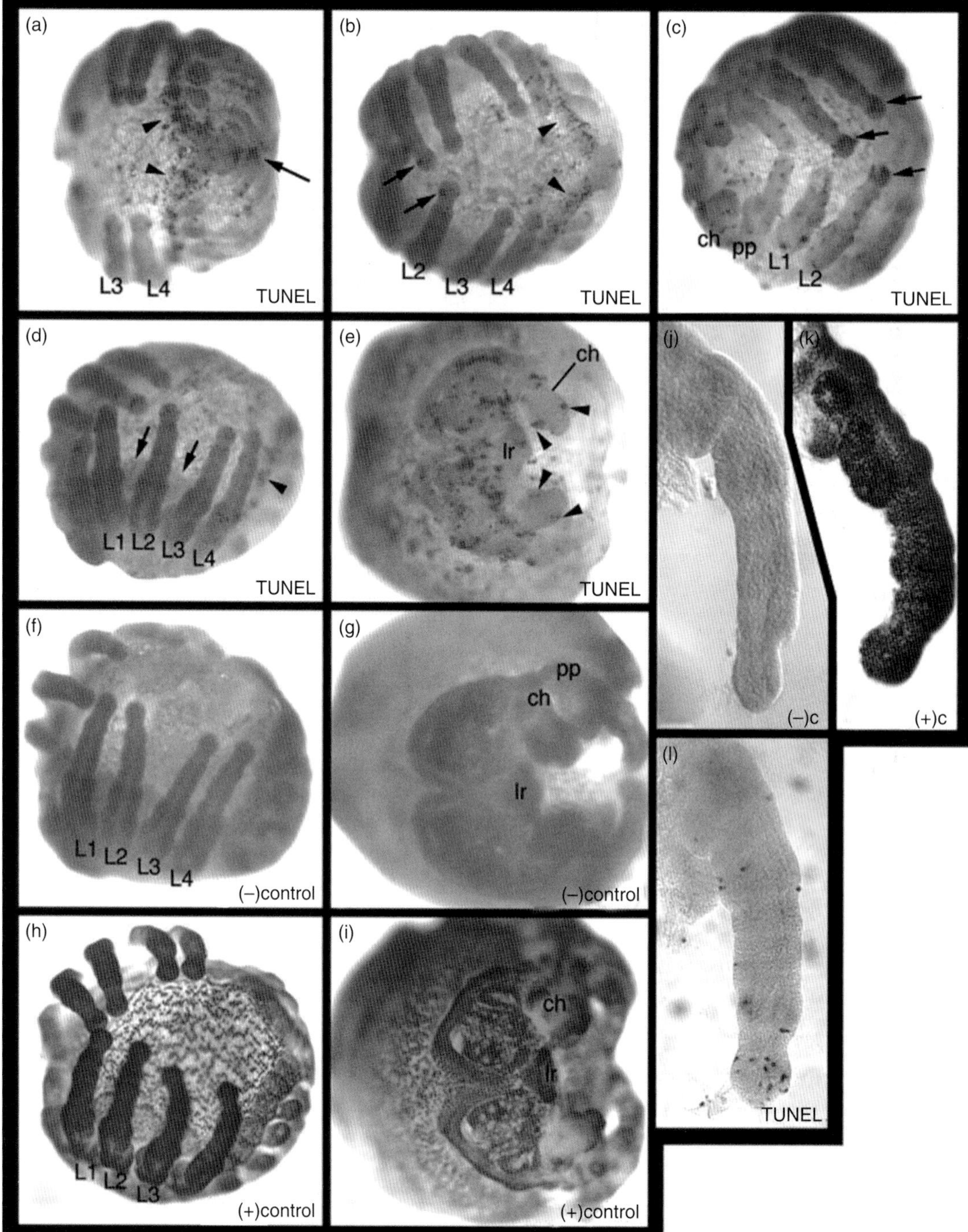

Fig. 10.1 TUNEL labeling of embryos of a species in the spider genus *Cupiennius*. (a–e) Whole mount embryos labeled with TUNEL at approximately 230 hr AEL. (a) View of the opisthosoma. (Source: Prpic *et al.* 2002. Reproduced with permission of John Wiley & Sons.)

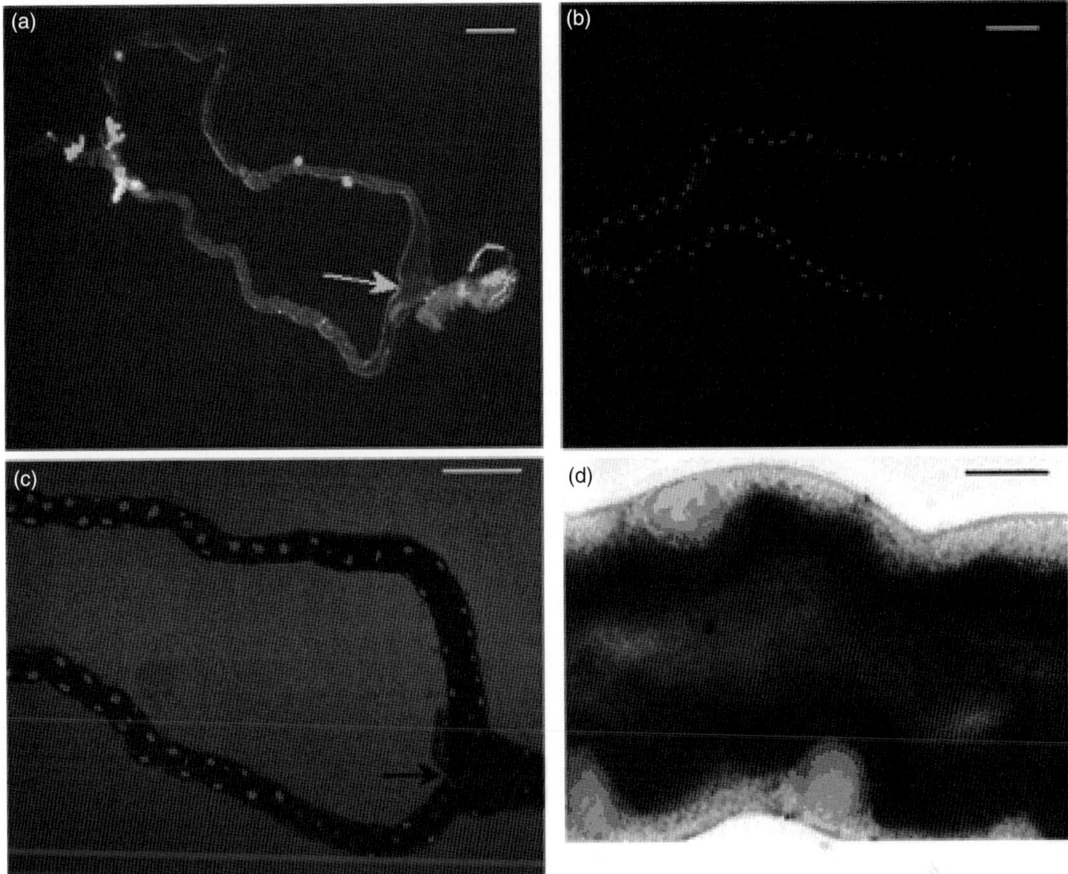

Fig. 10.2 Alkaline phosphatase domain in an intact adult *D. melanogaster* tubule pair: (a) control tubules subjected to alkaline phosphatase staining protocol without chromogenic substrate; the junction with the ureter is shown with an arrow; (b) DAPI-stained; (c) as in B but subjected to alkaline phosphatase staining protocol in the presence of NBT/BCIP chromogenic substrate; the junction with the ureter is shown with an arrow; (d) higher magnification to show predominantly apical localisation of staining. Scale bars, 100 μm for a–c, 10 μm for d. (Source: Cabrero *et al.* 2004. Reproduced with permission of the authors.) See plate section for the color version of the figure.

10.1.7 *Methylene blue [8]*

Procedure:

1. Immerse tissue in methylene blue overnight at 5°C.
2. Wash tissue with insect saline.
3. Post-fix for 30 min with picric acid adjusted to pH 7.0.
4. Rinse in saturated aqueous solution of ammonium molybdate.

10.1.8 *Peroxidase technic [9]*

Procedure:

1. Fix tissues overnight in 1% glutaraldehyde in 0.1M cacodylate buffer, pH 7.3, at room temperature.
2. Stain tissues in 2.5 mM DAB (3,3′-Diaminobenzidine tetrahydrochloride) in 0.1 mM sodium cacodylate buffer, pH 6.5 for 60 min at 37°C.
3. Add 0.03% hydrogen peroxide for 60 min.
4. Dehydrate, clear, and mount.

10.1.9 *Ruthenium red [10]*

Formula:

Solution A:

10% glutaraldehyde	2.5 pt
Ruthenium Red	0.1 pt
0.2M Cacodylate Buffer	5 pt
Distilled Water	2.4 pt

Solution B:

10% Osmium Tetraoxide	1 pt
0.2M Cacodylate Buffer	5 pt
Distilled Water	4 pt

Solution C:

10% Osmium Tetraoxide	1 pt
Ruthenium Red	0.1 pt
0.2M Cacodylate Buffer	5 pt
Distilled Water	3.9 pt

Procedure:

1. Fix tissues for 2 hr in solution A in the dark.
2. Post-fix for 1 hr in solution B.
3. Post-fix and stain for 1 hr in solution C in the dark.
4. Dehydrate in acetone and embed in resin.

10.1.10 *Modified Schafer and Sanchez silver staining technic [11, 12]*

Procedure:

1. Fix tissues in 2.5% glutaraldehyde in 0.1M Cacodylate buffer.
2. Wash in acetone, three times for 10 sec each.
3. Immerse in 0.1M $AgNO_3$ for 10 min.
4. Wash in distilled water.
5. Soak in developer for 10 min.
6. Rinse in 3% acetic acid.
7. Dehydrate, clear and mount in Permount.

10.1.11 *Modified Hansson's technic for carbonic anhydrase [13]*

Application:

Recommended for the detection of carbonic anhydrase in midguts of *Aedes aegypti, Aedes albopictus, Culex quinquefasciatus, Culex, nigripalpus, Ochlerotatus taeniorhynchus, Anopheles albimanus*, and *Anopheles quadrimaculatus*.

Formula:

Solution A:

100 mM $CoSO_4$	1 pt
500 mM H_2SO_4	6 pt
66 mM KH_2PO_4	10 pt

Solution B:

$NaHCO_3$	0.75 g
dH_2O	40 ml

Reaction Mix:

Solution A	17 pt
Solution B	40 pt

Procedure:

1. Cold anesthetize adult mosquitoes (males and females separately) and dissect 15–25 whole midguts in hemolymph substitute saline (see Appendix).
2. Fix overnight in 3% gluteraldehyde in phosphate buffered saline (PBS) at 4°C.
3. Rinse three times in PBS.
4. Incubate in Reaction Mix for 5 min.
5. Rinse in PBS.
6. Incubate in 0.5% $(NH_4)_2S$ for 2 min.
7. Rinse in distilled water.
8. Mount in depression well slides and image.

10.2 Detection of NAPDHd

Histological detection of NAPDHd is used as a marker of nitric oxide synthase (NOS). The enzymatic activity of NAPDHd is sensitive to fixation and that sensitivity varies from species to species within insects. [14] A series of methodologies for detecting NAPDHd activity are presented below with the tested species indicated. Some of the methodologies presented are contradictory and the reader is encouraged to test the procedures empirically.

10.2.1 *Ott and Elphick technic [15–17]*

Application:

Recommended for the detection of NAPDHd activity in the cockroach *Periplaneta americana* and in the locust *Schistocerca gregaria*.

Formula:

HEPES-buffered saline (HBS)

1M NaCl	140 pt
1M KCl	5 pt
1M $CaCl_2$	5 pt
1M $MgCl_2$	1 pt
1M Sucrose	100 pt
1M Glucose	20 pt
1M HEPES	5 pt
pH to 7.2	

Fixative

Formaldehyde	3.7%
Methanol	91–91.5%

Glyercol/Gelatin Mount

Gelatin	5 g
dH_2O	25 ml
Glycerol	35 ml
Sodium azide	0.03 g
Thimerosal	0.01 g

Protocol:

1. Dissect brains from cockroaches or locusts in HBS at room temperature and transfer to ice-cold HBS for temporary storage (up to 3 hr).
2. Fix in ice-cold fixative for 30–60 min on ice.
3. Rinse in several changes of 0.1M Tris pH 7.2 on ice.

4. Cryoprotect the tissue in 20% sucrose in phosphate buffered saline (PBS) with 0.005% sodium azide on ice until the tissues sink.
5. Embed tissue in Tissue-Tek OCT and freeze on the surface of liquid nitrogen.
6. Cut cyrosections at 30 µm, mount on chrome alum/gelatin-coated slides and air dry at room temperature.
7. Rehydrate sections in 0.1M Tris pH 8.0 with 0.2% Triton X-100 (Tris-Tx).
8. Treat sections with cold 0.1M sodium acetate/acetic acid buffer pH 4.0.
9. Wash in Tris-HCl.
10. Stain for NADPHd activity in 0.2 mM nitro blue tetrazolium (NBT) and 0.2–1.0 mM β-NADPH in Tris-Tx between 45 min and 5 hr at room temperature.
11. Wash thoroughly in distilled water, and either coverslip and mount in glycerol-gelatin or dehydrate in a graded ethanol series and embed in Entellan, (Merck®).

10.2.2 *Zayas technic [18]*

Application:
Recommended for the detection of NADPHd in the tobacco hornworm, *Manduca sexta*.

Formula:
Saline

1M NaCl	140 pt
1M KCl	5 pt
1M $CaCl_2$	4 pt
1M Glucose	28 pt
1M HEPES	5 pt
pH to 7.4	

Protocol:
1. Anesthetize larvae in CO_2 or by chilling for 30 min at 4°C.
2. Expose the nervous system by making a cut along the dorsal midline, removing the gut and pinning the larvae in an elastomer-lined dish containing cold saline.
3. Fix for 2 hr in 4% paraformaldehyde in phosphate buffered saline (PBS) at room temperature.
4. Wash two times in PBS.
5. Permeabilize tissues overnight in 0.5% saponin in PBS at 4°C.
6. Rinse two times in PBS.
7. Stain in 50 mM Tris-HCl pH 7.8 containing 0.1% Triton X-100, 0.1 mM *p*-nitroblue tetrazolium (NBT) and 0.1 mM NADPH at room temperature for 90 min.
8. Rinse in PBS and dissect the nerve cords away from the body.
9. Dehydrate cords through an ethanol series, clear in methyl salicylate and mount in Cytoseal™ (Stephans Scientific®) and image.

10.3 **SEM**

10.3.1 *Akiyama-Oda's spider egg technic for SEM [19]*

Application:
Recommended for the preparation of spider eggs for SEM.

Formula:
Fixative

Heptane	5 pt
40% paraformaldehyde	1 pt
10% glutaraldehyde	0.7 pt

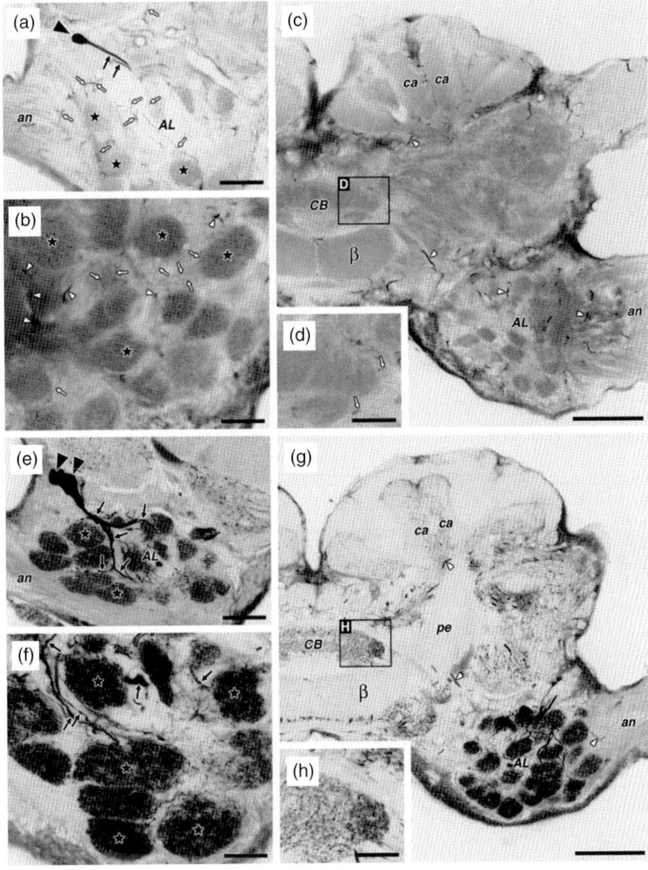

Fig. 10.3 NADPH diaphorase (NADPHd) staining in the brain of a cockroach after fixation in buffered formaldehyde (a–d) and in methanol/formalin (e–h); 30-m horizontal sections; anterior is top. *AL*, antennal lobe; *an*, antennal nerve; mushroom body -lobe; *ca*, mushroom body calyces; *CB*, central body; *pe*, mushroom body pedunculus. (a) After fixation in buffered formaldehyde at 4°C, only two neurons in the antennal lobe (*AL*) are detectable in the entire brain; one of the two cell bodies (black arrowhead) and both primary neurites (black arrows) are seen in this section. The antennal lobe neuropil contains no stained neuronal processes; all fibrous structures are weakly stained tracheae (white arrows). Weak, diffuse staining is present in the glomeruli (black stars). (b) Fixation on ice reveals faint meshworks of fine, beaded fibers against diffuse background in the glomeruli of the antennal lobes (some glomeruli are indicated by black stars). The weakly stained meshwork between the glomeruli is composed entirely of tracheae (white arrows). Tracheae that contain air show up sharp and dark (white arrowheads). (c) Next to no selective neuronal staining occurs in the remainder of the brain (4 hr fixation on ice). Strong staining is seen in certain regions of glia; diffuse staining is widespread in the neuropils. Air-filled tracheae show up strongly (some are indicated by white arrowheads). The frame indicates the region shown at higher magnification in d. (d) Central body and surrounding neuropil; note the absence of selective neuronal staining and the faint staining in tracheae (white arrows). (e) NADPHd staining in the antennal lobe (*AL*) after fixation in methanol/formalin (30 min on ice). Heavy staining occurs in the cell bodies (black arrowhead) and neurites (black arrows) of two antennal lobe interneurons. The glomeruli contain dense meshworks of strongly stained fibers (black stars). Tracheae are unstained. (f) Neurites (black arrows) of the interneurons in (e) innervate the antennal lobe neuropil and contribute to the dense fiber meshworks in the glomeruli (black stars). (g) Methanol-formalin fixation reveals selective fiber staining in many other brain neuropils including the central body (CB). Background staining is considerably lower than after fixation in buffered formaldehyde (cf. c). The frame indicates the region shown at higher magnification in h. (h) Intense NADPHd staining in fibers in the central body after fixation in methanol/formalin; tracheae are unstained. Scale bars: 100 m in a, e, 50 m in b, d, f, and h, 200 m in c, and g. Note: Air-filled tracheae that could be mistaken for axons are indicated by white arrowheads. (Source: Ott *et al.* 2002. Reproduced with permission of John Wiley & Sons.)

2X C&G's Balanced Saline	2.5 pt
Distilled Water	0.8 pt

Procedure:
1. Dechorionate spider eggs with 10% bleach.
2. Fix eggs in fixative solution for 4 hr to overnight at 4°C.
3. Rinse in phosphate buffered saline with 0.1% Tween-20 (PBSTw).
4. Manually remove the vitelline membrane.
5. If the inside of the egg is to be visualized, cut the eggs with a razor blade, remove the yolk and post-fix in 2.5% glutaraldehyde in PBS for 1 hr or longer.

10.3.2 *Koga technic [20]*

Application:
Recommended for the preparation of fragile organs such as the ptilinum and eye for SEM.

Procedure:
1. Fix tissue in 5% glutaraldehyde for 5 min in the microwave at a maximum temperature of 75°C.
2. Incubate in fixative for 24 hr at room temperature.
3. Dehydrate in a graded series of alcohols to 95% ethanol for 24 hr each.
4. Rinse in 100% ethanol three times for 1 hr each.
5. Immerse in the following ratios of t-butyl alcohol to ethanol (2:1, 1:1 and 1:2) for 24 hr each.
6. Immerse in 100% t-butyl alcohol, three times for 24 hr each.
7. Freeze-dry sample and proceed with SEM preparation.

10.3.3 *Rohrbough and Broadie IHC preparation [21]*

Application:
Recommended for the visualization of and accessibility to the internal embryo.

Procedure:
1. Dechorionate embryos in bleach.
2. Rinse embryos in water then in saline.
3. Glue embryos to coverslips, dissect dorsally, and glue flat.
4. Fix for 10–15 min in 4% paraformaldehyde.
5. Proceed with standard IHC techniques.

10.4 *In situ* hybridization

Application:
ISH procedure for the large eggs from spiders.

10.4.1 *Damen* In Situ *Hybridization (ISH) for spider eggs [22]*

Formula
PEMS Buffer

1 M PIPES	1 pt
10 mM EDTA	1 pt
10 mM $MgSO_4$	2 pt
Distilled Water	6 pt

Procedure:
1. Dechorionate embryos in 50% bleach and fix overnight in a mixture of equal parts heptane and 5.5% formaldehyde in PEMS buffer with gentle agitation.

2. Transfer embryos to methanol and remove the vitelline membranes manually using forceps.
3. Store embryos in methanol at −20°C or they can be used immediately.
4. After preparation, embryos can be stained using standard ISH techniques.

10.4.2 *Dearden* In Situ *Hybridization for locusts [23]*

Application:
Recommended for ISH of *Diptera* ovarioles, embryos, and early eggs.

Formula:

PEM-FA

0.2 M PIPES	5 pt
10mM EGTA	2 pt
1M $MgSO_4$	0.01 pt
37% Formaldehyde	1 pt
Distilled Water	1.99 pt

PBSTw

1X Phosphate Buffered Saline (PBS)	100 pt
Tween-20	0.1 pt

Hybridization Buffer

Formamide	5 pt
20X SSC	4 pt
10X Denhardt's Solution	1 pt
Blocking Solution	0.1 pt
100 mg/ml heparin	0.01 pt
100 mg/ml yeast	0.1 pt
Tween-20	0.001 pt
CHAPS	0.001 pt

Procedure:

1. Tissues are fixed in the following way:
   a. Embryos are dissected from eggs and fixed in PEM-FA for 45 min and washed in PBS.
   b. Ovary tissue is dissected from the ovary mass. The intima surrounding each ovariole is removed and the ovariole is fixed in 4% paraformaldehyde in PBS. The ovarioles are washed in PBSTw.
   c. Early eggs are pricked in 4% paraformaldehyde in PBS and fixed for 2 hr. The chorion is peeled from the egg and the eggs washed in PBS.
2. Ovaries and eggs are washed in phosphate buffered saline with 0.1% Tween-20.
3. Ovaries and eggs are treated with 5 µg/ml proteinase K for 1 hr at room temperature then rinsed in PBSTw.
4. Ovaries are then treated with 0.2N HCl for 15 min and rinsed in PBSTw.
5. Pre-hybridize all tissue types in hybridization buffer for 2–6 hr at room temperature and then for 10 min at 65°C.
6. Replace pre-hybridization solution with the probe in hybridization solution.
7. Hybridize overnight at 65°C.
8. Wash in hybridization solution for 10 min, 30 min, and 1.25 hr at 55°C.
9. Wash two times in 2X SSCP with 1% CHAPS Lysis buffer (3-[(3-Cholamidopropyl) dimethylammonio]-1-propanesulfonate).
10. Detect probe using appropriate technique for label used.

10.4.3 *Hughes* In Situ *Hybridization for firebrat (Order Thysanura) embryos [24]*

Formula:

RIPA Detergent mix

1M NaCl	150 pt
NP-40	10 pt
Sodium Deoxycholate	5 pt
SDS	1 pt
0.5M EDTA	2 pt
1M Tris-HCl pH 8.0	20 pt
Distilled Water	Up to 1000 pt

Procedure:

1. Soak embryos in 1:1 heptane/ethanol mix for 15 min.
2. Immerse embryos in RIPA detergent mix for 1 hr.
3. Treat embryos to proteinase digestion for 10 min.
4. Post-fix embryos for 20 min.
5. After preparation, embryos can be stained using standard ISH techniques.

10.4.4 *Liu and Kaufman* In Situ *Hybridization [25]*

Application:

Recommended for the preparation of blastoderm stage embryos.

Formula:

Fixative

Heptane	5 pt
40% paraformaldehyde	3.3 pt
2X Phosphate Buffered Saline with Tween	0.85 pt
Distilled Water	0.85 pt

Procedure:

1. Boil eggs for 1 min.
2. Remove chorion in fixative solution for 20 min with shaking.
3. Repeat shaking in a mixture of 1 pt heptane with 1 pt methanol for 20 min.
4. Fix embryos in 4% paraformaldehyde for 1 hr.
5. Proceed with standard ISH techniques.

10.4.5 *Lobbia's* In Situ *Hybridization for* Lepidoptera *wings [26]*

Formula:

FEA

Formalin	2 pt
45% Ethanol	6 pt
Acetic Acid	1 pt

Procedure:

1. Remove wings just after pupation and rinse in PBS.
2. Fix wings in FEA overnight.
3. Wash in 70% ethanol three times.
4. Slowly dehydrated in graded series of methanol and PBS with 0.1% Tween to 100% methanol.
5. Wings can be stored in methanol at −20°C or used immediately.
6. After preparation, wings can be stained using standard ISH techniques.

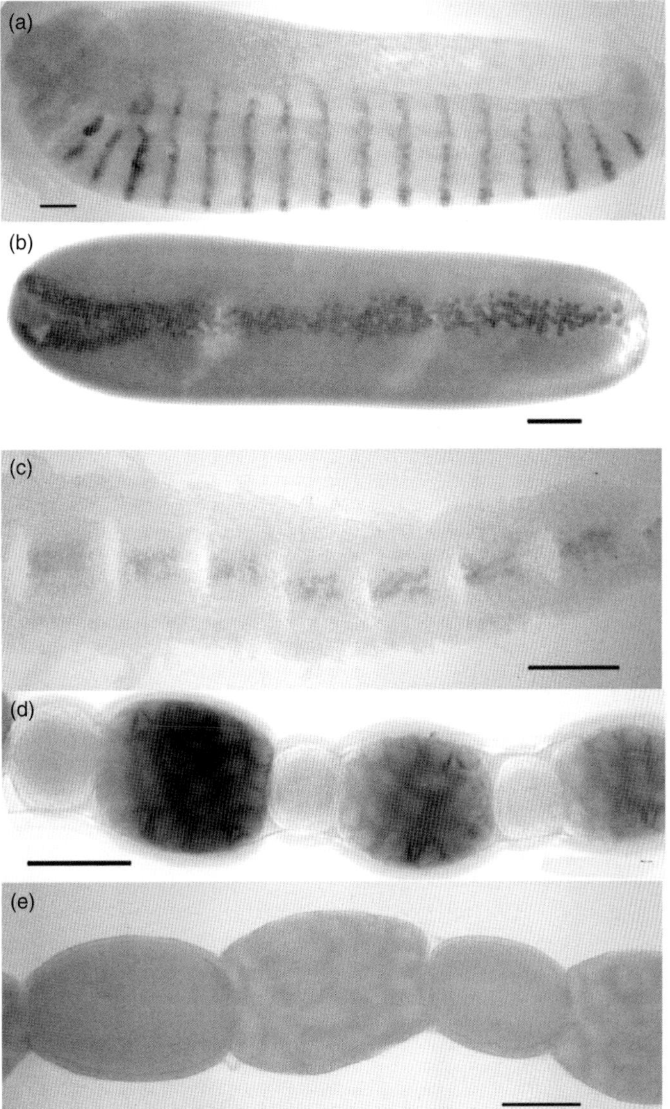

Fig. 10.4 Honeybee embryos (a, b and c) and ovaries (d and e) hybridized with a selection of DIG labelled RNA probes as described. Scale bars represent 100 μm. (a) Late stage embryo hybridized with a probe against e30, a homologue of the *Drosophila* engrailed gene, derived from a genomic clone. *Drosophila* engrailed is expressed in the posterior of each segment and acts to define and maintain the parasegment boundary. Similar to *Drosophila* engrailed, e30 is expressed in a stripe in the posterior of each segment from the gnathum posterior of embryos. Faint domains of expression are also present in the head. (b) Early stage embryo hybridized with a probe for a homologue of the Hox 3 gene, derived from a genomic clone. Hox 3 is expressed in the dorsal stripe, a pattern consistent with the function of zen in *Drosophila*, and Hox 3 in *Tribolium* and *Schistocerca*, where it regulates the formation of the extraembryonic membranes. (c) Dissected late stage central nervous system hybridized with a probe against a homologue of the slit gene derived from a cDNA clone. Slit is a conserved regulator of axon targeting, expressed in the midline glia in *Drosophila*. (d) Part of an ovariole hybridized with a probe for a homologue of the vasa gene, a conserved regulator of germ cell fate, derived from a genomic clone. (e) Sense control for d. (Source: Osborne *et al.* 2005. Reproduced with permission of EDP Sciences.) See plate section for the color version of the figure.

10.4.6 *Osborne and Deardon* In Situ *Hybridization [27]*

Application:
Recommended for the preparation of ovarioles.

Formula:
Fixative

Heptane	5 pt
40% paraformaldehyde	3.3 pt
2X Phosphate Buffered Saline	0.85 pt
Distilled Water	0.85 pt

Procedure:
1. Fix ovarioles in fixative overnight with shaking.
2. Proceed with standard ISH techniques.

References

1. Hall G.M. *et al.* Male Jeffrey pine beetle, *Dendroctonus jeffreyi*, synthesizes the pheromone component frontalin in anterior midgut tissue. *Ins. Biochem. and Mol. Biol.* **32**, 1525–1532 (2002).
2. Karr T.L. & Alberts B.M. Organization of the cytoskeleton in early *Drosophila* embryos. *J. Cell Biol.* **102**, 1494–1509 (1986).
3. Prpic N.M., Schoppmeier M. & Damen W.G. Detection of cell death in spider embryos using TUNEL. *Cold Spring Harbor Protoc.* (2008).
4. Prpic N.M. & Damen W.G. Cell death during germ band inversion, dorsal closure, and nervous system development in the spider *Cupiennius salei*. *Dev Dyn* **234**, 222–228 (2005).
5. McCall K., Peterson J.S. & Pritchett T.L. Detection of cell death in Drosophila. *Methods Mol. Biol.* **559**, 343–356 (2009).
6. Cabrero P., Pollock V.P., Davies S.A. & Dow J.A.T. A conserved domain of alkaline phosphatase expression in the Malpighian tubules of dipteran insects. *J. of Exp. Biol.* **207**, 3299–3305 (2004).
7. Fiorillo B.S., Lino-Neto J. & Bao S.N. Structural and ultrastructural characterization of male reproductive tracts and spermatozoa in fig wasps of the genus *Pegoscapus*. *Micron* **39**, 1271–1280 (2008).
8. Chiang R.G. Functional anatomy of the vagina muscles in the adult western conifer seed bug, *Leptoglossus occidentalis* (Heteroptera: Coreidae), and its implication for the egg laying behaviour in insects. *Arthropod Struct. & Dev.* **39**, 261–267 (2010).
9. Pascoa V. *et al. Aedes aegypti* peritrophic matrix and its interaction with heme during blood digestion. *Ins. Biochem. and Mol. Biol.* **32**, 517–523 (2002).
10. Mancini K. & Dolder H. Ultrastructure of apyrene and eupyrene spermatozoa from the seminal vesicle of *Euptoieta hegesia* (Lepidoptera : Nymphalidae). *Tissue & Cell* **33**, 301–308 (2001).
11. Spiegel C.N., Oliveira S.M.P., Brazil R.P. & Soares M.J. Structure and distribution of sensilla on maxillary palps and labella of *Lutzomyia longipalpis* (Diptera : Psychodidae) sand flies. *Microscope. Res Tech.* **66**, 321–330 (2005).
12. Schafer R. & Sanchez T.V. Nature and development of sex attractant specificity in cockroaches of genus *Periplaneta* .1. Sexual dimorphism in distribution of antennal sense-organs in 5 species. *J. Morph.* **149**, 139–157 (1976).
13. del Pilar Corena M. *et al.* Carbonic anhydrase in the adult mosquito midgut. *J. Exp. Biol.* **208**, 3263–3273 (2005).
14. Ott S.R. & Burrows M. NADPH diaphorase histochemistry in the thoracic ganglia of locusts, crickets, and cockroaches: Species differences and the impact of fixation. *J. Comp. Neurol.* **410**, 387–397 (1999).

15. Ott S.R. & Elphick M.R. Nitric oxide synthase histochemistry in insect nervous systems: Methanol/formalin fixation reveals the neuroarchitecture of formaldehyde-sensitive NADPH diaphorase in the cockroach *Periplaneta americana*. *J. Comp. Neurol.* **448**, 165–185 (2002).
16. Kurylas A.E. *et al.* Localization of nitric oxide synthase in the central complex and surrounding midbrain neuropils of the locust *Schistocerca gregaria*. *J. Comp. Neurol.* **484**, 206–223 (2005).
17. Herbert Z. *et al.* Developmental expression of neuromodulators in the central complex of the grasshopper *Schistocerca gregaria*. *J. Morph.* **271**, 1509–1526 (2010).
18. Zayas R.M., Qazi S., Morton D.B. & Trimmer B.A. Neurons involved in nitric oxide-mediated cGMP signaling in the tobacco hornworm, *Manduca sexta*. *J. Comp. Neurol.* **419**, 422–438 (2000).
19. Akiyama-Oda Y. & Oda H. Early patterning of the spider embryo: a cluster of mesenchymal cells at the cumulus produces Dpp signals received by germ disc epithelial cells. *Dev.* **130**, 1735–1747 (2003).
20. Koga D., Ueno M. & Yamashina S. SEM observation of the insect prepared by microwave irradiation. *J. Electron Micro.* **52**, 477–484 (2003).
21. Rohrbough J. & Broadie K. Anterograde Jelly belly ligand to Alk receptor signaling at developing synapses is regulated by Mind the gap. *Dev.* **137**, 3523–3533 (2010).
22. Damen W.G.M. & Tautz D. A Hox class 3 orthologue from the spider *Cupiennius salei* is expressed in a Hox-gene like fashion. *Dev. Genes and Evol.* **208**, 586–590 (1998).
23. Dearden P. & Akam M. A role for Fringe in segment morphogenesis but not segment formation in the grasshopper, *Schistocerca gregaria*. *Dev. Genes and Evol.* **210**, 329–336 (2000).
24. Hughes C.L., Liu P.Z. & Kaufman T.C. Expression patterns of the rogue Hox genes Hox3/zen and fushi tarazu in the apterygote insect *Thermobia domestica*. *Evol. & Dev.* **6**, 393–401 (2004).
25. Liu P.Z. & Kaufman T.C. Hunchback is required for suppression of abdominal identity, and for proper germband growth and segmentation in the intermediate germband insect *Oncopeltus fasciatus*. *Dev.* **131**, 1515–1527 (2004).
26. Lobbia S., Futahashi R. & Fujiwara H. Modulation of the ecdysteroid-induced cell death by juvenile hormone during pupal wing development of lepidoptera. *Arch. of Insect Biochem. and Physiol.* **65**, 152–163 (2007).
27. Osborne P. & Dearden P.K. Non-radioactive *in-situ* hybridisation to honeybee embryos and ovaries. *Apidologie* **36**, 113–118 (2005).

11 Preparation of genitalia, mouthparts and other body parts

11.1.1 *Peacock's technic [1]*

Application:
Recommended for the preparation of slide mounts of genitalia, mouthparts, and so on.

Procedure:
1. Boil specimen in potassium or sodium hydroxide (10%) for 5 min.
2. Dehydrate through 30, 50, 70, 90, and 100% alcohol for 2 min each.
3. Clear in xylene.
4. Mount.

11.1.2 *Leston's technic [2]*

Application:
Recommended for the preparation of slide mounts of male genitalia, spermathecae, and head capsules.

Procedure:
1. Boil specimen in caustic potash (10%) for 1 to 5 min.
2. Transfer to glacial acetic acid.
3. Remove excess materials and make fine dissections.
4. Add a grain of acid fuchsin to acetic acid, observe staining and repeat as needed.
5. Transfer to creosote for 15 to 20 min.
6. Place 1 to 2 drops of warmed balsam on a slide.
7. Transfer specimen to slide.
8. Apply warmed coverslip.

11.1.3 *Rowell's technic [3]*

Application:
1. Recommended for clearing and mounting of insect exoskeletons.
2. Recommended as particularly useful for the preparation of the head, mouthparts, legs, and whole mounts of bees.

Formula:
Solution A:

HCl (conc.)	3 pt
Alcohol (70%)	10 pt

Insect Histology: Practical Laboratory Techniques, First Edition.
Pedro Barbosa, Deborah L. Berry and Christina S. Kary.

Companion Website: www.wiley.com/go/barbosa/insecthistology

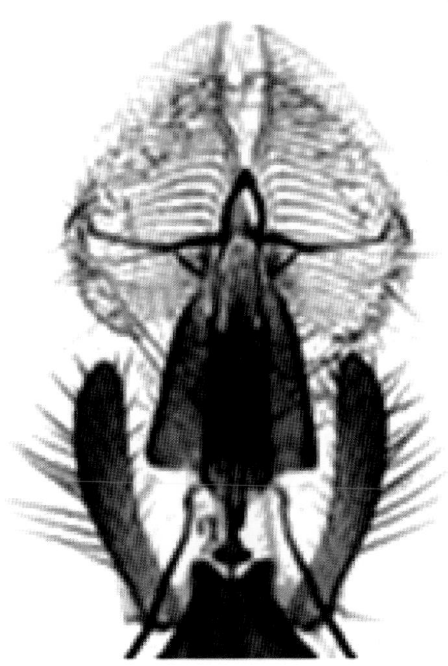

Fig. 11.1 Fly mouthparts. (Source: 3B Scientific.)

Procedure:

1. Remove desired body parts and place each group in small cheesecloth bags.
2. Place specimens in KOH (20%) in covered dish and boil for 15 to 30 min.
3. Wash in running water 12 hr.
4. Bleach by placing specimens in solution A with a few potassium chlorate crystals added.
5. Specimens can be left until light tan in color.
6. Wash in four to five changes of distilled water for 30 min.
7. Arrange body parts as desired on slide.
8. Place second slide on top and hold together with clip or rubber band.
9. Place in alcohol (80%) for 1 hr.
10. Drain and place in alcohol (95%) for 1 hr.
11. Transfer hardened body parts through two changes of absolute alcohol for 1 hr and 2 hr.
12. Clear in clove oil for 24 hr.
13. Mount in balsam.

Note:

1. In step 2, add more solution as evaporation occurs thus maintaining concentration.
2. Periodic pressing of bags at 2 to 3 hr intervals is suggested in step 3.
3. In step 4, when liberation of chlorine stops, more potassium chlorate may be added.
4. In bleaching process it is suggested that a few body parts be processed at a time.
5. Those body parts that do not need to be flat will be carried through steps 7 and 8.

11.1.4 *Bennett's technic [4]*

Application:

Recommended for the dissection and preparation of genitalia of Lepidoptera.

Procedure:

1. Remove the whole abdomen from the specimen carefully.
2. Soak the abdomen in alcohol (25%), to which 2 to 3 drops of a saturated potassium hydrate solution has been added, for 3 to 4 hr or overnight.
3. Heat abdomens in boiling water bath for 1 to 3 min (if soft) or 5 to 6 min (if hard).
4. Transfer to alcohol (50%) when soft enough for dissection.
5. Dissect out genitalia by using gentle pressure on small soft abdomens just below the base of the genitalia.
6. Or, by an incision in the terminal 2 to 3 segments of larger abdomens.
7. Carefully remove armature from segments.
8. Transfer body to alcohol (50%) for 1 hr.
9. Transfer to alcohol (90%) followed by absolute alcohol for 15 min each.
10. Transfer greasy specimens to xylol for several min.
11. Blot body dry and return to remainder of specimen.
12. Arrange genitalia on slide.
13. Mount in a small amount of balsam.

Note:

1. Rearrangement of genitalia when moved by addition of balsam (step 13) may be accomplished by using needles dipped in xylol.
2. After an incision is made (step 7) the connection between armature and body wall may be teased away carefully.

11.1.5 *Depieri technic [5]*

Application:
Appropriate for mouthparts containing chitin.

Procedure:

1. Insect heads are fixed in Bouin's for 24 hr and rinsed in phosphate buffer.
2. Immerse in 7% EDTA in an ice bath and microwave for 0.5 hr.
3. Dehydrate through a graded series of ethanols.
4. Clear in xylene and embed in paraplast.

11.1.6 *Modified Murakami and Imai technic [6]*

Application:
Appropriate for visualization of meiosis in testes.

Procedure:

1. Dissect testes in 0.075M KCl, incubate for 4–5 min.
2. Fix in 1:3 acetic acid/methanol for 30–40 sec.
3. Transfer testes to a drop of 45% acetic acid on glass slide.
4. Squash testis with cover glass.
5. Freeze on carbon dioxide ice to remove cover glass.
6. Immerse briefly in 100% ethanol.
7. Immerse in acetic acid for 25 sec.
8. Air dry slide and incubate overnight at 40°C before staining.

11.1.7 *Modified Yamashiki and Kawamura [7]*

Application:
Preparation of sperm bundles for IHC.

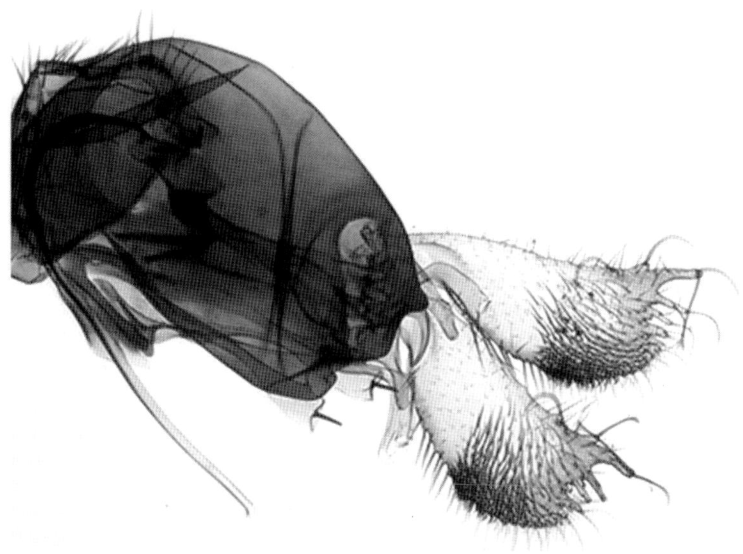

Fig. 11.2 Detail of genitalia. *Dolichopus ungulatus* a fly in the family Dolichopodidae. (Source: David Walker, Micscape. Reproduced with permission.)

Procedure:

1. Smear sperm bundles from excised testes onto a cover glass coated with 3-aminopropyl-triethoxy-silane.
2. Fix in 4% paraformaldehyde in phosphate buffered saline (PBS) for 1 hr.
3. Post-fix in cold acetone for 10 min.
4. Proceed with standard IHC techniques.

11.1.8 *Chawanji technic [8]*

Application:

For preparation of spermatozoa.

Procedure:

1. Prepare glass slide by coating with gelatin, chrome alum and 1 drop glycerol and baking overnight at 60°C.
2. Dissect testes and seminal vesicles from insect and release spermatozoa.
3. Spread spermatozoa on prepared slide and fix in 3:1 methanol/acetic acid for 4 min.
4. Wash in saline for 1 min.
5. Proceed with downstream staining protocols.

11.1.9 *Hanson's Clearing technic [9]*

Application:

Recommended as a simple and quick method for delicate insects or structures.

Procedure:

1. Treat specimen with KOH (5 to 10%) until internal organs are loosened but not completely dissolved.
2. Transfer to alcohol and squeeze lightly for a few seconds or leave in alcohol for 1 to 2 min.
3. Return specimen to hot KOH.

Note:
1. This method is designed to replace the usual method of rolling out internal tissues softened by KOH treatment.
2. Step 3 causes alcohol to boil quickly and carry out loose tissues through a hole in the integument made previously.

11.1.10 *Hazeltine's technic [10]*

Application:
1. Recommended for histological preparation for general morphological studies of external and internal chitinous structures.
2. Recommended as valuable for clearing mosquito larvae, bees, adult terminalia, etc.

Procedure:
1. Heat specimen in nitric acid (10%) until dark areas turn to a brownish color.
2. Transfer to hot hydrogen peroxide (6%) for a short time.
3. Briefly transfer to hot chlorine bleach.
4. Return specimen to heated peroxide until decolorized.

Note:
1. A large queen bumble bee can be cleared in about 10 to 15 min.
2. In this process hairs are completely cleared or lost, membranes remain intact, muscles and other tissues can be removed easily.
3. Even though nervous tissue and muscle remain in place, distortion and yellowing takes place.
4. The temperature and concentrations will vary depending upon the specimens being prepared.

11.1.11 *Clarke's technic [11]*

Application:
Recommended for the preparation of mounts of the genitalia of Lepidoptera.

Formula:
Solution A:
Mercurochrome (0.5%):

Mercurochrome	4.6 pt
Distilled Water	2 oz.

Procedure:
1. Remove the abdomen of the specimen carefully.
2. Transfer to test tubes with KOH (10%) for 2 to 3 hr (for most small Microlepidoptera), 24 hr (for Geometridae, Phalaenidae, and most Rhopalocera (butterflies)) and 48 hr (for Sphingidae, Bombycidae, etc.).
3. Transfer to dish with water.
4. Remove remaining internal tissues and scales.
5. Cut at the eight abdominal intersegmental membrane (if male) or seventh segment (if female).
6. Remove genitalia.
7. To stain add a few drops of solution A in watch glass with water until specimen is faint pink in color.
8. Transfer to alcohol (95%).
9. Spread and arrange genitalia.

10. Transfer to oil of cloves until cleared.
11. Transfer to large watch glass with xylol.
12. Arrange and mount in balsam.

Note:
1. In this procedure, most transfers are to small watch glasses with the appropriate substance.
2. In step 4, excess tissues may be removed with the aid of camel's hair brushes or small syringes.
3. In mounting large genitalia, rings may be used to support coverslip and provide adequate depth.
4. Drifting of genitalia may be prevented by not using thick balsam and by storing (drying) slides for several months in a horizontal position.

11.1.12 *Hardwick's technic [12]*

Application:
Recommended for the preparation of mounts of genitalia of large and medium sized Lepidoptera.

Procedure:
1. Place specimen in a KOH solution (10%) overnight.
2. Place abdomen in dish with alcohol (30%) and dissect out genitalia.
3. Transfer to alcohol (95%) for 20 min.
4. Transfer to clove oil for clearing.
5. Stain in a 1:100 mixture of saturated alcoholic safranin and clove oil for ½ hr.
6. Transfer to xylol.
7. Mount in synthetic mounting medium.
8. To hasten drying, place in oven at 55°C.

Note:
1. Hardwick gives detailed instructions on the dissection of genitalia.
2. Other stains may serve adequately.
3. Small glass chips may be used as supports for thick genitalia.
4. When slides are oven dried, bubbles may form and can be eliminated by filling these with xylol and returning to oven.

11.1.13 *Wickham's technic [13]*

Application:
Recommended for the preparation of slide mounts of Coleoptera body parts.

Procedure:
1. Remove legs, elytra, antennae, labrum, maxilla, and any other desired structures.
2. Place specimens in an aqueous caustic potash solution (15%) until internal tissues are loosened and chitin is fairly clear.
3. Transfer to water and squeeze out remaining internal tissues.
4. Transfer briefly to clean water.
5. Transfer to alcohol (90%) for several minutes.
6. Transfer to absolute alcohol for several hours.
7. Transfer to oil of cloves or a 1:1 mixture of turpentine and molten phenol for several hours.
8. Arrange and mount balsam.

Note:

1. Hind wing may be mounted but steps 1–4 should be omitted and they should be placed in alcohol (step 5).
2. Body parts of the same approximate size should be selected for each mount.
3. Thicker mounts should be supported by small glass chips.

11.1.14 *Marson's technic I [14]*

Application:
Recommended as a general procedure for the preparation of mounts of insect parts.

Procedure:

1. Treat alcohol preserved specimens with caustic soda (10%) and heat.
2. Remove from heat when boiling point of alcohol has been reached (84°C).
3. After 2 to 3 min, reheat until internal brown fluids exit from specimen.
4. Transfer specimen to water.
5. Place specimens on slide, dissect, arrange, place another slide on top and tie with thread.
6. Transfer to glacial acetic acid for about 24 hr.
7. Transfer to xylene for at least 24 hr.
8. Cut threads and place specimens in carbol xylene (xylene saturated with phenol crystals) for 10 min.
9. Transfer to xylene for 10 min.
10. Mount in balsam.
11. Leave slides in cool place for 24 hr.
12. Fill in any spaces that develop with balsam and place in warm (50°C) place until dry.

Note:

1. Specimens used in this procedure should be preserved in alcohol for at least 24 hr.
2. No fixatives containing formalin should be used.
3. When warming caustic soda the temperature should not reach the boiling point.
4. Although not necessary, if ringing is desired, ring with an alcoholic shellac (ringing) medium.

11.1.15 *Marson's technic II [15]*

Application:
Recommended for mounts of honeybee parts including wings, legs, cornea, ocelli, antennae, head, mouthparts, mandible, sting, abdomen, spiracles, and wax plates.

Procedure:

1. Specimen must be preserved in alcohol for at least 24 hr.
2. Kill in chloroform (if specimens are to be used for whole mount of head, undissected mouthparts or mouthparts without mandibles) and then place in alcohol.
3. Proceed as in steps 2 and 3 (in Technic I) for all parts.
4. Wash all parts well in water.
5. Repeat step 5 of Technic I except for wings and cornea in which case it is not necessary.
6. Transfer wings through glacial acetic acid, absolute alcohol and xylene for 10 min each.
7. No treatment is needed for cornea.

8. Remove sting from binding, transfer to alcoholic (70%) acid fuchsin (0.1%) for 1 to 3 min and then bind again for another 24 hr.
9. All other parts are treated as in steps 6 and 7 of Technic I.
10. For all parts except cornea the remaining procedure is the same as in Technic I.
11. Place cornea in 1:1 mixture of glycerine and water for 10 min.
12. Warm glyercine in water bath until melted and place drop on slide.
13. Transfer cornea to slide.
14. Warm cover glass and place on preparation.
15. After it sets, place drop of formalin around preparation for more permanency.
16. Wash off formalin with water and dry.

11.1.16 *Vizzi's technic [16]*

Application:
Recommended for the histological preparation of mosquito mouthparts.

Procedure:
1. Take specimen from alcohol (70%), remove antennae and cut through the middle of thorax.
2. Place specimen in a solution of acetic acid (20%) for 2 hr.
3. Dehydrate through a graded series of 75, 85, 95%, and absolute alcohol for 3 to 5 min each.
4. Place in solution of absolute alcohol-xylol (50:50) for 5 min.
5. Place in xylol for 5 min.
6. Place in xylol-paraffin (50:50) mixture for 5 min.
7. Embed in paraffin and section.
8. Stain with fuchsin.

Note:
1. Acetic acid (step 2) is designed to soften cuticle.

11.1.17 *Lovitt, Okumura, and Nelson technic [17]*

Application:
1. Recommended for the preparation of mounts of genitalia from fresh, dried, and alcohol-preserved specimens.

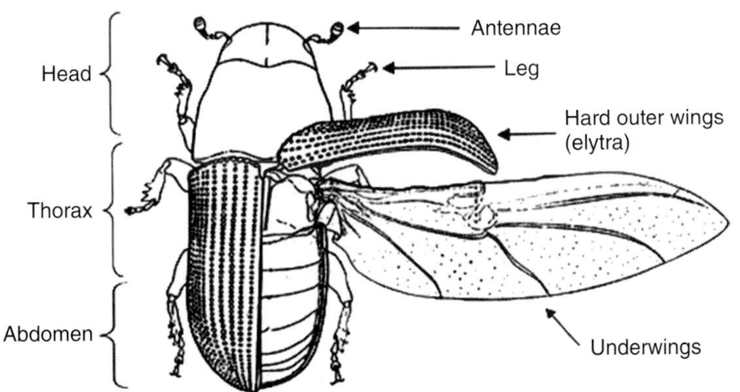

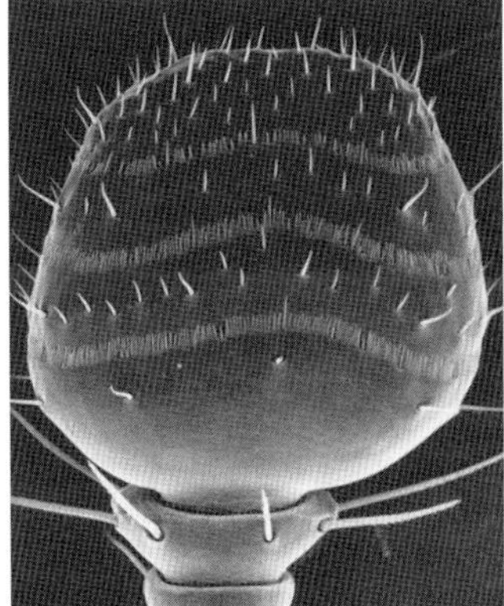

Fig. 11.3 Diagram of the mountain pine beetle and SEM of a representative antenna. (Source: Malcolm Furniss, The Forest History Society. Reproduced with permission.)

2. Recommended for the preparation of slide mounts of female genitalia of *Trogoderma granarium* (Khapra beetle) and seven other *Trogoderma* species.

Formula:

Solution A: Essig's Aphid Fluid

Lactic Acid	20 pt
Carbolic Acid	2 pt
Glacial Acetic Acid	4 pt
Distilled Water	1 pt

Procedure:

1. Fresh Specimens
   a. Remove abdomen and place in 5 pt of potassium hydroxide (10%).
   b. In a covered vessel, heat at 66°C for 15 min on a hot plate.
   c. Transfer to water.
   d. Evert the genitalia by applying gentle pressure.
   e. Split the abdomen along the pleural sclerite.
   f. Transfer to 5 pt of solution A with 3 drops of lignin pink stain (2%) added and heat at 66°C for 1 hr.
   g. Transfer to cellosolve (ethylene glycol monoethyl ester).
   h. Dissect out genitalia from abdomen.
   i. Place in a 1:1 mixture of xylene and Canada balsam for 5 min.
   j. Arrange and mount in medium of 5 pt balsam to 1 pt xylene.
   k. Dry at 32°C for 48 hr.
2. Alcohol-Preserved Specimens
   a. Place abdomen in 5 pt of solution A for 1.5 hr at 66°C.
   b. Remove abdomen and place in 5 pt of KOH (10%) for 45 min at 66°C.
   c. Transfer to absolute alcohol for 5 min.
   d. Transfer to solution A with 3 drops of lignin pink stain (2%) added for 1.5 hr at 66°C.
   e. After 30 min of the 1.5 hr, evert genitalia with dissecting needles.
   f. Transfer to cellosolve and dissect out genitalia.
   g. Repeat steps 9–11 in Procedure I.

Note:

1. The procedure for dried specimens is identical to those for alcohol-preserved specimens except that the time required for step 1 is 1 hr.
2. Bent dissecting needles are suggested for manipulations required in procedures.

11.1.18 *Bronskill's technic [18]*

Application:

1. Recommended for the preparation of slide mounts of the sex pheromone gland of female Lepidoptera.
2. Recommended as particularly applicable for comparative studies of several species of subfamily Tortricinae.

Formula:

Solution A: Carnoy's A

Absolute Ethanol	6 pt
Chloroform	3 pt
Glacial Acetic Acid	1 pt

Procedure:

1. Extrude genitalia by grasping anal papillae with fine forceps and pulling outwards.
2. Flood specimen with solution A.
3. When fixed enough to remain extended (30 sec) remove abdomen from remainder of specimen.
4. Prick with minuten pin (two to three times) and place in fresh solution A for 8 to 12 hr under vacuo (15 in. Hg.).
5. Hydrate through ethanol series (i.e., 95, 80, and 70%).
6. Cut caudal tip off from rest of abdomen (between 5th and 6th segments).
7. Transfer to alcohol (70%).
8. Hydrate through ethanol series (50 and 30%) down to distilled water.
9. Transfer to a 1:1 methanol-chloroform solution at 37°C overnight.
10. Rinse in cold HCl (1 N) for 1 min.
11. Transfer to HCl (1 N) at 60°C for 8 min.
12. Rinse in cold HCl (1 N) for 1 min.
13. Stain in Schiffs reagent for 1 hr.
14. Transfer through three changes of acid sulfite for 10 min each.
15. Dehydrate through ethanol series up to absolute ethanol for 15 min, each in vacuo (15 in. Hg).
16. Clean in benzene.
17. Rinse in xylene.
18. Mount in synthetic resin.

Note:

1. For a better preparation, hairs and scales should be scraped off the genital region at step 4.

11.1.19 *Arnett's technic [19]*

Application:

1. Recommended for the preparation of slide mounts of the male genitalia of beetles.
2. Recommended for the preparation of slide mounts using dry specimens.

Formula:

Solution A: Gage's Stain (Stock)

Acid Fuchsin	0.5 pt
HCl (10%)	25 pt
Distilled Water	300 pt

Solution B: Gage's Working Stain

Solution A	1 drop
Distilled Water	5 drop

Solution C: Acetic Alcohol

Ethanol (50%)	3 pt
Acetic Acid	1 pt

Solution D: Carbo-xylene

Carbolic Acid	1 pt
Xylene	1 pt

Procedure:
1. Place dry specimen in hot (not boiling) water until relaxed.
2. With fine insect pin, fine scissors or forceps, extract genitalia.
3. Transfer to clean water.
4. Treat with KOH (10%) and heat in covered vessel.
5. Transfer to solution C.
6. Transfer to solution B in covered vessel for 12 hr.
7. Transfer to water for 5 min.
8. Transfer to alcohol (95%) for 5 min except if specimen is very delicate.
9. Transfer to solution D for 5 min.
10. Transfer to xylene for 5 min.
11. Mount in balsam.

Note:
1. The time suggested for steps 6–10 are generally applicable, but for these steps and others, details are a function of the particular specimen.

11.1.20 *Jefferson, Rubin, McFarland, and Shorey technic I [20]*

Application:
Recommended for the preparation of mounts of antennae of moths (e.g., *Trichoplusia ni*).

Procedure:
1. Fix whole moths in Bouin's fluid or picroformol.
2. Wash in water.
3. Immerse whole insect in an aqueous crystal violet solution (0.5%) for 1 to 2 hr.
4. Rinse in distilled water.
5. Remove the antennae and dry on absorbent paper under heat lamp.
6. Transfer to xylene for 1 hr.
7. Mount in Permount.

11.1.21 *Jefferson, Rubin, McFarland, and Shorey technic II [20]*

Application:
Recommended for the preparation of mounts of antennae.

Formula:
Solution A: Fixative

Lactic Acid	1 pt
Ethanol (70%)	1 pt

Procedure:
1. Remove antennae, fix and clear in solution A for 1 hr in an oven at 61°C.
2. Let stand in solution A at room temperature for about 1 week.
3. Mount in C-M medium (see Chapter 4).
4. Ring with Zut's compound or euparal.

References

1. Peacock H.A. *Elementary Technique*. Edward Arnold Ltd., London (1966).
2. Leston D. A simple method for making stained mounts. *Entomologist* 86, 254 (1953).
3. Rowell L.S. Preparation of bee slides. *Science* 73, 320 (1931).

4. Bennett,N. The dissection and preparation of the genitalia of Lepidoptera. *Entomologist* **62**, 220–248 (1929).
5. Depieri R.A., Andrade C.G.T.J. & Panizzi A.R. A new methodology for processing mouthparts of hemipterans (Hemiptera) for microscopic studies. *Micron* **41**, 169–171 (2010).
6. Nokkala S. & Nokkala C. Achiasmatic male meiosis in 2 species of *Saldula* (Saldidae, Hemiptera). *Hereditas* **99**, 131–134 (1983).
7. Sahara K. & Kawamura N. Roles of actin networks in peristaltic squeezing of sperm bundles in *Bombyx mori*. *J. Morph.* **259**, 1–6 (2004).
8. Chawanji A.S., Hodgson A. & Villet M.H. Sperm morphology in four species of African platypleurine cicadas (Hemiptera : Cicadomorpha : Cicadidae). *Tissue & Cell* **37**, 257–267 (2005).
9. Hanson J.F. A simple technique for improving and accelerating KOH clearing of insects. *Bull. Brooklyn Entomol. Sec.* **49**, 21 (1954).
10. Hazeltine W.E. A new insect clearing technique. *J. Kansas Entomol. Soc.* **35**, 165–166 (1962).
11. Clarke J.F.G. The preparation of slides of the genitalia of Lepidoptera. *Bull. Brooklyn Entomol. Soc.* **36**, 149–154 (1941).
12. Hardwick D.F. Preparation of slide mounts of lepidopterous genitalia. *Canad. Entomol.* **82**, 231–235 (1950).
13. Wickham H.F. The preparation of beetles for the microscope. *Canad. Entomol.* **41**, 1–5 (1909).
14. Marson J.E. The preparation of microslides of the honeybee. Part I: General methods for all insects. *Bee World* **33**, 183–188 (1952).
15. Marson J.E. The preparation of microslides of the honeybee. Part II: Dissection and mounting of parts. *Bee World* **34**, 4–7 (1953).
16. Vizzi F.F. The mouthparts of the male mosquito *Anopheles quadrimaculatus*. *Say. Ann. Entomol. Soc. Amer.* **46**, 496–504 (1953).
17. Lovitt A.E., Okumra G.T. & Nelson H.D. Techniques for preparing slide mounts of female genitalia of the Khapra beetle *Trogoderma granarium* and related species. *Ann of the Entomol. Soc. of Amer.* **61**, 1623–1624 (1968).
18. Bronskill J.F. Permanent whole-mount preparation of lepidopterous genitalia for complete visibility of the female sex pheromone gland. *Ann. Entomol. Soc. Amer.* 898–900 (1970).
19. Arnett R.H. A technique for staining, dissecting and mounting the male genitalia of beetles. *Coleop. Bull.* **1**, 63–66 (1947).
20. Jefferson R.N., Rubin R.E., McFarland S.U. & Shorey H. Sex pheromones of noctuid moths XXII. The external morphology of the antennae of *Trichoplusia ni, Heliothis zea, Prodenia ornithogalli*, and *Spodoptera exigua*. *Ann. Entomol. Soc. Amer* **63**, 1227–1238 (1970).

12 Preparation of chromosomes

12.1.1 *Moorman's smear stain [1]*

Application:

1. Recommended as a modified Heidenhain's iron-hematoxylin technique.
2. Recommended for use in staining smears of *Drosophila* larval polytene chromosomes and neuroblast cells.

Procedure:

1. Dissect out salivary glands and ganglion in Ringer's solution.
2. Transfer to a drop of acetic acid (45%) on a (number 2) coverslip for 3 to 4 min.
3. Invert coverslip on an albumen coated slide.
4. Crush specimens by applying pressure.
5. Blot out excess liquid.
6. Place slip on edge in staining dish with absolute methyl (AM) alcohol for a minimum of 12 hr.
7. Place slide in a coplin jar, submerged in AM alcohol for a minimum of 12 to 24 hr.
8. Coverslip should fall off, if not, pry off with fine needle or insert pin.
9. Transfer slide to coplin jar with ferric ammonium sulfate in alcohol for minimum of 12 hr.
10. Transfer to coplin jar with a few crystals of hematoxylin (dissolved) for at least 12 hr.
11. Rinse slide in AM alcohol and place in saturated alcoholic solution of picric acid.
12. Destain until light color, usually 30 min.
13. Rinse in AM alcohol and place in saturated alcoholic solution of lithium carbonate for 5 to 10 min.
14. If counterstaining is desired, place in an alcoholic solution of eosin Y for a few seconds.
15. Mount in green euparal. (Euparal can be colored green by adding some copper salt).

Note:

1. In step 3, albumin smeared slide should be allowed to dry for several days.
2. In step 6, level of alcohol must be maintained so as to touch lower edge of coverslip.

Insect Histology: Practical Laboratory Techniques, First Edition.
Pedro Barbosa, Deborah L. Berry and Christina S. Kary.

Companion Website: www.wiley.com/go/barbosa/insecthistology

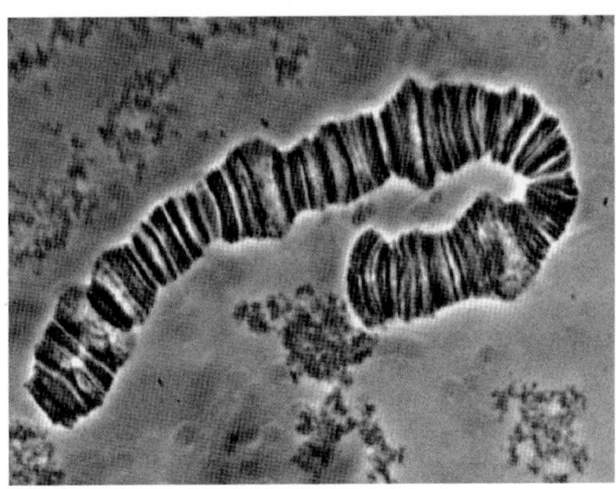

Fig. 12.1 Midge chromosome. (Source: © 1997 University of Massachusetts.)

3. In addition, petroleum jelly should be applied to edges between the lid of dish to prevent evaporation.
4. In step 9, after running first slide through, others should be dipped in alcohol briefly after the ferric ammonium sulfate.

12.1.2 *Frankston and Dambach technic [2]*

Application:
1. Recommended for chromosome preparations of salivary glands.
2. Recommended for *Drosophila* chromosome preparations.

Formula:
Solution A: Belling's Aceto-Carmine

Preparation:
1. Saturate acetic acid (45%) with carmine.
2. Boil for 10 min and filter when cold.

Solution B:

Hematoxylin (0.5%)	1 pt
Iron alum (4%)	1 pt

Solution C: Barrett's Iron-Hematoxylin

Solution A	1 pt
Alcohol (95%)	1 pt
Glacial Acetic Acid	2 pt

Preparation:
1. After solution C is dissolved, add small amount of sodium bicarbonate to hasten ripening.

Procedure:
1. Dissect out salivary glands.
2. Add a few drops of solution A to gland for 5 to 10 min.
3. Place coverslip on gland and make a smear by applying pressure.
4. Examine chromosomes.
5. If adequate, add more stain until coverslip rises up.

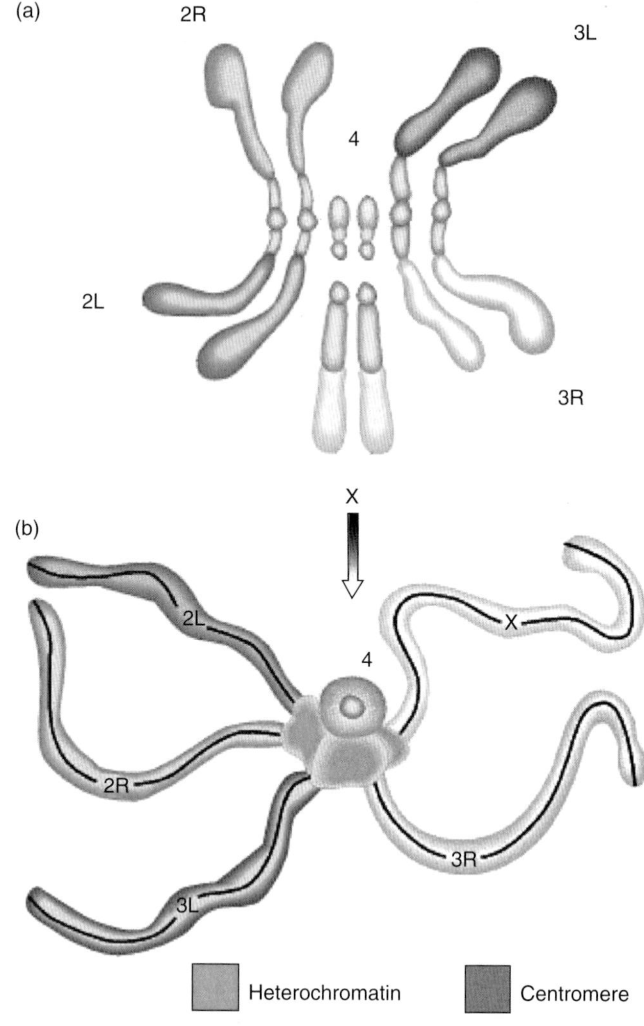

Fig. 12.2 The chromocenter of *Drosophila* salivary gland chromosomes is shown in figure, where L and R stand for arbitrarily assigned left and right arms.

6. Remove coverslip and add a few drops of solution C.
7. Pour off and add a few drops of alcohol (70% followed by 95%).
8. Clear and mount.

12.1.3 *LaCour's technic [3]*

Application:

1. Recommended generally for chromosomal preparations (e.g., of *Drosophila* salivary glands).
2. Recommended for quick chromosome counts.
3. Recommended as a good technic for differentiation of salivary gland chromosome structure.

Formula:

Solution A: Orcein Stain

Orcein	1 pt
Glacial Acetic Acid (45%)	45 pt
Distilled Water	55 pt

Preparation:

1. Dissolve orcein in hot acetic acid.
2. Cool and add water.
3. Shake well and filter.

Procedure:

I. General
   1. Tease tissues in drop of solution A or 0.5% stain (for large chromosomes) for 30 to 60 sec.
   2. Treat coverslip with Mayer's albumen, place over low flame (until gray smoke appears).
   3. Cool and place over tissues.
   4. Apply pressure under a few pieces of filter paper to coverslip.
   5. Place slide over low flame two or three times.
   6. Ring for temporary mounts or prepare permanent mounts.

II. Permanent Mounts
   1. Remove coverslips by transfer to 10% acetic acid for 3 to 10 min.
   2. Transfer through 80% alcohol and absolute alcohol for 2 min each.
   3. Transfer through two changes of cedarwood oil for 5 min each.
   4. Mount in thick oil or balsam and place coverslip on.
   5. Drain excess cedarwood oil.
   6. Dry on hot plate.

III. Salivary Glands (*Drosophila*)
   1. Use modified solution A by making it 2% orcein and increasing acetic acid content to 70%.
   2. Dissect out tissues in Ringer's or solution A.
   3. Let stand in stain for 5 to 10 min.
   4. Apply coverslip.
   5. Apply gentle pressure to flatten.

IV. Salivary Glands (*Sciara*; a genus of fungus gnats)
   1. Dissect out tissues in acetic acid (45%)
   2. Transfer to solution A containing 10% chloroform for 2 to 3 min.

Note:

1. Staining is good after acetic-alcohol fixation.
2. Prefixed materials should be rinsed in acetic acid (45%) before staining.
3. Materials previously stored in alcohol (70%) should be transferred to acetic acid for 1 hr +.
4. Modifications of solution A contain up to 70% acetic acid and from 0.5% to 2% orcein.

12.1.4 *Amirkhanian's technic [4]*

Application:

1. Recommended for use in contrasting type of dipteran spermatogenesis.
2. Recommended for testes, ovaries, and preparations of salivary glands.

Formula:

Solution A: Carnoy's

Glacial Acetic Acid	1 pt
Ethyl Alcohol (96%)	3 pt

Solution B: Battaglia's Hydrolyzing Fluid

	HCl (concentrated 37% gas)	1 pt
	Distilled Water	1 pt
Solution C:		
	Cresyl Violet solution (1%) in 50% acetic acid	
Solution D:		
	NaCl	0.65 pt
	Distilled Water	100 pt
Solution E:		
	Solution A	1 pt
	Distilled Water	19 pt
Solution F:		
	Solution A	1 pt
	Solution B	1 pt

Procedure:

1. Transfer larvae to solution D for 10 to 15 min.
2. Dissect desired organs on slide and drain.
3. Place drop of solution E on material for 30 sec.
4. Drain and place drop of solution F on material at 20–25°C for 2 to 6 min.
5. Rinse in solution A.
6. Apply drop of solution C and a coverslip treated with Mayer's albumen.
7. Differentiate with water by drawing fluids under and across coverslip with filter paper.
8. Gentle pressure with dissecting needle may be applied (prior to squashing) during washing.
9. Temporary preparations can be made by ringing with fingernail lacquer.
10. Mounting in Euparal requires dehydration in 95% and absolute alcohol.
11. Mounting in Canada Balsam requires additional transfer through xylene.

12.1.5 *Marshak's technic [5]*

Application:
Recommended for the preparation of *Drosophila* salivary gland chromosomes.

Formula:
Solution A: Aceto-Carmine

Preparation:

1. Boil saturated solution of carmine in solution of glacial acetic acid (45% aqueous) for several hours.
2. Attach reflux condenser to flask (with solution) to avoid concentrating solution.
3. By sedimentation of filtration a dark red solution is prepared.

Solution B: A saturated solution of carmine in glycerine.

Preparation:

1. Dissolve carmine in glycerine.
2. Filter.

Procedure:

1. Dissect out glands to be used.
2. Stain in well slide for 15 to 20 min.
3. Transfer to slide and wash with solution A.
4. Cover with coverslip and drain with filter paper.
5. Spread chromosomes by pressure applied with dissecting needle.
6. Place solution B around edges of coverslip and a piece of filter paper against one edge.
7. Allow the glycerine to replace all aceto-carmine overnight.
8. Transfer to alcohol (removes glycerine) and blot dry.
9. Seal with balsam or gum-mastic-paraffin.
10. If mounting in balsam is desired allow slides to stand with glycerine for 1 to 3 days.
11. Pry coverslip off or float off in absolute alcohol.
12. Clear in xylol or clove oil and mount.

12.1.6 *Pares' technic [6]*

Application:

1. Recommended for the study of salivary chromosomes.
2. Recommended for clearly differentiating chromosomal bands and structure.

Procedure:

1. Dissect out glands in acetic acid (45%).
2. Transfer to freshly prepared alcoholic hydrochloric acid (keep covered) for 1 to 2 min.
3. Wash material with water (two to three changes).
4. Place large drop of water on specimen and let stand for 5 min.
5. Drain carefully and completely.
6. Add drop of alcohol-soluble nigrosine and cover with coverslip for 10 to 15 min.
7. Squash under a piece of filter paper and place drop of glyercine on edge of coverslip.
8. Drain excess fluid and seal with wax.

12.1.7 *Schultz-Nicoletti technic [7]*

Application:

1. Recommended for salivary gland chromosome squash preparations.
2. Recommended for use with *Drosophila* larvae.

Formula:

Solution A: Shen's

NaCl	9 pt
KCl	0.42 pt
$CaCl_2$	0.25 pt
H_2O	1000 pt

Solution B: Lactic-Acetic Orcein

Powdered Orcein (2% by wt.)	1 pt
Lactic Acid	1 pt

Procedure:

1. Dissect specimen in drop of solution A on a siliconized slide.
2. Transfer organs to a drop of acetic acid (45%) for 1 min (on same slide).
3. Transfer organs to small drop of solution B on center of slide for 10 to 15 min.
4. Place siliconized coverslip over preparation.

5. Hold down with fingers of one hand while tapping several times with small rubber hammer over organs.
6. After folding piece of bulbous paper around slide, press with rolling movement of thumb.
7. Check preparation for effective spreading of chromosomes: if not sufficient, reapply pressure.
8. For permanent slides, smearing should be done by inverting coverslip on an unsiliconized slide.

Note:
1. Small toy hammers may be used for step 5.
2. Coverslip must not be allowed to move in step 6.
3. Siliconed slides may be made by dipping in undiluted G.E. SC-87 Dry-Film Silicone lubricant for several minutes (under a hood). This is followed by washing with hot soapy water.

12.1.8 *Crystal violet staining technic [8]*

Application:
Recommended for staining chromosomal preparations for maximum differentiation.

Procedure:

I. General
1. Stain in a 1% aqueous crystal violet solution (previously boiled, cooled, and filtered) for 3 to 10 min.
2. Rinse in water.
3. Mordant in 1% iodine and 1% KI (in 80% alcohol) solution for 30 to 45 sec.
4. Rinse in alcohol (95%).
5. Wash in absolute alcohol for 2 to 10 sec.
6. Differentiate (microscopically) in clove oil.
7. Transfer through three changes of xylol (to clear) for 5 min in each.
8. Transfer to new xylol for 30 min.
9. Mount in balsam.

II. Modification A
1. Stain in 1% aqueous crystal violet for 10 to 30 min.
2. Rinse in water.
3. Transfer to absolute alcohol for 2 to 3 sec.
4. Transfer to iodine for 1.5 to 2 min.
5. Transfer to absolute alcohol for 2 to 3 sec.
6. Transfer to 1% chromic acid for 15 sec.
7. Transfer to absolute alcohol for 5 sec.
8. Transfer to chromic acid for 15 sec.
9. Transfer to absolute alcohol for 5 to 15 sec.
10. Transfer to clove oil (to differentiate).
11. Clear in xylol and mount.

III. Modification B
1. Mordant in iodine for 10 to 20 min.
2. Rinse in water.
3. Transfer to stain for 5 to 20 min.
4. Rinse in water, iodine, then 95% alcohol.

5. Wash rapidly in absolute alcohol for several seconds.
6. Transfer to clove oil (differentiation).
7. Clear in xylol and mount.

Note:
1. Procedure II (Modification A) is suggested for insects having small chromosomes.
2. Stain is limited by its tendency to fade.

12.1.9 *French and Kitzmiller technic [9]*

Formula:

Solution A: Colchicine

Colchicine	0.1 pt
H_2O (Double Distilled)	100 pt

Solution B: Colchicine

Colchicine	0.1 pt
Distilled H_2O	50 pt
Honey	50 pt

Solution C: Hypotonic Sodium Citrate

$Na_3C_6H_5O_7$ $2H_2O$	1 pt
H_2O (Double Distilled)	100 pt

Solution D: Carnoy's

Glacial Acetic Acid (45%)	1 pt
Lactic Acid (85%)	1 pt

Procedure:
1. Incubate larvae or pupae for 3 to 24 hr in solution A.
2. Or, feed adults on solution B.
3. Dissect specimen in solution C on a silicone treated slide.
4. Fix in solution D and stain for 30 sec to 3 min.
5. Place untreated slide on top of coverslip, turn over, hold coverslip firmly with filter paper and tap gently.
6. Ring for temporary use or use liquid nitrogen for permanent preparations.

12.1.10 *Smith's Aceto-Carmine technic [8]*

Formula:

Solution A: Aceto Carmine Stain

Carmine	0.25 pt
Acetic Acid (45%)	50 pt

Preparation:
1. Add carmine to boiling acetic acid.
2. Simmer for 2 to 3 min.
3. Cool on ice and filter.

Procedure:

I. General
   1. Tease or squash tissues in drop of solution A.
   2. Cover with coverslip.

3. Seal with wax if satisfactory.
4. Store and ripen in refrigerator.
5. When stained as desired, replace stain with acetic acid (45%) by drawing across coverslip with filter paper.

II. Modification A
1. Prefix in alcohol-acetic (3:1) solution.
2. Fix for 10 to 15 min.
3. Mount in aceto-carmine or store in alcohol (70%).

III. Modification B
1. Heating at 60°C whole or after spreading improves the preparation (4 to 5 times).
2. Do not boil liquid.
3. Apply pressure between exposure to flame.

IV. Modification C
1. Add iron (as a ferric salt) to either stain or alcohol-acetic prefixer.

V. Modification D
1. Acetic-orein is sometimes substituted for aceto-carmine.

12.1.11 *Orcein staining of mitotic chromosomes [10]*

Applications:
Recommended for analysis of chromosome morphology, chromosome aberrations and degree of ploidy.

Protocol:
1. Dissect larval brains
   a. Place larvae in a drop of saline and use two pairs of forceps to grab the larval mouth parts and the larval body midway toward the posterior part and pull apart. Remove brain with forceps.
2. Wash in physiological saline for 1–5 min at room temperature.
3. Transfer brains to a petri dish with 2 ml of saline with a drop (~100 µl) of 10^{-3}M colchicine, cover and incubate for 1.5 hr at 25°C.
4. Transfer brains to a drop of hypotonic solution on a siliconized slide for 10 min at room temperature.
5. Fix the brains in 2 ml freshly prepared fixative (11:11:2 acetic acid/methanol/H_2O) until they become transparent (~10–20 sec).
6. Transfer brains individually to small drops (~2 µl) of 2% aceto-orcein for 1–2 min on a coverslip.
7. Lower a non-siliconized slide onto the coverslip.
8. Invert the sandwich and squash between two to three sheets of blotting paper.
9. Seal the edges of the coverslip with either melted depilatory wax or with nail polish.

12.1.12 *Smith's paraffin technic [8]*

Formula:
Solution A: 2BD

Chromic Acid (1%)	100 pt
Potassium Bichromate (1%)	100 pt
Saponin	0.1 pt
Osmic Acid (2%)	30 pt
Acetic Acid (5%)	30 pt

Solution B: B-15

Picric Acid (sat. aqueous solution)	1 pt
Formaldehyde	25 pt
Glacial Acetic Acid	5 pt
Chromic Acid	1.5 pt
Urea	2 pt

Solution C: Crafs

Chromic Acid	1 pt
Glacial Acetic Acid	7 pt
Distilled Water	92 pt
Formaldehyde	30 pt
Distilled Water	70 pt

Preparation:

1. Mix in equal parts just before using

Solution D: Kahle's

Alcohol (95%)	15 pt
Formaldehyde	6 pt
Glacial Acetic Acid	1 pt
Distilled Water	30 pt

Solution E: Modified Kahle's

Replace water with saturated aqueous solution of picric acid.

Procedure:

1. Fix depending on nature of material and stain to be used.
2. For ovaries, fix in B-15 of Picro-Kahle (solution E) before hematoxylin, solution E before Feulgen stain, solution A or C before crystal violet.
3. For testes, solution A and C before crystal violet and solution D before Feulgen.
4. For eggs (and embryos) solution E before Feulgen, hematoxylin or chromic-crystal violet.
5. Fixation times vary, for solutions A, B, and C (4 hr) and for solution D and E (2 hr).
6. Aid fixation by placing in vacuum pump for 3 min.
7. Wash materials from osmic fixatives in cold water (several times).
8. Soak in warm water for 15 min.
9. Transfer Kahle fixed material to alcohol (35%).
10. Transfer to solution B, D, and E fixed material to alcohol (70%).
11. Dehydration can be accomplished using substances such as ethyl alcohol, n-butyl alcohol, dioxane, etc. (see Chapter 3).
12. Infiltrate with paraffin of low melting point for 4 hr.
13. Transfer to pure wax for 2 hr.
14. Embed in wax and section (about 10 microns).

12.1.13 *Smith's smear technic [8]*

Application:

1. Recommended for quick and detailed chromosomal preparation.
2. Recommended as a limited technic that disturbs normal orientation of the cells.

Procedure:

1. Place dissected tissues on right hand side of slide.
2. Place scalpel, sharp edge down, on right side of tissues.
3. Sweep quickly and evenly across to the left end of slide.
4. Immediately transfer slide into fixative.
5. Remove lumps of tissue debris.
6. Fixatives which give good results include solutions A and C (previous Technic) for 1 to 2 hr.
   a. For solution C fixed material, transfer to alcohol (70%).
   b. For solution A fixed material, wash in running water for 15 min.
7. Then transfer through alcohol series (10, 30, 60, and 70) for 1 min each.
8. To destain osmic fixative transfer to water and 80% alcohol (1:2) solution for 20 to 30 min.
9. Transfer through alcohol series to water in preparation for staining.

12.1.14 *Smith's squash technic [8]*

Procedure:

1. Fix in modified Kahle's for 10 min.
2. Transfer to albuminized slide with drops of acetic acid (45%).
3. Cover with coverslip and heat over alcohol flame (4 to 5 times) without boiling.
4. Apply pressure using filter paper to drain excess acid.
5. Between repeated heatings, apply pressure to coverslip, directly.
6. Transfer to alcohol (70%) and pry off coverslip.
7. Let slide remain in alcohol until ready for staining.

12.1.15 *Heidenhain's hematoxylin technic [8]*

Formula:

Solution A: Mordant

Iron-alum crystals (ammonio-ferric sulphate)	4 pt
Distilled Water	100 pt

Solution B: Stain

Hematoxylin Crystals	0.5 pt
Alcohol (95%) small amount: diluted to make 0.5% aqueous solution	

Procedure:

I. General
   1. Mordant in solution A for 12 hr or overnight.
   2. Wash in running water for 5 to 15 min.
   3. Stain in solution B (time equal to step 1).
   4. Rinse in water.
   5. Transfer to 2% solution A (differentiate microscopically).
   6. Wash in water for 15 min.
   7. Dehydrate, clear (xylol) and mount (balsam).

II. Modification A
   1. Mordant in solution A for 30 min.
   2. Wash for 10 to 15 min.
   3. Transfer to 2% hematoxylin for 60 min.
   4. Differentiate in 0.5% iron-alum.

Note:
1. Time for steps in I are a function of the age of solutions.
2. Picric acid (saturated aqueous solution) can be substituted for iron-alum in differentiation (step 5 of I).

12.1.16 *Brazilian technic [8]*

Procedure:
1. Mordant in iron-alum (1% in 70% alcohol) for 24 hr.
2. Wash in alcohol (70%).
3. Soak in alcohol (70%) for 15 min to 3 hr.
4. Stain in 0.5% Brazilian in 70% alcohol (with 1–2 drops/50 pt alcoholic solution of iron-alum).
5. Wash in alcohol (70%) for 2 to 24 hr.
6. Transfer to 1% alcoholic iron-alum and differentiate for 1 min to 3 hr.
7. Wash in alcohol (70%) and differentiate in absolute alcohol.
8. Clear in alcohol-cedar oil and mount.

12.1.17 *Breland's technic I [11]*

Application:
1. Recommended generally for mosquito chromosome preparations.
2. Recommended as particularly useful for the following tissues:
   a. Brains of prepupae.
   b. Malphigian tubules of fourth instars.
   c. Ovaries and testes of newly emerged adults (less than 24 hr old).

Formula:
Modified Carnoy's:

Chloroform	1 pt
Absolute Alcohol	1 pt
Glacial Acetic Acid	1 pt

Procedure:
1. No prefixation is required for tissues (except brain) since aceto-orcein (1% solution) and aceto-carmine (1% solution) both fix and stain.
2. When using brain tissues, fix in modified Carnoy's fluid for 10 min.
3. Remove debris and excess water on slide with filter paper.
4. Apply drop(s) of stain to material (aceto-carmine suggested).
5. Place coverslip on tissues (make sure tissues are in center of coverslip) and let stand for several minutes.
6. Apply light pressure with dissecting needle.
7. Apply heat (alcohol lamp) to slide without allowing stain to boil and add stain when necessary.
8. Examine tissues microscopically to check stain uptake.
9. Place filter paper over coverslip and apply medium pressure with fingers.
10. Apply heat as before and add stain when necessary.
11. Examine microscopically and repeat above steps with more pressure and application of heat until appropriate spread and stain intensity is achieved.

Note:
1. Coverslip should not move during squashing and staining procedures.
2. During all sequences, preparations should not be allowed to dry.

3. All procedures including dissection should be preformed on the same slide.
4. The above procedures will give only temporary slides for study or photography.

12.1.18 *Breland's technic II [12]*

Application:
Recommended for mosquito chromosomal preparations using brains of prepupae.

Formula:
Solution A: Modified Carnoy's Fixative
Same as in Breland's (I).

Solution B: Aceto-Lactic–Orcein Stain

Distilled Water	33 pt
Lactic Acid	33 pt
Glacial Acetic Acid	33 pt
Orcein (synthetic)	2 pt

Preparation:
1. Heat (do not boil) with frequent stirring for 20 min.
2. Filter while hot.

Solution C: Balsam-Paraffin Sealer

Canada Balsam	60%
Paraffin	40%

Solution D: Belar's Saline

Sodium Chloride	6 pt
Potassium Chloride	0.2 pt
Calcium Chloride	0.2 pt
Sodium Bicarbonate	0.2 pt

Preparation:
1. Add distilled water to the above up to 1 liter.

Procedure:
1. Before fine dissection, head is placed in solution A for 45 sec to 1 min.
2. Place head on slide and remove excess fixative.
3. Place drop of solution D on head and dissect out brain.
4. Remove all debris and center brain.
5. Place drop of solution B on brain.
6. Place small metal ring around specimen and coverslip on top to prevent evaporation.
7. Let stand exposed to solution B for 20 to 45 min.
8. A few species require longer exposure of about 1.5 to 3 hr (e.g., *Anopheles* and *Toxorhynichites*).
9. Remove excess stain with filter paper.
10. Place a drop of acetic acid (50%) on brain and remove, with filter paper, excess stain and acetic acid.
11. Place another drop on brain and place a large coverslip (18 x 22 mm) on preparation.

12. Place piece of filter paper over coverslip and apply pressure with thumb.
13. Examine preparation, if not satisfactory add acetic acid to coverslip and repeat step 12.
14. When satisfactory, add acetic acid and seal with solution C using heated metal spatula.

Note:
1. During step 12, care should be taken to not move the coverslip.
2. Properly sealed slides can last from 1 month to 6 weeks.

12.1.19 *Rai's technic [13]*

Formula:
Solution A:

Methanol	6 pt
Chloroform	3 pt
Propionic Acid	2 pt

Procedure:
1. This procedure is basically the same as Breland's II except for several changes as follows.
2. Fix in modified Carnoy's solution (30 sec to 3 min): average time for most species is 30 to 45 sec.
3. An alternative solution (A) is suggested to both fix and preserve.
4. The brain is stained in aceto-lactic orcein for 10 to 30 min.
5. The preparation is warmed slightly and squashed.

Note:
1. If larvae are stored in solution A (in refrigerator) good preparations are attainable for as long as 6 months.

12.1.20 *French, Baker, and Kitzmiller technic [14]*

Application:
Recommended for preparation of mosquito chromosomal mounts.

Formula:
Solution A:

Colchicine Alkaloid	0.1 pt
Double Distilled Water	100 pt

Solution B: Modified Carnoy's

Glacial Acetic Acid (45%)	1 pt
Ethyl Alcohol (95%)	1 pt

Solution C:

Synthetic Orcein	2 pt
Lactic Acid (85%)	50 pt
Glacial Acetic Acid	50 pt

Solution D: Carnoy's

Glacial Acetic Acid	1 pt
Ethyl Alcohol (100%)	3 pt

Solution E:

Lactic Acid (85%)	1 pt
Glacial Acetic Acid (diluted 1:3 with 45% acetic acid)	1 pt

Solution F:

Orcein (2%) diluted to 0.5% with solution E

Procedure:

I. Pretreatment

1. Specimens are placed in solution A. The time varies in relation to the tissue as follows:
   i. Brain – place late fourth instars in solution A for 2 to 4 hr.
   ii. Testes – place male pupae (immediately after pupation) in solution A for 24 hr.
   iii. Ovary – place female pupae (24 hr old) in solution A for 24 hr.

II. Treatment

1. Transfer dissected tissues to drop of solution B on siliconized coverslip for 30 sec.
2. Transfer tissues to drop of solution C on same coverslip for 30 sec.
3. If phase microscopy is used, dilute solution C to 0.5% with lactic acid and glacial acetic acid (equal parts) and stain for 30 sec.
4. Place drop of acetic acid (45%) on tissues.
5. Remove drop of fixative (with filter paper) and place coverslip (non-siliconized) over preparation.
6. Apply light pressure.
7. Ring coverslip with dental wax with temporary mount (several weeks).
8. Permanent mounts can be prepared by following the steps listed below.
9. Dip slide (after above treatment) into liquid nitrogen for 30 sec.
10. Remove top coverslip with razor blade.
11. Transfer tissues to ethyl alcohol (95%) for 5 min.
12. Transfer to ethyl alcohol (100%) for 1 min.
13. Cover tissue with mounting medium.
14. Place coverslip on, after dipping in ethyl alcohol (100%) and drying off excess fluids with filter paper.
15. Place slide on slide warmer, cover with filter paper and place weight on top 50–100 gm of shot (i.e., tiny lead or steel pellets).
16. Let dry and maintain level for 48 hr.

III. Modification A

1. Transfer slide on piece of dry ice (coverslip up) after removal from liquid nitrogen for 30 to 45 sec, thus preventing loss of tissue.
2. To improve spreading and staining, dissection can be performed on siliconized slide on a cold surface.
3. Transfer to small drop of 0.5% stain on siliconized slide and destain with 4 drops of acetic acid (45 pt) and distilled water (55 pt).
4. Place on non-siliconized slide and destain with 4 drops of acetic acid (45 pt) and distilled water (55 pt).

Note:

1. Zeiss-Einschulussmittell L-15 is recommended by authors as a mounting medium (step 13).
2. A suggested cold surface (III-step 2) consists of a frozen wet sponge wrapped in plastic.

IV. Salivary Chromosomes

1. Transfer dissected glands to solution D for 1 to 2 min.
2. Treat (dip) slides and coverslip in G.E. SC-87 Dri-Film siliconization.
3. Place treated coverslip with drop of solution E (during fixation) in center of slide.
4. Transfer glands to stain after fixation for 30 sec to 2 min.
5. Apply very light pressure using non-treated slide over coverslip, then turn over.
6. Ring with wax (temporary) or dip in liquid nitrogen (see above).

Note:

1. Following all procedures a cold surface gives better results.
2. Better staining can be attained by slower penetration of stain.
3. In step 5 (Proc. 4) tapping of coverslip is sufficient pressure for mosquito chromosomes.

12.1.21 *Crozier's technic [15]*

Application:

1. Recommended for use in chromosomal preparations of cerebral ganglia, testes, etc., of *Drosophila* and ants.
2. Recommended for species with small and numerous chromosomes.

Formula:

Solution A: Insect Ringer

NaCl	14 pt/liter
$CaCl_2$	0.4 pt/liter
KCl	0.2 pt/liter
$NaHO_3$	0.2 pt/liter

Solution B: Colcemid-Ringer

Consisting of 0.05% colcemid in solution A

Solution C:

Acetic Acid	1 pt
Methanol	3 pt

Solution D:

Acetic Acid	1 pt
Ethanol	3 pt

Solution E: Aceto-Lactic Orcein

Orcein	1 pt
Lactic Acid (85%)	28 pt
Glacial Acetic Acid	22 pt

Procedure:

I. Pretreatment

1. Enclose larvae (*Drosophila*) in facial tissue envelope with yeast and soak in solution B.
2. Cover and let stand at 20–25°C for 5 hr.
3. For ants, puncture prepupae or early male pupae and transfer to solution B.
4. Cover and let stand at 20–25°C for 15 hr.

II. Treatment

1. Dissect out tissues in sodium citrate (1%), transfer to new citrate and let stand for 10 to 20 min.
2. Transfer to solution C for 30 min.
3. Transfer to drop of acetic acid (50% aqueous) on warmed clean slide.
4. Macerate with minuten pins, if tissues do not dissociate.
5. Add a drop of solution C spread evenly over tissue and let evaporate.
6. Transfer slide to solution D for 4 hr.
7. Rinse in stain solvent and drain.
8. Transfer to solution E.
9. Place coverslip on and put in covered container in oven (or warming plate) at 50°C for 12 hr (or until adequate staining occurs).
10. Transfer to solution D (vertically) to loosen coverslip.
11. Dehydrate through ethanol (70% with several drops of acetic acid), 95% and absolute ethanol.
12. Mount in Euparal or clear in xylene and mount in balsam.

Note:

1. In step I (pretreatment), solution B is introduced by feeding of larvae.

Formula:

Solution A: Modified Ringer's

KCl	0.1 pt
$CaCl_2$	0.0135 pt
$NaHCO_3$	0.012 pt
NaCl	0.75 pt
Distilled Water	add up to 100 pt

Solution B:

Sodium Citrate	1 pt
Distilled Water	100 pt

Solution C: Modified Carnoy's

Ethyl Alcohol (95%)	1 pt
Acetic Acid (45%)	1 pt

Procedure:

1. Dissect out organs in solution A.
2. Transfer to solution B at 40°F for 10 min.
3. Transfer to solution C at about 8°F for 10 to 15 min.
4. Transfer to clean slide and add a drop of acetic acid (45%).
5. Treat a coverslip in silicone solution.

6. Wash with running water, pass through several changes of ethyl alcohol (95%) and dry with lint-free cloth.
7. Place coverslip on tissues.
8. Place filter paper around slide.
9. Apply firm pressure over coverslip.
10. Using heavy metal rod, tap coverslip through filter paper several times.
11. Place slide on piece of "Dry Ice" for 30 min.
12. Remove siliconized coverslip with chilled razor blade.
13. Transfer slide to ethyl alcohol (95%) for 5 min.
14. Drain and add drop of orcein-fast green stain overnight at freezing temperature.
15. Place coverslip (siliconized and rinsed in absolute ethanol) over the stain drop.
16. Place filter paper over coverslip, press gently.
17. Seal with balsam-paraffin sealant for temporary slides.
18. If permanent mounts desired, after staining, place on piece of "Dry Ice' for 1 hr.
19. Transfer to ethyl alcohol (95%) for 5 min.
20. Transfer to absolute alcohol for 1 min and drain.
21. Place drop of euparal on slide and place siliconized coverslip (rinsed in absolute alcohol) on slide, let dry for 24 hr.

Note:
1. When applying pressure, do not move coverslip.
2. Silicone treatment prevents tissues from adhering to coverslip.
3. Preparations can be observed with phase microscopy.
4. Staining (step 15 or 18) can be enhanced by storage at freezing temperatures overnight.

12.1.22 *Preparation of meiotic chromosomes from larval and pupal* Drosophila *testes [16]*

Procedure:
1. Dissect larval or pupal testes in saline.
2. Transfer testes to a drop of 45% acetic acid on a siliconized coverslip for 2 min.
3. Place a clean slide over the coverslip and invert the sandwich; squash gently between two sheets of blotting paper.
4. Freeze the slide either on dry ice or in liquid nitrogen.
5. Remove the coverslip with a razor blade and let the slide air-dry.

12.1.23 *Preparation of meiotic chromosomes from larval and pupal* Drosophila *testes – aceto-orcein preparation [16]*

Procedure:
1. Dissect larval or pupal testes in saline.
2. Transfer testes to a drop of 2% aceto-orcein placed on a coverslip for 2 min.
3. Place a clean slide over the coverslip and invert the sandwich; squash gently between two sheets of blotting paper.
4. Seal with clear nail polish.

12.1.24 *Preparation of meiotic chromosomes from adult* Drosophila *testes [17]*

Procedure:
1. Dissect testes in Ringer's solution for *Drosophila*.
2. Fix testes in 45% acetic acid for 1 min.
3. Stain testes in 3% orcein dissolved in 60% acetic acid for 5 min.
4. Quickly rinse the testes in 60% acetic acid and transfer them into a small drop of 60% acetic acid placed on a slide.

5. Cut testes under dissecting microscope to release testis contents. Add a drop of lacto-aceto-orcein.
6. Gently lower a coverslip over the specimen. Remove excess liquid from the edges of the coverslip with a piece of blotting paper.
7. Seal with clear nail polish.

12.1.25 *LaFountain technic [18]*

Application:
Recommended for the visualization of chromosomes from testes.

Procedure:
1. Fix testes in paraformaldehyde with 2 μM Hoschst 33258 (nucleic acid stain).
2. Transfer to a coverslip with paraformaldehyde with 2 μM Hoechst 22358 and rupture testes. Incubate for 20 min.
3. Remove excess fixative and add a drop of oil.
4. Spread spermatocytes at the glass-oil interface and image.

12.1.26 *Della Torre technic [19]*

Application:
Recommended for the preparation of nurse cells for chromosome banding.

Procedure:
1. Pre-fix tissues in Carnoy's.
2. Dissect ovaries in 50% propionic acid.
3. Stain with 2% lacto-orcein solution.
4. Wash cells in 50% propionic acid to remove excess stain.
5. Cover with coverglass and squash to reveal chromosomes.

12.1.27 *Dionne and Spicer's technic [20]*

Application:
Recommended as an effective technic for chromosomal preparations of aphids.

Formula:
Solution A: Ringer's

Sodium Chloride	0.65 pt
Potassium Chloride	0.025 pt
Calcium Chloride	0.03 pt
Sodium Bicarbonate	0.02 pt
Distilled Water	100 pt

Solution B:

Glacial Acetic Acid	2 pt
Ethyl Alcohol	2 pt
Chloroform	0.125 pt

Procedure:
1. Transfer embryos to 25% of solution A for 15 min.
2. Transfer to solution B for 1 hr.
3. Hydrolyze embryos in normal hydrochloric acid at 60°C for 5 min.
4. Wash in distilled water.
5. Transfer to Gomori's hematoxylin at 60°C for 30 min.

6. Transfer to acetic acid (45%) for 30 min or more.
7. Transfer five to six stained embryos to slide in minimal acetic acid (45%).
8. Place coverslip and tap with blunt instrument.
9. Flood under coverslip with acetic acid (45%).
10. Drain excess with filter paper and press slide between filter paper.

12.1.28 *Guthrie, Dollinger, and Stetson's technic [21]*

Application:
1. Recommended for chromosomal preparations of the European corn borer, smartweed borer and lotus borer.
2. Recommended for the preparation of tissues including testes and brain.

Formula:
Solution A:

Glacial Acetic Acid	1 pt
Ethyl Alcohol (95%)	3 pt

Procedure:
1. Dissect out tissues in Ringer's.
2. Transfer to solution A for 24 hr.
3. Transfer to normal HCl for 6 to 7 min.
4. Transfer to solution A for 5 min.
5. Transfer to slide with drop of lactic-aceto-orcein stain (and cover) for 2.5 to 3 hr.
6. Drain excess stain with filter paper and add drop of glacial acetic acid (50%).
7. Drain excess stain and acid and repeat process.
8. Apply gentle pressure with thumb on coverslip covered with filter paper.
9. Add drop of glacial acetic acid and repeat pressure if preparation unsatisfactory.
10. When satisfactory, place over flame for an instant.
11. Seal or dehydrate in alcohol series for permanent mount.

Note:
1. When applying pressure, care must be taken not to move coverslip.
2. A shorter staining time is recommended for brain tissues, i.e., 2 hr.

12.1.29 *Warren, Gassner, and Breland's technic [22]*

Application:
1. Recommended for chromosomal preparations of insects with simple metamorphosis.
2. Recommended as particularly useful for chromosomal brain preparations of nymphal praying mantis.

Formula:
Solution A: Modified Carnoy's

Absolute Alcohol	1 pt
Glacial Acetic Acid	1 pt
Chloroform	1 pt

Procedure:
1. Transfer brain tissue to solution A for 1 min.
2. Place drop of lacto-aceto-orcein on tissues (avoid evaporation) for 40 to 45 min.

3. Drain excess stain with filter paper.
4. Place drop of acetic acid (50%) to partially destain. Drain and repeat.
5. Place a coverslip on, place filter paper over it and apply pressure.
6. Examine microscopically and repeat above as necessary.
7. Seal with balsam-paraffin sealant.

Note:
1. The steps in this procedure are very similar to Breland's technic.

12.1.30 *Schmuck and Metz technic [23]*

Application:
1. Recommended as a technic that reveals the chromosomes in insect eggs distinctly.
2. Recommended for eggs of the fungus gnat *Sciara coprophila* Lint.

Formula:
Solution A: Modified Carnoy's (as above)

Procedure:
1. Fix eggs in solution A for ½ to 2 hr.
2. Wash in absolute alcohol (several changes) for 2 hr.
3. Transfer through higher alcohols (95, 85, and 70%) for 1 hr each.
4. Transfer through lower alcohols (50, 30, and 15%) for ½ hr each.
5. Transfer through several changes of water.
6. Transfer to cold normal HCl for 15 min.
7. Transfer to normal HCl at 60°C for 8 to 10 min.
8. Rinse in cold HCl.
9. Transfer to SO_2 water for 2 to 5 min.
10. Stain in fuchsin sulfuric acid for 1 hr.
11. Wash in SO_2 water (two changes) for 15 min each.
12. Rinse in water (several changes).
13. Dehydrate quickly in alcohol up to 85% for 5 min each.
14. Clear in xylol and mount in balsam.

12.1.31 *Whiting's modified Schmuck-Metz technic [24]*

Application:
Recommended for chromosomal preparations of insect eggs, embryos, and insect ovaries.

Formula:
Solution A: Kahle's

Distilled Water	30 pt
Alcohol (95%)	15 pt
Formalin	6 pt
Glacial Acetic Acid	1 pt

Solution B: Leuco Basic Fuchsin

Preparation:
1. Saturate distilled water with sulphur dioxide gas.

Procedure:

1. Fix in solution A for ½ to 24 hr.
2. Transfer to distilled water (two changes) for 1 hr each.
3. Transfer to normal HCl at room temperature for 10 min.
4. Transfer to HCl at 60°C for 10 min.
5. Transfer to solution B for 2 hr.
6. Transfer to solution C for an instant.
7. Transfer through two changes of distilled water for 10 min each.
8. Transfer through two changes of triethyl phosphate for 15 min each.
9. Transfer to mixture of triethyl phosphate and xylene (equal parts) for 15 min.
10. Transfer to xylene and mount in balsam.

Note:

1. It is suggested that movement of embryos be accomplished using lens paper sacks.
2. Triethyl phosphate is used instead of ethyl alcohol method to prevent reduction in staining intensity.

12.1.32 *VonBorstel and Lindsley technic [25]*

Application:

Recommended for the display of insect embryo chromosomes.

Formula:

Solution A: Modified Kahle's

Distilled Water	37 pt
Alcohol (95%)	32 pt
Formalin	27 pt
Glacial Acetic Acid	4 pt

Solution B: Leuco Basic Fuchsin

Preparation:

1. Bubble SO_2 gas into a 0.5% basic fuchsin until decolorized.

Solution C: Sulfonated Azure A

Azure A (aqueous solution)	0.25%
Thionyl Chloride	2–3 drops/10–12 pt of dye

Preparation:

1. Prepare 2 to 10 min before staining.

Solution D: Sulfurous acid bath.

Preparation:

1. Bubble SO_2 into distilled water until saturated.

Procedure:

1. Fix in solution A for ½ hr to several days.
2. Transfer through 70%, 95%, and absolute alcohol for 2 to 5 min each.
3. Transfer to methanol-chloroform mixture (1:1) at 37°C for 1 to 16 hr (or overnight).

4. Transfer through alcohol series to water for 2 to 5 min each.
5. Transfer to one normal HCl at 60°C for 14 to 16 min.
6. Rinse in water.
7. Transfer to solution B or C for 30 min to 2 hr.
8. Transfer to solution D (if stain solution B is used) for 3 to 5 min.
9. Or, transfer through five rinses in water for 2 to 3 min.
10. Transfer through alcohol series to absolute for 2 to 5 min each.
11. Transfer to (1:1) absolute alcohol and xylene solution.
12. Transfer to xylene and mount in Permount™ or balsam.

Note:
1. Eggs with heavy chorion can be soaked in 5% sodium hypochlorite or Clorox for 2 to 5 min.

12.1.33 *Oster Balaban technic [26]*

Application:
1. Recommended as a method for histological preparations of somatic chromosomes.
2. Recommended as a particularly useful method for slides of ganglion and imaginal disc chromosomes of larval and pupal *Drosphila* and house flies.

Formula:
Solution A:

Sodium Citrate	1 pt
Distilled Water	100 pt

Solution B:

Natural Orcein	2%
Fast Green (in 50% glacial acetic acid)	0.25%
Lactic Acid (85%)	50%

Procedure:
1. Dissect out dorsal and ventral portions of cerebral ganglion and imaginal discs in solution A.
2. Check that no extraneous material is attached to specimens.
3. Transfer to a drop of solution A on a siliconized slide for 10 min.
4. Transfer to a drop of acetic acid (50%) on a siliconized slide, for 15 min.
5. Transfer with siliconized pipette to a non-siliconed slide.
6. Cover specimens with siliconed coverslip which is covered by a piece of filter paper.
7. Tap with blunt instrument and apply pressure with thumb.
8. Examine for degree of spreading.
9. Place slide on piece of "Dry Ice" for at least 30 min.
10. Flip off coverslip with single-edged razor blade.
11. Immerse in absolute alcohol for 5 min or more.
12. Although not necessary, leave in alcohol 12 hr or overnight to improve quality of preparation.
13. Apply a drop of solution B on tissues and apply a non-siliconized coverslip.
14. To store slide, ring with nail polish.

Note:

1. The best preparations are made with well-nourished larvae.
2. Water used in preparation of solution A should be double glass distilled.
3. Siliconed slides are made by rubbing slides with siliconed tissues.
4. Siliconed pipettes are prepared by dipping a fine pipette into a silicone solution and rinsing off with alcohol.
5. Fine forceps may be used for transfers before step 5 but not afterwards.
6. Solution B is prepared without heating and should be filtered periodically.

12.1.34 *Stephen's technic [27]*

Application:
Preparation of chromosomes for FISH (fluorescent *in situ* hybridization)

Formula:
Cohen Buffer:

1M $MgCl_2$	10 pt
1M Sodium Glycerol 3-phosphate	25 pt
1M $CaCl_2$	3 pt
1M KH_2PO_4	10 pt
NP40	5 pt
1M KCl	30 pt
1M Sucrose	160 pt
Distilled Water	To 1000 pt

Formaldehyde Fixative:

1M NaCl	100 pt
1M KCl	2 pt
1M NaH_2PO_4	10 pt
NP40	20 pt
40% Formaldehyde	50 pt
Distilled Water	Up to 100 pt

Procedure:

1. Dissect out salivary glands and incubate in Cohen buffer for 8–10 min.
2. Fix glands in formaldehyde fixative for 3–25 min.
3. Transfer glands to 45% acetic for 3–60 min.
4. Transfer glands to a drop of 45% acetic acid on cover glass.
5. Position a glass slide over the cover glass and break open the salivary gland nuclei by picking up the glass slide and moving the cover glass gently back and forth.
6. Freeze slide in liquid nitrogen and remove cover glass with a razor blade.
7. Rinse in tris buffered saline and proceed with standard FISH staining techniques.

12.1.35 *Preparation of slides for FISH on mitotic chromosomes of mosquitoes [28]*

Procedure:

1. Hatch mosquito eggs at 28°C and after 2–3 days, transfer 2nd to 3rd instar larvae to 16°C for *Ae. aegypti* and *C. quinquefasciatus* and to 22°C for *An. gambiae*.
2. Place 4th instars on ice for several minutes for immobilization.
3. Transfer larvae to a drop of cold hypotonic solution (0.5% sodium citrate or 0.075 M potassium chloride) on a slide.

4. Under a stereomicroscope, select larvae with oval imaginal discs (IDs) for further dissection.
5. Decapitate larva, and cut the cuticle from the ventral side of the larval thorax using dissecting scissors. Make an additional cut in the second or third abdominal segment to dissect the gut from the larva.
6. Open the cuticle and remove the gut and fat body from the larva. Remove the hypotonic solution using filter paper, replace with a fresh drop of hypotonic solution and incubate for 10 min at room temperature.
7. Replace hypotonic solution with Carnoy's (3:1 ethanol/acetic acid). IDs will become white and easily visible in the fixative.
8. Remove IDs from the body using dissecting needles and transfer them to a drop of 50% propionic acid.
9. Remove any other tissues (e.g., gut and fat body) and cover IDs with a 22x22 coverslip for 10 min at room temperature.
10. Cover the slide with filter paper and squash the tissue by tapping the eraser of a pencil on the perimeter of the coverslip.
11. Check the quality of the slide and select slides with >50 chromosome spreads.
12. Dip and hold the slide in liquid nitrogen until it stops bubbling.
13. Remove the coverslip from the slide using a razor blade and transfer the slide to a container of 70% ethanol chilled at −20°C.
14. Store at 4°C for at least 1 hr or up to several days.
15. Dehydrate slides through ascending concentrations of ethanol (70%, 80%, 100%) at 4°C for 5 min each, and air dry at room temperature.
16. Store dry slides at −20°C until ready for FISH staining.
17. Proceed with FISH staining as per standard protocols (see Chapter 7).

12.1.36 *Preparation of slides for FISH on mitotic chromosomes [29]*

Procedure:
1. Prepare fresh 60% acetic acid and pre-warm 1 ml to 40°C.
2. Prepare a Gene Frame® from ABgene®. Remove the full-size protective film from a Gene Frame®. Attach the blue frame to a clean slide. Place slides on a dry heating block at 40°C.
3. Dissect two to three brains per slide from large, feeding *Drosophila* third-instars in 0.7% NaCl.
4. Transfer brains to a 100 µl drop of 0.5% sodium citrate for 15 min at room temperature.
5. Transfer the brains to a 1 ml drop of 75% ethanol:25% acetic acid and fix for 10 min at room temperature.
6. Place a pre-warmed drop of 60% acetic acid in the middle of a Gene Frame on the heated slide and transfer brains into the drop.
7. Immediately remove the slide from the heating block and spread the liquid over the surface within the frame for at least 1 min.
8. Air-dry the slides at room temperature.
9. Store the slides for at least 2 days, up to 2 months, in a dark and dry environment.
10. Proceed with FISH staining as per standard protocols (see Chapter 7).

12.1.37 *High pressure preparation of polytene chromosomes [30]*

Application:
Recommended for preparation of chromosome preparations for *in situ* hybridization.

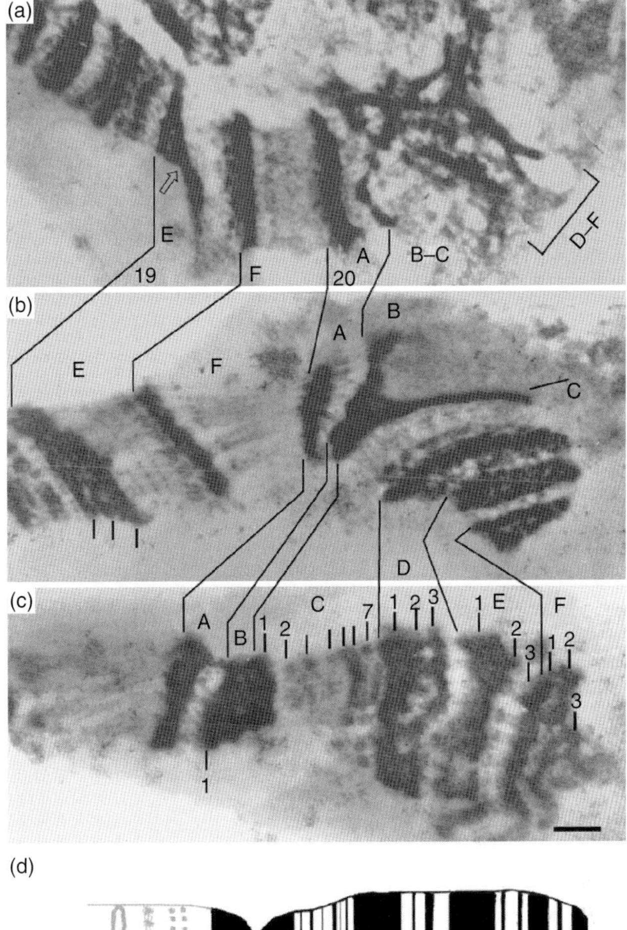

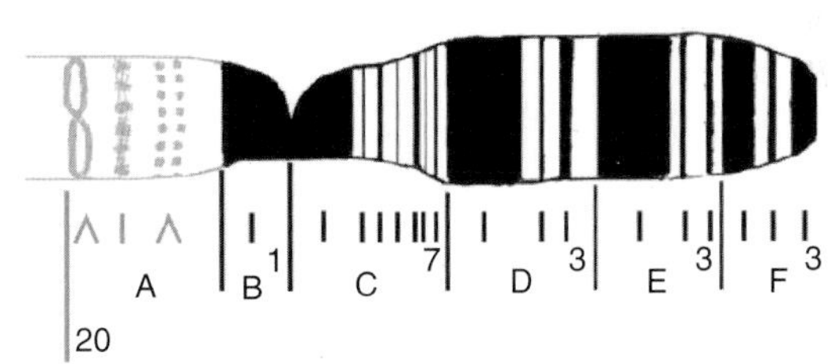

Fig. 12.3 The 19E–20F pericentric region of the X chromosome. (a) Strain Batumi L; numerous ectopic contracts disrupt the banding pattern in the 20BC region. (b, c) Strain In(1)scv2Su(UR): suppression of DNA underreplication makes the banding pattern clearly visible. The weak spot in the 19E region (open arrows in a) disappears. (d) Map drawn for the 20B-F region; the new bands appearing in the mutant strain are in black; the gray bridges correspond to those on Bridges' map. Bar represents 1um. (Figure 1 a–d). (Source: Semeshin *et al.* 2001. Reproduced with permission of Springer.)

Formula:

Modified Carnoy's solution with methanol:

100% Methanol	3 pt
Glacial acetic acid	1 pt

Modified Carnoy's solution with ethanol:

100% Ethanol	3 pt
Glacial acetic acid	1 pt

Procedure:

1. Dissect half-gravid *Anopheles* females at 25 hr post-blood feeding under a dissection microscope and fix ovaries from five females in 500 ml fresh modified Carnoy's fixative with methanol for 24 hr at room temperature. Transfer to −20°C for long-term shortage.

2. Place on pair of ovaries in a drop of fresh modified Carnoy's solution with ethanol on a slide.
3. Split ovaries into approximately six sections with dissecting needles and transfer them to a drop of 50% propionic acid on a slide.
4. Separate follicles using dissecting needles, removing any excess tissue.
5. Add a new drop of 50% propionic acid to the follicles and leave in acid for 3–5 min at room temperature. Place a coverslip on top and let slide stand for 1 min.
6. Wrap the slide in filter paper and plastic. Using a Dremel 200 rotary tool with a Flex-Shaft attachment and soft plastic tip set between 3000–5000 RPM, express the follicles by swirling the tip in circles and lightly pressing the coverslip to evenly spread the nuclei for approximately 1 min.
7. Prepare a sandwich by placing an additional coverslip next to the coverslip covering the chromosomes and cover them with a second microscope slide to reduce the chance of crushing the slide in the vise. Wrap the slides in the plastic sheet that comes in glass slide containers and filter paper to hold the sandwich and to protect the slide from scratching due to the vise.
8. Apply pressure of 85–120 inch-lbs to the slides via mechanical vise to flatten the chromosomes.
9. Remove the second microscope slide and the additional coverslip.
10. Heat the slide at 55°C on a slide denaturation/hybridization system for 10–15 min to further flatten the chromosomes.
11. Dip the slide in liquid nitrogen for at least 15 sec and when bubbling has stopped, proceed to quickly remove the coverslip with a razor blade.
12. Place slides in cold 50% ethanol for 5 min.
13. Dehydrate slides in ascending ethanol series (70%, 90%, 100%) for 5 min each.
14. Air dry slides.

12.1.38 *S Phase indices [31]*

Application:
Recommended for the quantification of S phase indices (a measure of cell growth and viability).

Protocol:
1. Prepare reagents from Click-It® reaction kit from Invitrogen™ (Cat. #C10337).
2. Dissect *Drosophila* brains in fresh Grace's unsupplemented cell culture medium.
3. Add an equal volume of 200 μM (EdU 5-ethynyl-2'-deoxyuridine) solution in DMSO and incubate brains in the dark for 30 min at room temperature.
4. Remove the EdU and rinse brains in phosphate buffered saline (PBS).
5. Swell brains in 0.5% sodium citrate solution for 10 min.
6. Fix brains in an 11:11:2 acetic acid:methanol:water solution for 30 sec.
7. Transfer brains to a dot of PBS on a slide and cover with a coverslip.
8. Sandwich the slide and coverslip between filter paper and an additional microscope slide. Place sandwich into a mechanist vise and apply pressure using a torque wrench to 15 Nm for 2 min.
9. Remove and place slide and coverslip into liquid nitrogen.
10. Pop off coverslip and wash slide in 3% BSA (Bovine Serum Albumin) in PBS.
11. Incubate slide in PBS with 0.5% Triton X-100 for 20 min.
12. Wash two times in PBS with 3% BSA.
13. Incubate in Click-It® reaction mixture in the dark for 30 min.

14. Wash two times in PBS with 3% BSA.
15. Counterstain nuclei with 5 μg/ml Hoescht 33342 in PBS for 20 min in the dark.
16. Rinse two times in PBS and mount in Vectashield® with DAPI (Vector Labs®).

12.1.39 *Narukawa technic [32]*

Application:
Preparation of chromosomes for Atomic Force Microscopy.

Formula:
Hypotonic Solution:

1M KCl	75 pt
1M NaCl	15 pt

Farmer's Fixative:

Ethanol	3 pt
Acetic Acid	1 pt

Procedure:
1. Dissect out tissues and immerse in hypotonic solution.
2. Fix in Carnoy's fixative.
3. Fix in Farmer's fixative.
4. Disrupt tissues by vortex, then centrifuge.
5. Remove supernatant and re-suspend cells in 60% acetic acid.
6. Spread an aliquot of the cell suspension on a glass slide and dry on a 50°C heat block.

12.1.40 *Semeshin technic [33]*

Application:
Recommended for the preparation of chromosomes for TEM.

Procedure:
1. Fix tissues in cold 3:1 ethanol/acetic acid for 2–4 hr.
2. Squash tissue in 45% acetic under cover glass.
3. Freeze the slides in liquid nitrogen and remove cover glass.
4. Stain with 1.5–2% uranyl acetate in 70% ethanol for 12–24 hr.
5. Dehydrate and embed in resin.

12.1.41 *Matsumoto technic [34]*

Application:
Recommended for the detection of negatively supercoiled regions of DNA.

Procedure:
1. Treat salivary glands with 0.01% digitonin in dissection buffer for 10 min.
2. Rinse in dissection buffer.
3. Soak in 0.2 ng/ml biotinylated psoralen in dissection buffer for 10 min.
4. Illuminate glands with long wave (365) UV lamp for 10 min to cross link psoralen.
5. Fix salivary glands with 40% acetic acid and squash.
6. Detect biotinylated psoralen with labeled streptavidin.

12.1.42 *Sumner's technic [35]*

Application:
Recommended for the preparation of mitotic chromosomes.

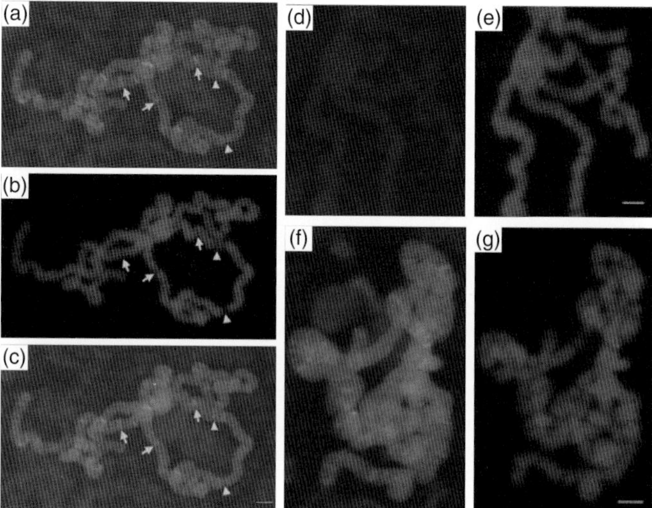

Fig. 12.4 Visualization of psoralen signals on polytene chromosomes. (a) Biotinylated psoralen signals were detected with Alexa 488-labeled streptavidin (light green color). (b) DAPI signals (blue color). (c) Merged image. Arrows indicate representative interbands and puffs. In some loci, intense psoralen signals were observed only on the side margins of an interband or puff (arrowheads). (d) Biotinylated psoralen signals disappeared when the photo-crosslinking step was omitted. (e) DAPI image of (d). (f) Biotinylated psoralen signals were resistant to RNase treatment. (g) DAPI image of (f). Bars: 10 μm. (Source: Matsumoto & Hirose 2004. Reproduced with permission of Journal of Cell Science.) See plate section for the color version of the figure.

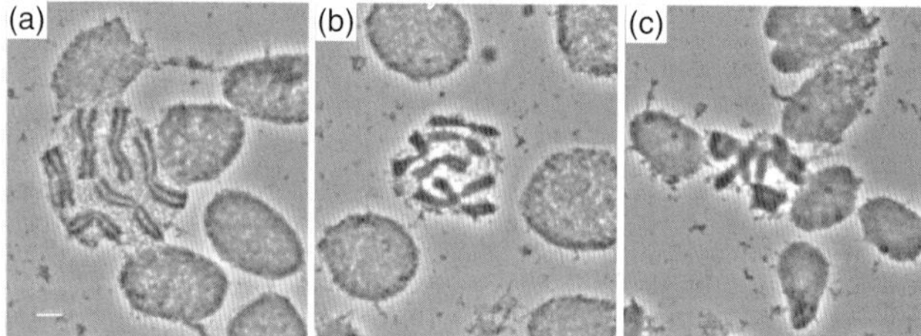

Fig. 12.5 Different qualities of chromosome spreads: a) a perfect chromosome spread - round shape of the cells demonstrates sufficient treatment of the IDs in hypotonic solution; b) a perfect hypotonic treatment - chromosomes are slightly under squashed; c) a poor chromosome spread - the result of insufficient hypotonic treatment is indicated by oval shape of the cells. (Source: Timoshevskiy *et al.* 2012. Reproduced with permission of the authors.)

Procedure:

1. Insect brains are dissected 0.7% sodium chloride.
2. Incubate in 0.7% sodium chloride with 0.00025% colchicine for 90 min at 25°C.
3. Disrupt nucleus by incubating in a 1% trisodium citrate hypotonic solution for 5 min.
4. Fix in 3:1 acetic acid/methanol for 1 hr with four changes at intervals of 15 min at 25°C.
5. Transfer to slide, add 1 drop 45% acetic acid and squash under cover glass.
6. Freeze slides in liquid nitrogen and remove cover glass.
7. Post-fix slides in 3:1 ethyl alcohol/acetic acid for 1 hr.
8. Air dry slide.
9. Incubate in 0.2N chloride acid for 10 min at 25°C.
10. Transfer to barium hydroxide saturated 50°C for 9–12 min.

11. Rinse in acid water.
12. Immerse slides in 2X SSC (Saline-sodium citrate buffer) for 30 min at 60°C.
13. Wash in distilled water.
14. Stain in 4% Giemsa for 90 min.
15. Wash in tap water.

12.1.43 *Howell and Black technic [36]*

Application:
Nucleolus organizer regions are stained black and chromosome arms are yellow.

Formula:
Solution A: Colloidal Developer Solution

Powdered gelatin	2 pt
Formic Acid	1 pt
Distilled Water	99 pt

Preparation:
1. Stir for at least 10 min to fully dissolve gelatin.
2. Solution is stable for two weeks.

Solution B: Aqueous Silver Nitrate Solution

$AgNO_3$	4 pt
Distilled Water	8 pt

Procedure:
1. Place chromosome preparation on glass slide.
2. Add two drops colloidal developer solution and four drops aqueous silver nitrate solution.
3. Mix solutions and cover with cover glass.
4. Place slide on 70°C slide warmer for 2 min.
5. Rinse off stain and cover glass with running distilled water.
6. Blot dry and examine preparation.

12.1.44 *Van't Hof technic [37]*

Procedure:
1. Dissect tissues and fix in Carnoy's for 20 min.
2. Macerate tissues in 60% acetic acid.
3. Spread suspension on a slide at 45°C.
4. Dehydrate through graded alcohols to 96% ethanol for 30 sec each.
5. Dry at room temperature.

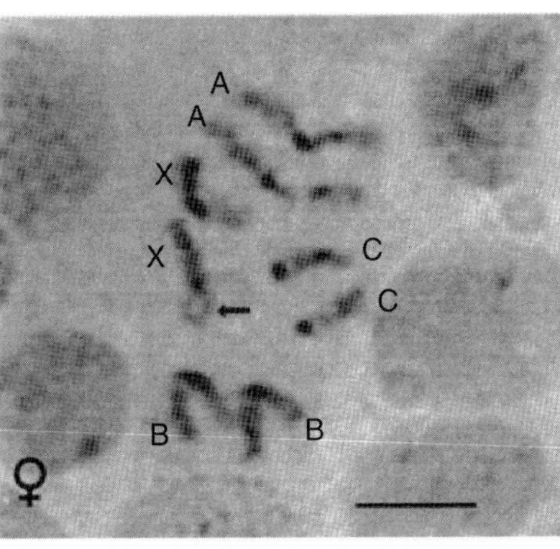

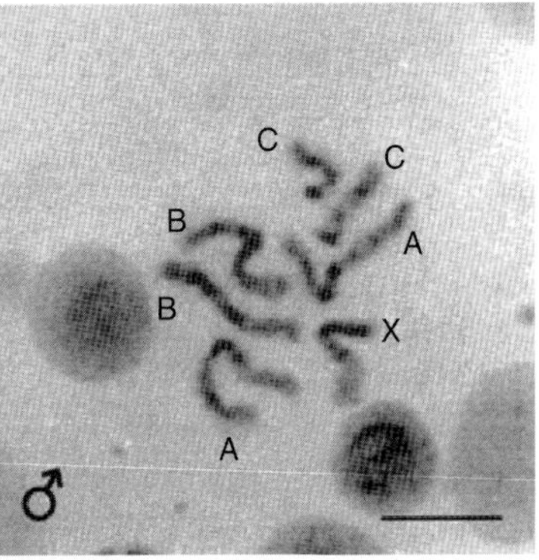

Fig. 12.6 C-band in *Bradysia hygida* larvae neuroblast chromosomes. The arrow indicates the ring shaped structure in one of the X chromosomes. The bar represents 5 μm. (Source: Gaspar *et al.* 2002. Reproduced with permission of Genetica.)

12.1.45 *Zacharopoulou technic [38]*

Procedure:

1. Dissect tissue in Ringer's solution.
2. Transfer to hypotonic solution (1% sodium citrate) for 10–15 min.
3. Fix for 3 min in 3:1 methanol/acetic acid.
4. Transfer to a small drop of 66% acetic acid.
5. Macerate tissue by pipetting up and down several times.
6. Lay cell suspension on a clean slide and place on a 40–45°C hot plate to dry.
7. Stain chromosomes with 5% Giemsa in phosphate buffer.

References

1. Moorman A.E. A modification of Heidenhain's iron-haematoxylin technique, as used to stain smears of *Drosophila* larval polytene chromosomes and neuroblast cells. *Canad. J. Zool.* **49**, 132–133 (1971).
2. Frankston J.E. & Dambach G.J. A rapid method for making permanent smears of salivary chromosomes. *Proc. Penn. Acad. Sci.* **10**, 112–113 (1936).
3. LaCour L. Acetic-orcein: a new stain-fixative for chromosomes. *Stain Technol.* **16**, 169–174 (1941).
4. Amirkhanian J.D. A combined HCL-acetic-alcohol fixation and hydrolysis followed by cresyl violet staining for mosquito chromosome spreads. *Stain Technol.* **43**, 167–170 (1968).
5. Marshak A. A rapid method for making permanent mounts of *Drosophila* salivary gland chromosomes. *Amer. Nat.* **70**, 406–407 (1936).
6. Pares R. Application of nigrosine to the study of the salivary chromosomes. *Science* **172**, 1151–1152 (1953).
7. Nicoletti B. An efficient method for salivary gland chromosome preparations. *Dros. Inform. Serv.* **33**, 181–182 (1959).
8. Smith S.G. Techniques for the study of insect chromosomes. *Canad. Entomol.* **75**, 21–34 (1943).
9. French W.L. & Kitzmiller J.B. Mosquito chromosome preparations using colchicine pre-treatment (abstract). *Amer. Zool.* **1**, 449 (1961).
10. Pimpinelli S., Bonaccorsi S., Fanti L. & Gatti M. Preparation and orcein staining of mitotic chromosomes from *Drosophila* larval brain. *Cold Spring Harbor Protoc.* (2010).
11. Breland O.P. Preliminary observations on the use of the squash technique for the study of the chrosomomes of mosquitos. *Texas J. Sic.* **11**, 183–190 (1959).
12. Breland O.P. Studies on the chromosomes of mosquitoes. *Ann. Entomol. Soc. Amer.* **54**, 360–375 (1961).
13. Rai K.S. A comparative study of mosquito karyotypes. *Ann. Entomol. Soc. Amer.* **56**, 160–170 (1963).
14. French W.L., Baker R.H. & Kitzmiller J.B. Preparation of mosquito chromosomes. *Mosq. News* **22**, 377–383 (1962).
15. Crozier R.H. An acetic acid dissociation, air-drying technique for insect chromosomes, with aceto-lactic orcein staining. *Stain Technol.* **43**, 171–173 (1968).
16. Bonaccorsi S., Giansanti M.G., Cenci G. & Gatti M. Preparation of meiotic chromosomes from larval and pupal *Drosophila* testes. *Cold Spring Harbor Protoc.* 3, (2011).
17. Bonaccorsi S., Giansanti M.G., Cenci G. & Gatti M. Preparation of meiotic chromosomes from adult *Drosophila* testes. *Cold Spring Harbor Protoc.* 3, (2011).
18. LaFountain J.R., Hard R. & Siegel A.J. Visualization of kinetochores and analysis of their refractility in crane-fly spermatocytes after aldehyde fixation. *Cell mot. and Cytoskel.* **40**, 147–159 (1998).
19. Sharakhov I.V., Sharakhova M.V., Mbogo C.M., Koekemoer L.L. & Yan G. Linear and spatial organization of polytene chromosomes of the African malaria mosquito *Anopheles funestus*. *Genetics* **159**, 211–218 (2001).

20. Dionne L.A. & Spicer P.B. A squash method for somatic chromosomes of aphids. *Canad. J. Zool.* **35**, 711–713 (1957).
21. Guthrie W.D., Dollinger E.J. & Stetson J.F. Chromosome studies of the European corn borer, smart-weed borer and lotus borer. *Ann. Entomol. Soc. Amer.* **58**, 100–105 (1965).
22. Warren M.E., Gassner G. & Breland O.P. Somatic chromosomes of praying mantids as shown by the squash technique. *Texas Repts. Biol. Med.* **18**, 674–679 (1960).
23. Schmuck M.L. & Metz C.W. A method for the study of chromosomes in entire insect eggs. *Science* **74**, 600–601 (1931).
24. Whiting A.R. A modification of the Schmuck-Metz whole-mount technique for chromosome study. *Stain Technol.* **25**, 21–22 (1950).
25. VonBorstel R.C. & Lindsley D.L. Insect embryo chromosome techniques. *Stain Technol.* **34**, 23–26 (1959).
26. Oster I.I. & Balaban G. A modified method for preparing somatic chromosomes. *Dros. Inform. Serv.* **37**, 142–144 (1963).
27. Stephens G.E., Craig C.A., Li Y., Wallrath L.L. & Elgin S.C. Immunofluorescent staining of polytene chromosomes: exploiting genetic tools. *Meth. Enzymol.* **376**, 372–393 (2004).
28. Timoshevskiy V.A., Sharma A., Sharakhov I.V. & Sharakhova M.V. Fluorescent *in situ* hybridization on mitotic chromosomes of mosquitoes. *J Vis. Exp* 67, (2012).
29. Blattes R. & Kas E. Fluorescent *in situ* hybridization (FISH) on diploid nuclei and mitotic chromosomes from *Drosophila melanogaster* larval tissues. *Cold Spring Harbor Protoc.* 9, (2009).
30. George P., Sharakhova M.V. & Sharakhov I.V. High-throughput physical mapping of chromosomes using automated *in situ* hybridization. *J Vis. Exp* 64, (2012).
31. Gouge C.A. & Christensen T.W. *Drosophila* Sld5 is essential for normal cell cycle progression and maintenance of genomic integrity. *Biochem. Biophys. Res. Comm.* **400**, 145–150 (2010).
32. Narukawa J., Yamamoto K., Ohtani T. & Sugiyama S. Imaging of silkworm meiotic chromosome by atomic force microscopy. *Scanning* **29**, 123–127 (2007).
33. Semeshin V.F., Belyaeva E.S. & Zhimulev I.F. Electron microscope mapping of the pericentric and intercalary heterochromatic regions of the polytene chromosomes of the mutant Suppressor of underreplication in *Drosphila melanogaster*. *Chromosoma* **110**, 487–500 (2001).
34. Matsumoto K. & Hirose S. Visualization of unconstrained negative supercoils of DNA on polytene chromosomes of *Drosophila*. *J. Cell Sci.* **117**, 3797–3805 (2004).
35. Gaspar V.P., Borges A.R. & Fernandez M.A. NOR sites detected by Ag-dAPI staining of an unusual autosome chromosome of *Bradysia hygida* (Diptera:Sciaridae) colocalize with C-banded heterochromatic region. *Genetica* **114**, 57–61 (2002).
36. Howell W.M. & Black D.A. Controlled silver-staining of nucleolus organizer regions with a protective colloidal developer: a 1-step method. *Experientia* **36**, 1014–1015 (1980).
37. Van't Hof A.E., Marec F., Saccheri I.J., Brakefield P.M. & Zwaan B.J. Cytogenetic characterization and AFLP-based genetic linkage mapping for the butterfly *Bicyclus anynana*, covering all 28 karyotyped chromosomes. *PLOS One* 3, (2008).
38. Gariou-Papalexiou A., Yannopoulos G., Robinson A.S. & Zacharopoulou A. Polytene chromosome maps in four species of tsetse flies *Glossina austeni, G-pallidipes, G-morsitans morsitans* and *G-m. submorsitans* (Diptera : Glossinidae): a comparative analysis. *Genetica* **129**, 243–251 (2007).

13 Preparation of other specific insect organs and tissues

13.1 Introduction

Most general procedures for the histological preparation of insect tissues are the result of the combination of various appropriate sub-procedures. Thus, depending upon the type of insect, stage, size, time available, and so on, a selection is made of the most suitable fixation, dehydration, clearing, embedding, and mounting regimes. However, some procedures have been reported for the preparation of specific tissues. Most of these procedures deal with highly specialized tissues, organs or stages of development such as nervous tissue, gonads, and eggs. Included in this section are some of these techniques. The other sections in this book should be a basic source from which other procedure can be combined to fit specific needs.

13.1.1 *Brazil technic I [1]*

Application:
Recommended for the preparation of Sand Flies.

Procedure:
1. Fix specimens in Bouin's for 4 hr.
2. Post-fix specimens in Carnoy's for 2 hr.
3. Rinse in distilled water until no yellow color remains.
4. Transfer to 70% ethanol and dehydrate through graded series of alcohols.
5. Immerse specimens in a mixture of 1 pt 95% ethanol and 1 pt Historesin overnight at 4°C.
6. Infiltrate specimen in pure resin for at least a week at 5°C.
7. Embed in resin.

13.1.2 *Brazil technic II [1]*

Application:
Recommended for the preparation of Sand flies for protein detection.

Procedure:
1. Fix specimens in calcium formaldehyde for 24 hr.
2. Rinse in distilled water quickly, then for three times, 15 min each.
3. Transfer to 70% ethanol and dehydrate through a graded series of alcohols.
4. Immerse specimens in a mixture of 1 pt 95% ethanol and 1 pt Historesin overnight at 4°C.
5. Infiltrate specimen in pure resin for at least a week at 5°C.
6. Embed in resin.

Insect Histology: Practical Laboratory Techniques, First Edition.
Pedro Barbosa, Deborah L. Berry and Christina S. Kary.

Companion Website: www.wiley.com/go/barbosa/insecthistology

13.1.3 *Squash method [2]*

Application:
Recommended for the preparation of eggs with thick chorions and abundant yolk mass.

Procedure:
1. Place eggs in a drop of 0.9% NaCl on a cover glass.
2. Cut the posterior end of the egg with a sharp knife.
3. Gently squeeze out egg contents.
4. Remove excess NaCl and spread and affix egg contents on cover glass.
5. Fix in absolute methanol for 1 hr and proceed with standard staining techniques.

13.1.4 *Anderson's technic [3]*

Application:
Recommended for the preparation of refractory (yolky) eggs of locust and other insects.

Formula:
Solution A: Dioxane Fixative

Dioxane (saturated with picric acid)	85 pt
Formaldehyde (40%)	10 pt
Formic Acid	5 pt

Procedure:
1. Fix in solution A for 12 hr.
2. Wash in dioxin (three changes) to remove excess picric acid.
3. Transfer through four changes of cellosolve (to dehydrate) for 2 hr each.
4. Transfer to a 2% solution of celloidin in cellosolve at 30°C for 12 hr.
5. Transfer through three changes of benzene (to clear) for 45 min.
6. Transfer to mixture of paraffin and benzene (equal parts) for 15 min.
7. Transfer to paraffin-ceresin mixture (m.p. 55°C) and change every 15 min (total – 2 hr.).
8. Embed, block, and trim block to exposure tissue.
9. Transfer block to a 5% tergitol ethane-diol solution for 12 hr.
10. Drain excess fluid and cool block on ice.
11. Section, stain, and mount.

Note:
1. Tergitol ethane-diol is a wetting agent and other wetting agents were tried successfully.
2. Sections can be stained with many stains including Masson's iron hematoxylin, Ehrlich's hematoxylin and eosin, Biebrich scarlet, etc.

13.1.5 *Slifer and King's technic [4]*

Application:
1. Recommended for cytological and embryological studies of insect eggs.
2. Recommended for use in the study of maturation and cleavage in grasshopper eggs.

Procedure:
1. Fix eggs in Carnoy-Lebrun.
2. Wash in iodized alcohol.
3. Cut in half and store in alcohol (70–80%).

4. Transfer to phenol (4%) in 80% alcohol solution for 24 hr.
5. Dehydrate in 95% alcohol and clear in carbol-xylol.
6. Infiltrate with paraffin and block with cut surface (of egg) facing out.
7. Trim block to expose tissues.
8. Soak block in water for 24 to 48 hr or, after step 4 do the following.
9. Dehydrate in alcohol (95%) and clear in aniline oil.
10. Wash in chloroform.
11. Embed in paraffin, block and continue with steps 7 to 8.
12. Section and mount.

Note:

1. Eggs may be stored in alcohol (step 3) for about 3 months until needed for histological preparation.
2. Feulgen's stain is recommended since it does not stain the yolk.

13.1.6 *Ewen's technic for refractory material [5]*

Application:

Recommended for highly refractory material such as insect eggs which are hard, yolky, and brittle.

Formula:

Solution A: Carnoy's Fluid

Absolute Ethanol	6 pt
Chloroform	3 pt
Glacial Acetic Acid	1 pt

Procedure:

1. Fix in solution A at 0 to 4°C (with chorion punctured) for 16 hr to overnight.
2. Or, fix in solution A at room temperature (with chorion punctured) for 6 hr.
3. Transfer to absolute alcohol (briefly) and remove chorion.
4. Transfer to dioxane for 3 to 4 hr.
5. Transfer through two changes of benzene (to clear) for 1 to 2 hr.
6. Infiltrate in two changes of Paraplast (or alternative) in vacuum oven for 1 hr each.
7. Embed and block.
8. Cut block to expose tissues.
9. Transfer to a 5% tergitol ethane-diol solution for 4 to 6 hr.
10. Wash the block.
11. Swollen exposed tissues may be re-embedded by dipping in molten wax.
12. Section.

Note:

1. Dioxane used in procedure was used after storage over anhydrous calcium oxide for 14 hr.
2. The following stains were found satisfactory for staining eggs: Ehrlich's acid heamatoxylin counterstained with eosin, Heidenhain's iron hematoxylin and Delafield's hematoxylin.

13.1.7 *Beckel's technic [6]*

Application:
Recommended for removing pigment and softening the chorion of mosquito eggs prior to sectioning.

Formula:
Solution A: Mayer's Chloride

Potassium Chlorate	0.1 pt
HCl	0.1 pt
Water	100 pt

Preparation:
1. See Craig's Technic.

Procedure:
1. Transfer from alcohol (70%) to solution A for 30 min.
2. Wash in alcohol (70%).
3. Dehydrate in absolute alcohol.
4. Clear in xylol.

Note:
1. Double embedding is suggested for brittle specimens that are to be sectioned serially.

13.1.8 *Ewen's technic for neurosecretory components [7]*

Application:
Recommended for the histological display of neurosecretory components in insects.

Formula:
Solution A: Aldehyde Fuchsin Stock

Basic Fuchsin	1 pt
Water	200 pt
HCl (concentrated)	2 pt
Paraldehyde	2 pt

Preparation:
1. Dissolve fuchsin in boiling water for 1 min.
2. Cool, filter, add HCl and paraldehyde.
3. Keep in stoppered jar at room temperature.
4. After red color is gone (4 days at 20°C) filter.
5. Dry precipitate on filter paper (in paraffin oven).
6. Store crystals in stoppered bottle.
7. Stock solution = 0.75 pt of dye crystal/100 pt ethanol (70%).

Solution B: Stain

Stock Solution	25 pt
Ethanol (70%)	75 pt
Glacial Acetic Acid	1 pt

Solution C:

$KMnO_4$	0.15 pt
H_2SO_4	0.10 pt
Distilled Water	50 pt

Solution D:

Ethanol (absolute)	100 pt
HCl (concentrated)	0.5 pt

Solution E:

Phosphotungstic Acid	4 pt
Phosphomolybdic Acid	1 pt
Distilled Water	100 pt

Solution F:

Distilled Water	100 pt
Light green SF yellowish	0.4 pt
Orange C	1 pt
Chromotrope 2R	0.5 pt
Glacial Acetic Acid	1 pt

Procedure:

1. Fix material in modified Bouin's, Helly's, or Susa's.
2. Embed, section, remove paraffin and hydrate as usual.
3. Transfer to solution C.
4. Rinse in distilled water.
5. Transfer to 2.5% sodium bisulfite to decolorize.
6. Dehydrate through ethanol series, from several water rinses to 30, 70% ethanol and then solution B for 2 to 10 min.
7. Wash with ethanol (95%).
8. Differentiate in solution D for 10 to 30 sec.
9. Hydrate through ethanol series from 70% to 30% ethanol and then distilled water.
10. Transfer to solution E (mordant) for 10 min.
11. Rinse in distilled water.
12. Counterstain with solution F for 1 hr.
13. Rinse in 0.2% acetic acid (in 95% ethanol).
14. Quickly dehydrate up to absolute ethanol.
15. Clear in xylol and mount in Canada Balsam.

Note:

1. Solution F will keep indefinitely.
2. The use of 0.5% to 1.0% trichloroacetic acid rather than acetic acid improves fixation by Bouin's (usually containing 2% to 5% acetic acid).
3. Staining should occur soon after sectioning.

Result:

1. Cytoplasm appears light green and nuclei orange.
2. Neurosecretions appear dark purple.

13.1.9 *Blest's technic [8]*

Application:

1. Recommended for the histological display of insect central nervous system.
2. Originally recommended for preparations of the desert locust, honeybee, and Lepidoptera.

Formula:

Solution A:

Boric Acid (0.2M)	27.5 pt
Borax (0.05M)	22.5 pt
Silver Nitrate (1% aqueous)	5–10 pt
Lutidine-Water (2:6) or Pyridine	2–6 pt
Distilled Water	250 pt

Solution B:

Hydroquinone	3 pt
Sodium Sulfate	30 pt
Distilled Water	300 pt

Procedure:

1. Fix material in alcoholic Bouin's for 24 hr.
2. Wash in alcohol (50%).
3. Dehydrate and clear through graded series of alcohol, cedarwood oil, and benzene.
4. Embed in wax (56°C) and cut sections.
5. Remove wax and hydrate quickly.
6. Transfer to 20% silver nitrate for 2 to 3 hr.
7. Rinse in distilled water for 5 min.
8. Transfer to solution A at 37°C for 16 to 20 hr.
9. Transfer to solution B at 55–60°C for 3 min.
10. Wash in running water for 3 min.
11. Rinse in distilled water for 3 min.
12. Transfer to 0.2% sodium gold chloride for 1 to 3 min.
13. Rinse in distilled water for 1 min.
14. Transfer to 2% oxalic acid for 5 min.
15. Rinse in distilled water for 3 min.
16. Transfer to 5% sodium thiosulfate for 1 min.
17. Wash well in distilled water.
18. Dehydrate and mount.

Note:

1. In the preparation of solution A, if lutidine is used, the following is recommended: 6 pt lutidine-water with 10 pt silver nitrate in 250 pt buffer.

13.1.10 *Holmes' technic [9]*

Application:

1. Recommended for the histological display of nerve tracts of embryonic, larval, and adult tissues.
2. The procedure has been used with blowflies, mantids, grasshoppers, houseflies, and fruit flies.

Formula:

Solution A:

Boric Acid (H_3BO_3)	12.4 pt
Distilled Water	1000 pt

Solution B:

Borax (NaB_4O_7 $10H_2O$)	19 pt
Distilled Water	1000 pt

Solution C:

Solution A	55 pt
Solution B	45 pt
$AgNO_3$ (1% aqueous)	1 pt
Pyridine (10% aqueous)	5 pt
Distilled Water	494 pt

Solution D:

Hydroquinone	1 pt
$Na_25O_3 - 7H_2O$	10 pt
Distilled Water	100 pt

Procedure:

1. Fix in alcoholic Bouin's or Carnoy's (with acetic acid) for 24 hr.
2. Dehydrate and embed in paraffin or other paraffin-type medium.
3. Cut sections (10–12 μm) and adhere to slides with albumen.
4. Remove paraffin with xylene and hydrate through graded alcohol series to distilled water.
5. Transfer to $AgNO_3$ (20% aqueous) at 25–27°C in darkness at 1 hr.
6. Wash in three changes of distilled water for 3 to 3.5 min each.
7. Transfer to solution C at 37°C for 24 hr.
8. Transfer to solution D for at least 2 min.
9. Wash in running water and rinse in distilled water.
10. Transfer to 0.2% gold chloride (aqueous) for 3 min.
11. Rinse in distilled water.
12. Transfer to 2% oxalic acid for 3 to 10 min; observe changes under microscope.
13. Rinse in distilled water and transfer to 5% $Na_2S_2O_3$ for 5 min.
14. Wash in water, dehydrate, and mount.

Note:

1. The $AgNO_3$ (20%) solution should not be used when black precipitate occurs.
2. The $AgNO_3$ (1% aqueous) should be used only once.
3. Any cloudy silver solution should not be used.

13.1.11 *Rowell's technic [10]*

Application:

Recommended as a synthesis of Blest's and Holmes' techniques for the histological display of invertebrate central nervous system.

Formula:

Solution A:

Boric Acid (0.2M)	Enough to give appropriate pH 7–9

Borax (0.05M)	Enough to give appropriate pH 7–9
Silver Nitrate (1%)	20 pt
Lutidine Water (2:6)	10 pt
Distilled Water	250 pt

Solution B:

Silver Nitrate (5%)	9 pt
Sodium Sulfite (9%)	300 pt
Hydroquinone (0.5%)	20 pt

Procedure:

1. Fix material (e.g., brains) in Carnoy's, alcoholic Bouin's or formalin.
2. Embed in paraffin, celloidin or ester wax.
3. Cut sections anywhere from 8 to 25 μm.
4. Transfer to 20% silver nitrate in darkness for 1 hr.
5. Rinse in distilled water.
6. Transfer to solution A at 30–70°C for 16 hr (see note).
7. Rinse well in distilled water.
8. Transfer to 2% sodium sulfate for 2 min.
9. Rinse well in distilled water.
10. Transfer to solution B at 20°C for 5 to 10 min.
11. Rinse in distilled water.
12. Wash in running water for 5 min.
13. Rinse in distilled water.
14. Transfer to 0.2% gold chloride (with a little acetic acid) for 5 min.
15. Rapidly rinse in distilled water.
16. Transfer to 2% oxalic acid for 5 min.
17. Rinse in distilled water.
18. Transfer to 5% sodium thiosulfate for 2 min.
19. Wash in running water.
20. Dehydrate and mount.

Note:

1. For step 6 different tissues respond best to different temperatures, pH, duration of treatment, etc.
2. For step 6, a suggested starting point for treatment is a pH of 7.8 and a temperature of 50°C for 16 hr.
3. An increase in temperature and pH will increase the number of axons stained and decrease selectivity.

13.1.12 *Debauche's technic [11]*

Application:

Recommended for the histological display of insect nerve cells and neurofibrillae.

Formula:

Solution A: Modified Duboscq-Brazil

Ethanol (80%)	100 pt
Picric Acid	2 pt
Acetic Acid	10 pt
Formaldehyde	40 pt

Solution B:

Alcohol (90%)	50 pt
Ammonia (concentrated)	15 drops

Solution C:

Distilled Water	100 pt
Ammonia	52 drops

Solution D:

Distilled Water	20 pt
Silver Nitrate	4 pt

Solution E:

Formaldehyde	3 pt
Distilled Water	100 pt

Procedure:
1. Fix material in solution A for as long as possible (see note 2).
2. Place material in solution B for 24 hr.
3. Transfer to absolute alcohol for 1.5 hr.
4. Transfer to Toluol for 20 min.
5. Infiltrate in paraffin with 20% beeswax at 58°C for 6 hr.
6. Embed in same mixture.
7. Section as usual but transfer ribbon to xylol (freeing sections).
8. Transfer sections to absolute alcohol.
9. Slowly add distilled water drop-wise until sections are hydrated.
10. Transfer to solution C for 15 min.
11. Transfer sections to solution D.
12. Heat (45°C) and agitate solution until sections turn light brown.
13. While agitating, add pure ammonia drop-wise until precipitate forms.
14. Redissolve precipitate by adding more ammonia until opalescent.
15. Pour off solution and sections into dish with a little distilled water.
16. Quickly transfer to solution E.
17. Dip in distilled water and transfer to 1% aqueous gold chloride.
18. Transfer to 5% aqueous sodium thiosulfate for 10 min.
19. Wash well in tap water.
20. Dehydrate and mount in Canada Balsam.

Note:
1. It is suggested that the purest chemicals be used.
2. A fixation period of 100 days gives clear preparations.
3. Good treatment with solution E (step 16) is indicated by uniform brown coloration of sections.

Results:
1. Nerve cells and neurofibers are dark blue on a pinkish background.
2. Nuclei are transparent but well defined.

13.1.13 *Gregory's Bodian [12]*

Application:

1. Recommended for the histological preparation of insect central nervous system.
2. Recommended for use in the preparation of nerve cord ganglia and brain tissues of *Periplaneta* sp. and *Schistocerca gregaria*.

Procedure:

1. Fix in aged alcoholic Bouin's for 16 hr to 24 hr at 20°–25°C.
2. Transfer to 10% ammonium acetate (in 80% ethanol) until picric acid is removed.
3. Color specimen with eosin in absolute alcohol for orientation purposes.
4. Clear with xylene, embed and section.
5. Impregnate with 2% protargol containing clean metallic copper for 18 hr to 24 hr at 37°C.
6. Dip in distilled water three times.
7. Transfer to hydroquinone-sodium sulfite solution for 10 to 15 min.
8. Wash three times in distilled water for 2 min each.
9. Repeat steps 1 to 8 using a fresh protargol-copper solution.
10. Transfer to 1% gold sodium chloride ($NaAuCl_4$ $2H_20$) for 15 min.
11. Quickly rinse in distilled water.
12. Transfer to 2% oxalic acid for 10 min.
13. Wash quickly in distilled water.
14. Transfer to 5% sodium thiosulfate for 10 to 15 min.
15. Wash in distilled water three times for 3 min each.
16. Dehydrate, clear, and mount.

Note:

1. It is suggested that slides be cleaned with chromic acid.
2. The fixative (Bouin's) may be aged by storing at 60°C for 40 days and may be kept thereafter at room temperature.
3. In general, for
   a. 10 µm sections, 0.5 pt to 1 pt of copper/65 pt of 2% protargol and 10% sodium sulfite in 0.25% to 0.5% hydroquinone are suggested.
   b. 20 µm sections, 4 to 5 pt of copper/65 pt of protargol and 5% to 10% sodium sulfite in 1% hydroquinone are suggested.

13.1.14 *Rogoff's Bodian [13]*

Application:

Recommended for the histological preparation of the mosquito nervous system.

Formula:

Solution A: Modified Petrunkevitch

Alcohol (50%)	100 pt
Nitric Acid	3 pt
Ether	5 pt
Cupric Nitrate	2 pt
Paranitro-phenol	5 pt
Formalin (20% – just before use)	12 pt

Solution B:

Protargol	100 pt
Metallic Copper	4–6 pt

Solution C: Reducer

Hydroquinone	5 pt
Sodium Sulfite	5 pt
Water	100 pt

Solution D: Toner

Gold Chloride	1% containing
Glacial Acetic Acid	3 drops/100 pt of solution

Procedure:

1. Fix in solution A.
2. Dehydrate with graded series of alcohols or with dioxane.
3. Clear with xylene.
4. Embed in Tissue-mat or paraffin.
5. Section and adhere to clean slide.
6. Transfer to solution B for 24 to 48 hr at 37°C.
7. Wash in distilled water.
8. Transfer to solution C for 5 to 10 min.
9. Wash in three changes of distilled water.
10. Transfer to solution D.
11. Rapidly dip in distilled water.
12. Place in 2% oxalic acid until faint bluish-purple tinge appears (2 to 5 min).
13. Wash in distilled water.
14. Transfer to 5% sodium thiosulfate for 5 to 10 min.
15. Wash in distilled water.
16. Dehydrate and mount in damar or balsam.

Note:

1. A fixative with 25% formalin is also appropriate.
2. Ten micron sections are suggested.
3. Clean slides are essential; a K_2CrO_4-H_2SO_4 mixture is suggested followed by a water rinse.
4. The protargol solution may be prepared with copper wire shavings or sheets and used only once.
5. It is suggested that the copper be cleaned with nitric acid before use.

13.1.15 *Modified Power's [14]*

Application:

1. Recommended for the histological preparations of neural elements.
2. Recommended as a procedure for use on cockroach and grasshopper material.

Formula:

Solution A: Modified Dubosq-Brazil

Picric Acid (0.67% in 80% alcohol)	5 pt
Concentrated Formalin (37% HCHO)	2 pt
Glacial Acetic Acid	0.5 pt

Procedure:

1. Fix in solution A for 12 to 18 hr.
2. Transfer specimen to 10% ammonium acetate (in 80% alcohol until picric acid is removed).
3. Dehydrate in alcohol series.
4. Embed in Paraplast.
5. Cut sections, adhere to slides and hydrate.
6. Infiltrate with 2% protargol in Coplin jars with at least 40 pt of bright metallic copper for 16 hr at 37°C.
7. Rapidly wash three times with distilled water.
8. Reduce by placing in 1% hydroquinone (in 5% Na_2SO_3) for 15 min.
9. Rapidly wash three times in distilled water.
10. Repeat step 6 but for not more than 8 to 10 hr.
11. Rapidly wash three times in distilled water.
12. Reduce with 1% hydroquinone (in 5% Na_2SO_4) for 15 min.
13. Rapidly wash three times in distilled water.
14. Transfer to 2% oxalic acid for 1.5 min.
15. Wash in distilled water three times.
16. Transfer to 1% gold chloride for 5 min.
17. Repeat steps 13 to 14 but for at least 6 min each (observe reaction under microscope).
18. Wash in distilled water three times for 2 min.
19. Transfer to 5% $Na_2S_2O_3$ for 10 min.
20. Wash well in distilled water.
21. Dehydrate, clear, and mount.

Note:

1. Any paraffin-type material may be used for embedding.
2. Ten micron sections are recommended.
3. Reduction in step 12 is done with Na_2SO_4, not Na_2SO_3.
4. For embryonic nerve tissues repeat with steps 7, 8, 9, and 10 but for 16 hr.
5. Freshly prepared solutions of protargol, hydroquinone, and oxalic acid should be used.
6. A 150 pt solution of gold chloride should not be used for more than 40 slides.
7. In step 17 the sections should be kept in oxalic acid for 3 min. beyond the point where fibers show against pink background.

Result:

All size nerve fibers stain in distinct colors ranging from purple to shades of red.

13.1.16 *Cajal's No. V [15]*

Application:

Recommended for the histological preparation of insect motor nerve ending.

Formula:

Solution A: Ammoniacal Alcohol

Alcohol (96%)	50 pt
Ammonia	4 drops

Solution B: Reducer

Hydroquinone	2 pt
Formol (40%)	10 pt
Distilled Water	100 pt

Procedure:

1. Fix in a 10% chloral hydrate solution for 24 hr.
2. Rinse in distilled water.
3. Transfer to solution A for 24 hr.
4. Transfer to a 1.5% silver nitrate solution for 4 days at 35°C.
5. Wash quickly in distilled water.
6. Transfer to solution B for 24 hr.
7. Dehydrate and embed quickly in paraffin.
8. Cut sections, remove wax and wash in absolute alcohol.
9. Transfer to xylol and mount.

Note:

1. Silver nitrate will stain fat a black color thus all such tissues should be removed.

13.1.17 *DeFano's technic [15]*

Application:
Recommended for the preparation of insect brain tissues.

Formula:
Solution A: DeFano's Fixative

Cobalt Nitrate	1 pt
Formol (40%)	15 pt
Distilled Water	100 pt

Solution B: Reducing Fluid

Hydroquinone	2 pt
Formol (40%)	15 pt
Sodium Sulfite	0.5 pt
Distilled Water	100 pt

Solution C:

Sodium Hydrosulfite	3 pt
Ammonium Sulfocyanide	3 pt
Distilled Water	100 pt

Solution D:

Gold Chloride	1 pt
Distilled Water	100 pt

Solution E:

Solution C	1 pt
Solution D	1 pt

Procedure:

1. Fix fresh materials in solution A for 3 to 8 hr.
2. Rinse in distilled water.

3. Place in aqueous 1.5% silver nitrate solution at room temperature for 36 to 48 hr.
4. Rinse in distilled water.
5. Place in solution B.
6. Wash in running water for 5 min.
7. Dehydrate with alcohol (e.g., 25, 50, 75, and 96% for 25 min and absolute for a little over 25 min).
8. Clear in cedarwood oil and embed.
9. Cut sections and mount.
10. Remove wax with xylol and hydrate.
11. Place in solution E for 5 to 10 min.
12. Wash in running water.
13. Dehydrate and transfer to xylol.
14. Mount.

13.1.18 *Bielschowski's technic [15]*

Application:
Recommended for the histological preparation of nerve tissue in insects.

Formula:
Solution A: Ammoniacal Silver

Silver Nitrate (10%)	4 pt
Sodium Hydrate (40%)	3 drops
Ammonia (strong)	as needed
Distilled Water	as needed

Preparation:
1. Add sodium hydrate to silver nitrate.
2. Add ammonia until precipitate is dissolved.
3. Add water up to 20 pt.

Solution B: Toner

Gold Chloride (2%)	3 drops
Distilled Water	20 pt

Procedure:
1. Fix in formol (6%) for 7 days.
2. Wash in running water for 3 hr.
3. Fix in pyridine for 2 days.
4. Wash in water for 24 hr.
5. Transfer through changes of distilled water over 24 hr.
6. Transfer to silver nitrate (3%) for 4 days in darkness.
7. Rinse in distilled water.
8. Transfer to solution A for about 5 hr.
9. Rinse in distilled water.
10. Transfer to formol (8%) for 12 hr.
11. Wash in distilled water.
12. Dehydrate, clear in xylol and embed in paraffin.
13. Cut and mount specimens.
14. Dissolve wax in xylol and hydrate.

15. Transfer to solution B.
16. Fix in sodium hyposulfite (4%) for 5 min.
17. Wash in distilled water for 30 min.
18. Dehydrate and mount in xylol balsam.

13.1.19 *Bretschneider's [15]*

Application:
1. Recommended as a procedure for the display of brain tissues.
2. Recommended for use on cockroach and beetle brain tissues.

Formula:
Solution A: Mallory's Mixture

Aniline Blue	0.5 pt
Orange G	2 pt
Oxalic Acid	2 pt
Distilled Water	100 pt

Procedure:
1. Fix in 10% formol for 4 to 6 hr.
2. Dehydrate and embed in paraffin.
3. Section and mount sections.
4. Stain in aqueous eosin solution (about 1%) for 20 to 30 min.
5. Wash in distilled water.
6. Stain in Delafield's hematoxylin for 30 sec.
7. Wash in water.
8. Place in 1% phosphomolybdic acid for 2 to 3 min.
9. Wash in water.
10. Stain in solution A for 6 sec.
11. Wash in water.
12. Dehydrate and mount.

13.1.20 *Buxton's technic [16]*

Application:
Recommended for the histological preparation of brain tissue.

Procedure:
1. Dissect specimen.
2. Place material in 1% aqueous silver nitrate (in darkness) for 10 days.
3. Wash in distilled water.
4. Dehydrate and embed in paraffin.
5. Section and attach to slides.
6. Remove wax and hydrate.
7. Place in a 1.5% aqueous silver nitrate solution and exposed to sunlight for 10 min.
8. Wash in distilled water.
9. Place in a 1% gold chloride solution in bright light for 2 min.
10. Wash and reduce in an aqueous pyrogallic acid solution (about 2%).
11. Dehydrate and mount.

Note:
1. It is suggested that this method not be used with tissues containing pigment cells.

13.1.21 *Modified Gomori's AF technic [17]*

Application:
Recommended as a procedure applicable for the preparation of slides of neurocrine organs of insects.

Formula:
Solution A: Oxidant

$KMnO_4$ (in 100 pt water)	0.3 pt
H_2SO_4 (concentrated)	0.30 pt

Solution B: Stain

Distilled Water	200 pt
Dextrine	0.5 pt
Victoria Blue 4R	2 pt
Resorcin	4 pt

Solution C: Halmi's Mixture

Distilled Water	100 pt
Light green SF yellowish	0.2 pt
Orange G	1 pt
Chromotrope 2R (may be omitted)	0.5 pt
Phosphotungstic Acid	0.5 pt
Glacial Acetic Acid	1 pt

Procedure:
1. Expose tissues and fix in Bouin's for 12 to 24 hr.
2. Wash well in alcohol (70%).
3. Hydrate down to distilled water.
4. Transfer to solution A for 2 to 3 min.
5. Blot excess fluid with filter paper.
6. Transfer through two to three changes of sodium bisulfite (4%) for 1 to 10 min. or until bleached (white).
7. Wash in distilled water for 5 to 10 min.
8. Transfer through alcohol series (30 to 70%).
9. Transfer to solution B for 2 to 10 min.
10. Blot off excess fluid quickly with filter paper.
11. Differentiate in alcohol (95%), two or more times, until no stain is noted in fluid.
12. Dehydrate.
13. Clear in cedarwood oil for 2 to 4 hr.
14. Transfer to xylene for 2 min.
15. Mount intact specimen, or,
16. Dissect components, embed in paraffin wax, cut sections and mount in Canada balsam as usual.
17. Counterstain in solution C for about 45 sec.

Note:
1. In step 11 if a precipitate forms around specimen transfer to alcohol (70%) until it dissolves.
2. The bleaching in step 6 should be observed under a microscope.

13.1.22 *Humberstone's performic-Victoria blue technic [17]*

Application:

1. Recommended for the histological display of neuorsecretory material in paraffin sections of insect brains.
2. Recommended as particularly useful for the preparation of insect brain tissue.

Formula:

Solution A: Oxidant (Performic Acid)

Formic Acid (98%)	20 pt
Hydrogen Peroxide (30%)	2 pt

Solution B: Stain

Distilled Water	200 pt
Dextrine	0.5 pt
Victoria Blue 4R	2 pt
Resorcin	4 pt
Ferric Chloride (29%)	25 pt
Alcohol (70%)	400 pt
HCl (concentrated)	4 pt
Phenol Crystals	4 pt

Preparation:

1. Mix first four ingredients and boil.
2. Add ferric chloride.
3. Boil for 3 min.
4. Cool and filter.
5. Dry precipitate in an oven at 50°C.
6. Dissolve precipitate in alcohol.
7. Add HCl and phenol.
8. Let stand for 2 weeks before use.

Procedure:

1. Fix brain in formaldehyde-saline (10%).
2. Embed in paraffin, section and affix to slide.
3. Transfer to xylene.
4. Transfer to alcohol.
5. Allow sections to dry.
6. Add solution A on sections drop-wise for 5 min.
7. Wash in distilled water for 15 min.
8. Rinse in alcohol (70%).
9. Transfer to solution B for 12 hr.
10. Wash in alcohol (70%).
11. Rinse in water.
12. Counterstain in 0.1% safranin (in 1% acetic acid) for 5 min.
13. Rinse in water.
14. Dehydrate, clear, and mount.

Note:

1. Do not keep in H_2O_2 (used in preparing the oxidant) for more than 3 weeks after opening.
2. Performic acid should be used within 24 hr but must stand for 1 min before use.

13.1.23 *Modified Humberstone technic [17]*

Application:
Recommended as a bulk-processing technique for the display of the neuroendrocrine system of insects.

Formula:
Solution A: Formaldehyde-saline

Physiological Saline	90 pt
Formalin	10 pt

Solution B: Oxidant

Formic Acid (98%)	20 pt
Hydrogen Peroxide	2 pt

Solution C: Stain

Distilled Water	200 pt
Dextrine	0.5 pt
Victoria Blue RN 275	2 pt
Resorcin	4 pt

Procedure:
1. Expose brain or other organ of anesthetized or live specimen to saline.
2. Fix in place in solution A for 2 to 3 hr.
3. Dissect out organs and place in fixative for 24 to 36 hr.
4. Wash well in tap water for 2 to 3 hr.
5. Wash in distilled water for 10 to 20 min.
6. Blot excess water with filter paper.
7. Transfer to solution B for 5 min or until transparent.
8. Blot excess fluid with filter paper.
9. Wash several times with distilled water for 20 to 30 min.
10. Dehydrate through series of alcohol solutions (30 to 70%).
11. Transfer to solution C for 12 to 18 hr.
12. Blot excess fluid quickly with filter paper.
13. Differentiate in several changes of alcohol (70%) until stain is not evident.
14. Dehydrate.
15. Clear in cedarwood oil for 2 to 4 hr.
16. Transfer to xylene for 2 min.
17. Mount intact specimen, or dissect components, embed in paraffin wax, cut sections and mount in Canada balsam as usual.
18. Counterstain if needed with safranin.

13.1.24 *Ittycheriah and Marks' technic [18]*

Application:
1. Recommended as a method for the *in situ* demonstration of neurosecretory material in insects.
2. Recommended as particularly useful for tissues of Lepidoptera, Dictyoptera, Diptera, and Hymenoptera.

Formula:
Solution A: Modified Bouin's

Picric Acid (saturated aqueous)	75 pt
Formalin	25 pt
Trichloroacetic Acid	0.5 pt

Solution B: Performic Acid (Oxidant)

Formic Acid (98%)	20 pt
Hydrogen Peroxide (30%)	2 pt

Solution C: Stain

Resorcin Fuchsin	1 pt
Ethanol (70%)	98 pt
HCl (concentrated 12 N)	2 pt

Procedure:

1. Dissect out brain or other tissues in saline.
2. Fix in solution A for 4 to 25 hr.
3. Wash with water containing a few crystals of lithium carbonate.
4. Transfer to solution B for 5 min or more.
5. Blot excess fluid with filter paper.
6. Wash several times in water for 20 to 30 min.
7. Transfer through alcohol series (30 to 70%).
8. Transfer to solution C for 1 to 30 min.
9. Blot excess fluid with filter paper.
10. Differentiate in several changes of alcohol (70%) until stain is not evident in fluid.
11. Dehydrate in alcohol series.
12. Clear in cedar oil for 4 hr or more.
13. Transfer to xylene.
14. Mount whole tissue in Permount or Euparal, or embed in paraffin wax and section as usual.
15. Counterstain with Ehrlich's hematoxylin and light green.

Note:

1. Solution B should stain for 1 hr and should be used within 24 hr.
2. Solution C should be kept stoppered at room temperature and it remains stable for at least 20 days.
3. Amount of time for staining is dependent on the size of the specimen.
4. If further differentiation is required (step 10) use acid alcohol and wash well in alcohol (70%).

13.1.25 *Paget's Aldehyde-Thionin technic [17]*

Application:

Recommended for the histological preparation of the neurocrine organs of insects.

Formula:

Solution A: Oxidant

H_2SO_4 (concentrated)	0.5 pt
$KMnO_4$ (0.5%)	100 pt

Solution B: Stain

Distilled Water	100 pt
Light green SF yellowish	0.2 pt

Orange G	1 pt
Chromotrope 2R (may be omitted)	0.5 pt
Phosphotungstic Acid	0.5 pt
Glacial Acetic Acid	1 pt

Procedure:

1. Expose tissues and fix in Bouin's for 12 to 24 hr.
2. Wash well in alcohol (70%).
3. Hydrate down to distilled water.
4. Transfer to solution A for 2 to 3 min.
5. Blot excess fluid with filter paper.
6. Transfer through two to three changes of potassium metabisulfite for 2 to 10 min, or until completely white.
7. Wash in distilled water.
8. Transfer to solution B for 15 min to 2 hr.
9. Differentiate in distilled water for 2 min.
10. Dehydrate.
11. Clear in cedarwood oil for 2 to 4 hr.
12. Transfer to xylene for 2 min.
13. Mount intact specimen, or dissect components, embed in paraffin wax, cut sections and mount in Canada balsam as usual.
14. Counterstain in solution C for about 45 sec.

Note:

1. To display the secretory neurons and neurosecretary pathways in a whole mount remove brain sheath by pressing gently with forceps, loosening sheath and peeling sheath off.
2. The sheath may be removed during differentiation in cedarwood oil.

13.1.26 *Wilson's technic [19]*

Application:

Recommended for the histological display of nerve endings in insect tissues.

Formula:

Solution A:

Formic Acid (25%)	100–200 pt

Solution B:

Gold Chloride (1% aqueous)	10–20 pt

Procedure:

1. Separate muscle from surrounding tissues.
2. Place in a solution A for 15 min.
3. Blot dry and place in solution B for 10 min.
4. Blot dry and place in solution A for 40 to 50 hr in darkness.
5. Blot dry and transfer to pure glycerol.

Note:

1. When specimen is transferred to solution B do not use metal forceps.

Result:
Nerve endings appear blue-black or gray.

13.1.27 *Methylene blue technic [19]*

Application:
Recommended for nerve endings preparations of *Periplaneta americana*, *Blaberus cranifer*, *Photinus pyralis*, *Photuris* sp. and *Melanoplus differentialis*.

Formula:
Solution A: Stock

Methylene Blue Chloride	0.5 pt
Distilled Water	100 pt

Preparation:
1. Heat the above solution.
2. Stir until dissolved.
3. Filter, cool, and store.

Solution B: Unna's

HCl (concentrated)	6 drops
Solution A	30 pt
Rongalit (12% – sodium formaldehyde sulfoxalate)	6 pt

Solution C: Modified Unna's

Solution A	30 pt
Rongalit™	6 pt

Preparation:
1. Add Rongalit to solution A.
2. Warm gently and stir until it starts to turn deep dirty green.
3. Remove heat and stir until colorless.
4. Cool, filter through fine filter paper and store in dark in air-tight container.

Solution D: Dithionite-reduced

$Na_2S_2O_4$ (Sodium Dithionite – 0.25M)	1 pt
Solution A	5 pt

Preparation:
1. Add freshly made NaS_2O_4 to solution A.
2. Stir until blue color disappears, filter and use immediately.

Solution E:

Ammonium Picrate

Preparation:
1. Add 3N NH_4OH drop by drop to saturated aqueous picric acid solution.
2. Stir continuously until pH reaches 7.
3. Store in refrigerator and use cold.

Solution F:

Ammonium Molybdate

Preparation:

1. Either an 8% solution or a saturated solution (saturated at 5°C) may be used.
2. Store in refrigerator and use cold.

Procedure:

1. Inject solution B into abdomen (for posterior ganglia) or into dorsal neck area (for anterior ganglia).
2. Introduce enough to slightly distend abdomen.
3. Dissect out viscera to expose nerve cord ½ to 2 hr after injection.
4. Fill the body cavity with solution E and let stand for 10 min.
5. Replace with solution F to cover whole specimen and store at 5°C for 12 to 48 hr.
6. Remove all fatty tissue attached to nerve cord.
7. Remove the nerve cord and wash well several times with distilled water.
8. Orient on albumenized slide, draw off water and let dry until firmly held on slide.
9. Dehydrate in two changes of t-butyl alcohol for 1 to 2 hr each.
10. Clear in two changes of xylene for 15 min each.
11. Mount.

Note:

1. For small species, several injections may be more effective.
2. All three leuco solutions (B, C and D) can be used but quality of staining depends on a) solution used, b) volume injected, c) time between injection and dissection, and d) the insect.

13.1.28 *Gaengel eye technic [20]*

Application:

Recommended for SEM of intact eyes and head structures in segmented insects.

Procedure:

1. Decapitate insect and cut away one of the two eyes to allow for fixative penetration into the head.
2. Fix heads in 2% glutaraldehyde in phosphate buffered saline (PBS) on ice for 10–30 min in centrifuge tube.
3. Spin tube gently to bring heads to base of tube.
4. Add an equal volume of 4% OsO_4 in PBS and incubate on ice for 30–60 min.
5. Remove solutions and replace with 1 pt 4% OsO_4 for 1 hr on ice.
6. Dehydrate on ice through a graded series of ethanols to 100% ethanol.
7. Wash in propylene oxide two times for 10 min at room temperature.
8. Immerse in a mixture of 1 pt propylene oxide and 1 pt resin overnight at room temperature.
9. Immerse heads in pure resin for 2–4 hr at room temperature.
10. Embed heads so that intact eye is oriented upwards for sectioning.

13.1.29 *Vademecum for standardized fixation, embedding and sectioning of the* Drosophila *facet eye [21]*

Formula:

Optimal fixation solution

Glutaraldehyde	2.5%
Household detergent	0.1%
Sodium Cacodylate buffer pH 7.4	0.1M

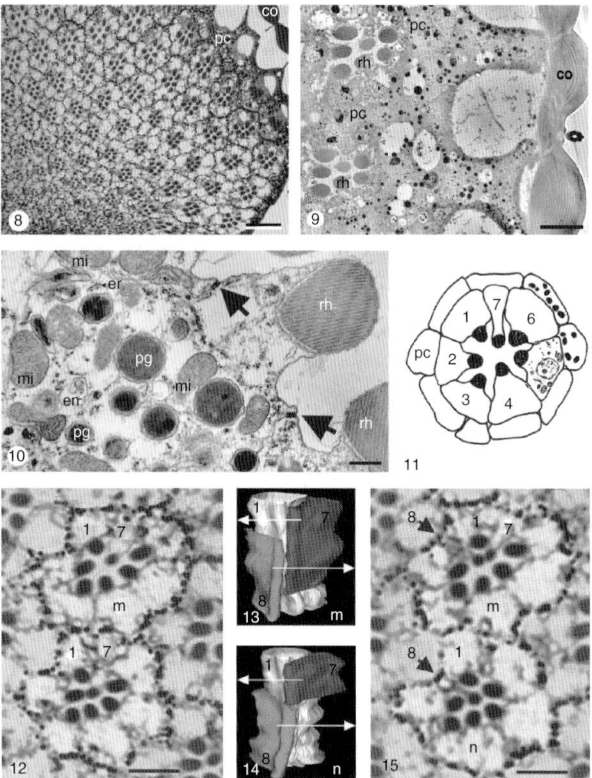

Fig. 13.1 LM-overview on 0.5-lm thick semi-thin cross section through p14-depleted *Drosophila* facet eye showing the cornea (co), pigment cells (pc), and the hexagonal unit eyes, the ommatidia. Toluidine blue staining, scale bar ¼ 10 lm. Fig. 9 and 10. TEM of ultrathin sections reveal generally acceptable, but not perfect ultrastructure preservation of the *Drosophila* compound eye after fixation with glutaraldehyde followed by OsO4; section post-staining with uranyl and lead salts. Fig. 9. Overview showing cornea (co), pigment cells (pc), and cross-sectioned ommatidia (of which the rhabdomere of photoreceptor cell 4 is marked by rh); scale bar ¼ 5 lm. Fig. 10. Subcellular architecture of a photoreceptor cell typically comprising mitochondria (mi), pigment granules (pg), multivesicular endosomes (en), endoplasmic reticulum (er), the apical rhabdomere (rh), and cell-to-cell contacts (marked by arrows; nucleus and Golgi apparatus are not hit in this section); scale bar ¼ 200 nm. Fig. 11. Morphology of *Drosophila* adult ommatidium as seen in apical cross section (schematized from TEM-micrographs and in accordance with Wolff and Ready, 1993: Fig. 2, p 1280). A honeycomb-like mesh of pigment cells (pc) surrounds the photoreceptor cells R1–R7 (R8 locates between R1 and R2, but below the virtual section plane). Each photoreceptor cell comprises a cell body and an apical, pseudocrystalline dense microvillous structure, the rhabdomere. Figs. 12 and 15. Semi-thin cross sections through the lower part of a normal (n) vs. a structurally modified (m) ommatidium in p14 depleted *Drosophila* eye. Morphological preservation and resolution are good enough to permit clear distinction of the characteristic contours of photoreceptor cells R7 and R8 that differ from each other, and from their neighbor R1 (as well as from R2 to R6). In normal ommatidia (n) the rhabdomeres of R7 and R8 do not occur within the same section, as it is the case in structurally aberrant ommatidia (m). The sections are out of a series used for the 3D reconstructions shown in Figures 13 and 14, where their respective positions are indicated by arrows. LM, section thickness ¼ 0.6 lm, Toluidine blue staining; scale bar ¼ 5 lm. Figs. 13 and 14. Three-dimensional modeling of p14-depleted *Drosophila* ommatidia: the segment of the photoreceptor cells R1, R7, and R8, shown here, is from the area where the central position is occupied by R7 and/or R8. Arrows roughly indicate the positions of the semi-thin sections shown in Figures 12 and 15. Normal ommatidia (Fig. 14) show an abrupt change from R7 to R8. By contrast, in structurally aberrant ommatidia (Fig. 13) R7 is slightly enlarged that leads to an 5-lm long overlap of R7 with R8. (Fig. 8). (Source: Hess *et al.* 2002. Reproduced with permission of John Wiley & Sons.) See plate section for the color version of the figure.

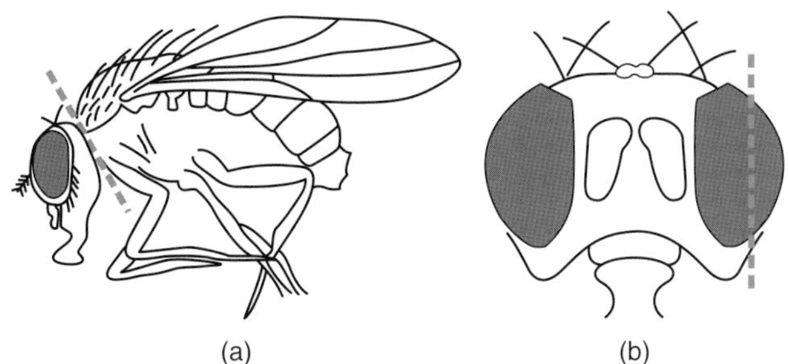

Fig. 13.2 Demonstration of cutting techniques for preparing *Drosophila* retinas for semithin plastic sections. (Source: Gaengel & Mlodzik 2002. Reproduced with permission of Springer.)

Procedure:

1. Anesthetize adult flies with CO_2 and immerse into fixative.
2. Under a dissecting scope, hold the abdomen with tweezers and open the cuticle of the fly's face between the eyes with a fine sharp tip of a tuberculin syringe needle.
3. Cut the head off with a scalpel and continue fixation for at least 4 hr at room temperature or overnight at 4°C.
4. Rinse heads thoroughly in several changes of buffer over 30–60 min.
5. Post fix in 1% unbuffered aqueous OsO_4 for 1–2 hr at 4°C.
6. Dehydrate in graded ethanol series.
7. Rinse three times in anhydrous acetone.
8. Infiltrate heads with graded series of Epon epoxy resin mixed with anhydrous acetone for 3 days at room temperature with continuous agitation.
9. Place heads into silicone flat embedding molds with their occipital plane at the bottom.
10. Polymerize the resin overnight at 60°C.
11. Glue the flat embedded head upside-down onto a cylindrical or square resin block.
12. The combined block is mounted into motor-driven trimming device.
13. Tilt and rotate the specimen to orient the occipital plane of the head to almost parallel to the milling and sectioning plane. Verify orientation by semi-thin sectioning through the fly's occiput.
14. Glue another cylindrical or square resin block to sandwich the original block.
15. Trim the sandwiched block and mount into an ultramicrotome with the head's midline running strictly parallel to the horizontal or vertical axis of the microtome. Verify orientation by semi-thin sectioning and adjust tilt as necessary.
16. Section 0.6-μm semithin sections and/or 80-nm ultrathin sections.

13.1.30 *Preparation of cuticles from uneclosed first-instar* Drosophila *larvae [22]*

Procedure:

1. Prepare egg laying chamber, collect eggs on juice-agar plates for 3 hr, remove adult flies and age eggs for 18 hr at 25°C.
2. Add tap water to cover the eggs, use a paint brush to dislodge the eggs and transfer to an egg sieve.
3. Wash eggs thoroughly in tap water.
4. Place sieve in dish with 12% sodium hypochlorite for 2–3 min.
5. Wash eggs with tap water.
6. Mix 500 μl each of heptane and methanol in 1.5 ml eppendorf.
7. Dry the mesh of the egg sieve by blotting the bottom of the sieve on a paper towel.

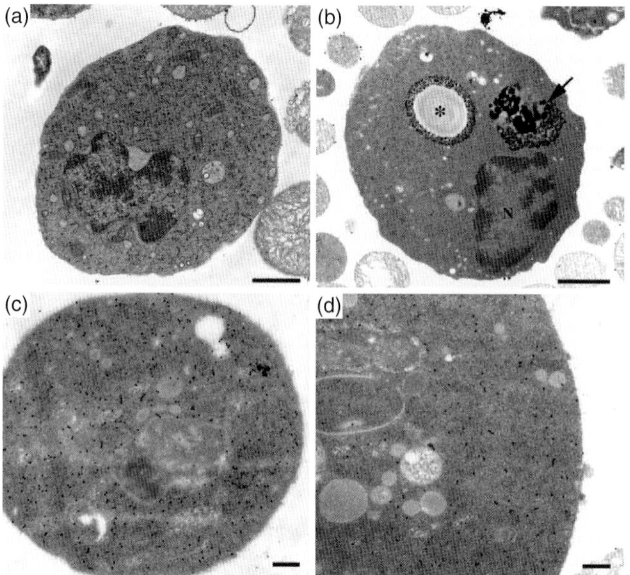

Fig. 13.3 (a) Oenocytoid, characterized by having an eccentric nucleus, no prominent nucleolus, and a homogeneous cytoplasm. (b), Oenocytoid with crystalline-like inclusion (arrow) and a lipid droplet (asterisk). The crystalline-like inclusion and the dark material surrounding the lipid droplet are probably products of a phenoloxidase reaction inside the cell. (c) Colloidal gold labeling of prophenoloxidase in the cytoplasm of an oenocytoid. (d) Colloidal gold labeling of peroxidase in the cytoplasm of an oenocytoid. N Nucleus. Bars: (a) 1 μm; (b) 2 μm; (c, d) 200 nm. (Source: Hillyer & Christensen 2002. Reproduced with permission of Springer.)

8. Swirl a dry paintbrush around the bottom of the egg sieve to collect the eggs on the paintbrush.
9. Shake the brush in the heptane top phase of the heptane:methanol mix to release the eggs from the paintbrush. Repeat steps 7 and 8 until all eggs have been transferred to the heptane.
10. Close the eppendorf tube and shake vigorously for 30 sec.
11. Let embryos settle to the bottom of the tube and remove heptane and methanol.
12. Wash three times with fresh methanol, 1 ml each.
13. Wash two times in methanol with 0.1% Triton X-100.
14. Use a glass pipette to transfer embryos to a clean slide, use a tip of filter paper or paper towel to wick away most of the liquid. Do not let the embryos dry out.
15. Add a drop of Hoyer's medium (see Chapter 8, or 3:1 lactic acid:H_2O) to embryos.
16. Spread and arrange the embryos with fine forceps and cover with coverslip.
17. Incubate for 1 hr at 60°C.
18. Seal the edges of the coverslip with nail polish.

13.1.31 *Preparations of cuticle from feeding* Drosophila *larvae – needle prick method [23]*

Procedure:

1. Prepare an egg laying chamber, allow females to deposit eggs for 24 hr, remove females and allow larvae to grow to desired stage. Transfer larvae to a 1.5 ml eppendorf tube with 0.5 ml phosphate buffered saline.
2. Incubate the larvae at 4°C for 15 min.
3. Transfer larvae to 70% glycerol. Prick the larvae with fine tungsten needle.

4. Transfer larvae to 1.5 ml eppendorf tube with fixative (4:1 glycerol:acetic acid). Incubate at 60°C overnight.
5. Transfer larvae to a drop of Hoyer's medium (see Chapter 8 or 3:1 lactic acid/H_20) on a clean slide.
6. Spread out and arrange the larvae, then place coverslip over preparation.
7. Incubate at 60°C overnight.
8. Seal edges of the coverslip with nail polish.

13.1.32 *Preparations of cuticle from feeding* Drosophila *larvae – pressure method [23]*

Formula:

Perenyi Fixative:

dH_2O	51 pt
1% Chromic Acid	15 pt
Nitric Acid	4 pt
95% Ethanol	30 pt

Procedure:

1. Prepare an egg laying chamber, allow females to deposit eggs for 24 hr, remove females and allow larvae to grow to desired stage. Transfer larvae to a 1.5 ml eppendorf tube with 0.5 ml phosphate buffered saline.
2. Incubate the larvae at 4°C for 10 min.
3. Transfer larvae to Perenyi fixative for 20 min.
4. Wash the larvae in dH_2O two times.
5. Transfer larvae to a drop of dH_2O on a clean slide and remove excess liquid with the tip of a piece of filter paper or paper towel.
6. Add a drop of Hoyer's medium (see Chapter 8 or 3:1 lactic acid/H_20) over the larvae.
7. Spread out and arrange the larvae, then place coverslip over the preparation.
8. Incubate at 60°C overnight or until clear.
9. Seal edges of the coverslip with nail polish.

13.1.33 *Millares' technic [24]*

Application:

1. Recommended for the simultaneous fixation and staining of the internal organs of Culicidae.
2. Recommended as a rapid method of preparation of insect tissue slide mounts.

Procedure:

1. Dissect the internal organs in a saline solution.
2. Fix and stain in a Giemsa solution for 30 min.
3. Destain with absolute methyl alcohol.
4. Clear and dehydrate with terpineol.
5. Clear in cedarwood oil.
6. Mount.

Note:

1. Giemsa solution is prepared with absolute methyl alcohol in a ratio of 2 drops of Giemsa for every pt of alcohol.
2. The Giemsa solution should be freshly made and no more than 5 days old.
3. Stain in covered dishes will undergo color changes and precipitation after 15 min.
4. Use non-blood fed individuals and dissect soon after killing.

13.1.34 *Schneider and Rudinsky's technic [25]*

Application:

1. Recommended for histological preparations of insect malphigian tubules, salivary glands, oviduct and ovarioles, digestive organs, fat bodies, and muscles.
2. Recommended as particularly useful for several species of ambrosia beetles.

Procedure:

1. Fix insect in hot Schandinn's (which consists of two parts saturated aqueous solution of corrosive sublimate (i.e. mercuric chloride) to one part absolute alcohol).
2. Wash in specimens in weak solution of iodine in alcohol (70%) for 24 hr.
3. Dehydrate in ethanol (70 and 96%), isopropanol and xylol for 24 hr each.
4. Infiltrate in paraffin (52°C) and then in paraffin (60°C) and section.
5. Embed in paraffin (60°C) and section.
6. Stain with Weigert's hematoxylin.
7. Differentiate in ferrichloride.
8. Dehydrate and mount in Canada Balsam.

13.1.35 *Cameron's technic [26]*

Application:

1. Recommended for the general display of insect tissues including heart, brain, nerve cord, digestive system, malphigian tubules, fat bodies, etc.
2. Recommended as a procedure well demonstrated in the American cockroach, *Periplaneta americana*.

Procedure:

1. Fix in Duboscq-Brazil (see Chapter 2) for 24 to 48 hr.
2. Transfer through several changes of alcohol (90%).
3. Transfer through two changes of absolute alcohol for 24 hr.
4. Transfer through alcohol (95%) and then into terpineol for several weeks, or,
5. Transfer to benzoate-celloidin for several weeks.
6. Embed in paraffin.
7. Section, stain, and mount.

Note:

1. In step 6, when using terpineol, several changes of paraffin are needed or until its characteristic odor is absent.
2. Any increase in celloidin above 1% does not improve sectioning.

13.1.36 *Hazelton Malpighian tubule technic [27]*

Application:

Recommended for the preparation of Malpighian Tubules (Mt).

Procedure:

1. As Mt are highly O_2-sensitive, dissect Mt in insect Ringer solution that is constantly oxygenated with water saturated carbogen (also known as Meduna's Mixture).
2. Remove the basal lamina by enzymatic digestion in 0.4% elastase in Ringer's for 30 min at room temperature.
3. Fix in 2.5% glutaraldehyde in 0.1M sodium cocadylate buffer for 18–19 hr at 4°C.
4. Dehydrate in graded alcohol series.
5. Dry in hexamethyldisilazane (also known as Bis(trimethylsilyl)amine) and proceed with SEM prep.

13.1.37 *Modified Shute's technic [28]*

Application:
Recommended for the preparation of mosquito midguts.

Formula:
Solution A:

Hematoxylin Crystals	1 pt
Alcohol (90%)	100 pt

Solution B:

Liquid Ferri Sesquichloride (sic)	4 pt
HCl	1 pt
Distilled Water	95 pt

Solution C:

Solution A	0.5 pt
Solution B	0.5 pt
Alcohol (50%)	25 pt

Procedure:
1. Draw the fixative (Bles' – See Chap. 2 Fixation) under coverslip until specimen becomes cloudy (a few seconds).
2. Cover entire slide with fixative for 5 min.
3. Lift coverslip, flush specimen with alcohol (70%) and then cover with alcohol (70%) for 5 min.
4. Replace with alcohol (50%) for 5 min.
5. Place slide in solution C for 30 sec (or drop solution on to slide).
6. Wash specimen with alcohol (50%).
7. Cover with alcohol (70%) for 5 min.
8. Counterstain with weak Bordeaux red alcohol solution.
9. Wash with absolute alcohol.
10. Cover with absolute alcohol for 5 min.
11. Draw off alcohol and mount with Euparal.

13.1.38 *Saxena's technic [29]*

Application:
Recommended for the histological display of salivary sheaths of feeding aphids.

Formula:
Solution A: Stain

Ferric Sulfate (1%)	75 pt
Potassium Ferricyanide (0.1%)	25 pt

Procedure:
1. Use 5 to 10 mm pieces of leaves with feeding aphids.
2. Fix specimen in Carnoy's for 12 hr at 4°C.
3. Dehydrate in three changes of absolute acetone for 30 min each at room temperature.
4. Place in chloroform for 1 hr.
5. Infiltrate in two changes of paraffin at 52–54°C for 30 min each.

6. Section and attach to slides.
7. Remove paraffin with petroleum ether and hydrate through a series of 90, 70, 50, and 30% ethanol.
8. Stain in solution A for 10 to 15 min.
9. Counterstain with NaOH (0.5%) if needed.
10. Dehydrate in series up to absolute ethanol, clear in xylene and mount in neutral Canada Balsam.

Note:
1. In step 5, temperatures higher than those indicated should be avoided.
2. The most appropriate staining time (step 8) can be found by running four slides through at 5, 10, 15, and 20 min.

Result:
The salivary sheaths stain dark blue against a light blue background.

13.1.39 *Guthrie, Campbell, and Bacon technic [30]*

Application:
Recommended for the histological display of aphid feeding sites.

Procedure:
1. Aphid infested leaf pieces are quickly cut and dropped into hot Farmer's Fluid (see Chapter 2) for 15 min.
2. Decant green liquid and quickly replace with hot Farmer's Fluid for 10 min.
3. Decant fluid and wash pieces with water.
4. Clear in sodium hydroxide solution (1%) at 110°C for 30 min.
5. Slowly decant and transfer through five changes of water over 24 hr.
6. Dehydrate up to absolute ethanol.
7. Place in xylene.
8. Stain with Safranin (in 50% ethanol and 50% xylene) for 2 min.
9. Transfer up through graded xylene-alcohol series to pure xylene.
10. Mount in Canada Balsam.

13.1.40 *Kislow, Barbosa, and Edwards technic [31]*

Application:
Recommended for the histological display of aphid feeding.

Procedure:
1. Fix specimen in freshly prepared cold aqueous solution of acrolein (10%) for 12 to 24 hr.
2. Transfer through two changes of 2 methoxy-ethanol at 0°C in 4 to 24 hr period (= dehydration).
3. Transfer through absolute ethanol, n-propanol and n-butanol at 0°C for 4 to 24 hr each.
4. Infiltrate in three changes of paraffin at 56–58°C over three-day period.
5. Embed in paraffin or polyester wax.

Note:
1. The recommended fixative is prepared with distilled water at 0°C.
2. Since acrolein is a tear gas, care should be exercised when it is handled.

3. An alternative but slower penetrating fixative is glutaraldehyde in 0.025M phosphate buffer, pH 6.8 (3%).
4. The size and specimen permeability should determine timing in various steps.

13.1.41 *Morgan's technic [32]*

Application:
Recommended for whole mount preparations of gonads of house flies.

Procedure:
1. Dissect out gonads in modified Ringer's solution.
2. Fix in Kahle's fixative for 12 to 24 hr.
3. Transfer through two changes of distilled water for 1 hr each.
4. Hydrolyze with normal HCl at 26°C for 10 min.
5. Transfer to HCl at 60°C for 10 min.
6. Stain with Schiff's reagent for 2 hr.
7. Destain with a sulfurous acid rinse for 10 min.
8. Transfer through two changes of distilled water for 10 min each.
9. Transfer through three changes of triethyl phosphate for 15 min each, to clear.
10. Transfer through two changes of a 1:1 solution of triethyl phosphate and xylene for 15 min each, to clear.
11. Transfer through two changes of xylene for 24 hr each.
12. Tease apart or leave intact and mount.

13.1.42 *Kramer's technic [33]*

Application:
1. Recommended for highlighting muscles of dipteran larvae.
2. Recommended for adult hemipteran insects.

Procedure:
1. Fix in Bouin's for 8 to 10 hr at 30°C.
2. Place in ethyl alcohol (50%) for 10 min.
3. Place in ethyl alcohol (70%) for 1 hr.
4. Place in ethyl alcohol (95%) for 10 min.
5. Place in eosin alcohol (0.5%) for 6 to 8 hr.
6. Place in ethyl alcohol (95%) and add oil of wintergreen drop by drop hourly over a 4 to 5 day period.
7. Transfer to oil of wintergreen.

Note:
1. Bouin's should be prepared with formalin solution neutralized with magnesium sulfate.
2. If step 6 is not done gradually, the larvae will collapse and shrivel.

13.1.43 *Hillyer hemolymph technic [34]*

Application:
Recommended for the histological preparation of hemolymph.

Procedure:
1. Perfuse hemolymph with phosphate buffer directly into microcentrifuge tube. For each sample, a pool of hemolymph from 200–300 insects is necessary for adequate volume.

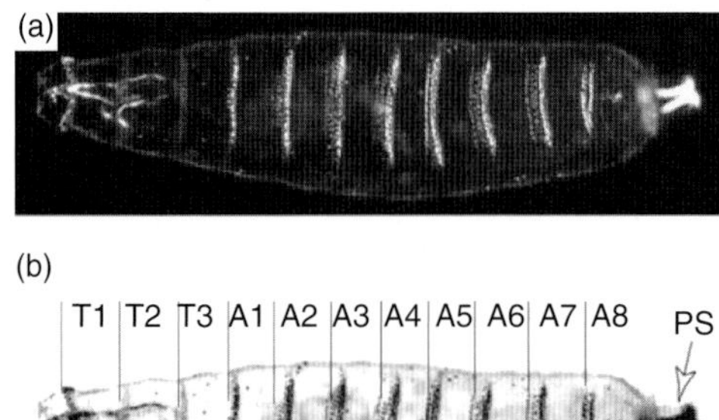

Fig. 13.4 (a) Dark-field image of the ventral cuticle of a first instar *D. melanogaster*. (b) Segments are labeled on a digitally inverted version of this image: (T1–T3) first, second, and third thoracic segments; (A1–A8) the eight abdominal segments; (PS) posterior spiracles. (Source: Stern & Sucena 2011. Reproduced with permission of Cold Spring Harbor Protocols.)

2. Fix in 1% glutaraldehyde and 0.67% formaldehyde in phosphate buffer for 20–30 min.
3. Centrifuge sample for 20 min at 800 rpm to form a cellular pellet.
4. Wash pellet and post-fix in 2% OsO_4.
5. Rinse pellet and dehydrate through graded ethanol series.
6. Embed in resin.

13.1.44 *Townsend technic [35]*

Application:
Recommended for the preservation of invertebrate membranes for SEM.

Procedure:
1. Fix specimens in 2.5% glutaraldehyde in 0.1M sodium cacodylate buffer for 18 hr at 4°C.
2. Wash specimens in cacodylate buffer.
3. Post fix in 1% OsO_4 for 1 hr and rinse with buffer.
4. Immerse specimen in 0.1% OsO_4 for 3–6 days to extract cytosol.
5. Wash specimens in cacodylate buffer.
6. Stain in 1% tannic acid for 1 hr.
7. Wash specimens in cacodylate buffer.
8. Dehydrate through graded ethanol series.
9. Immerse in hexamethyldisilizane and allow evaporation to dry sample.
10. Proceed with SEM.

13.1.45 *Trancik silk technic [36]*

Procedure:
1. Wind extracted silk gently around a 175 μm thick polyester film with grooved notches at each end to hold the silk in place and prevent sliding.
2. Wash the silk on the support film by soaking in 100% ethanol for 2–3 min.
3. Embed the silk and its support film in epoxy resin.
4. The resin block is trimmed in appropriate orientations to obtain transverse or cross sections of silk fibers.

13.1.46 *Kilgour's technic [37]*

Procedure:
1. Fix eggs in Carnoy-Lebrun (see Chapter 2) for 2 hr.

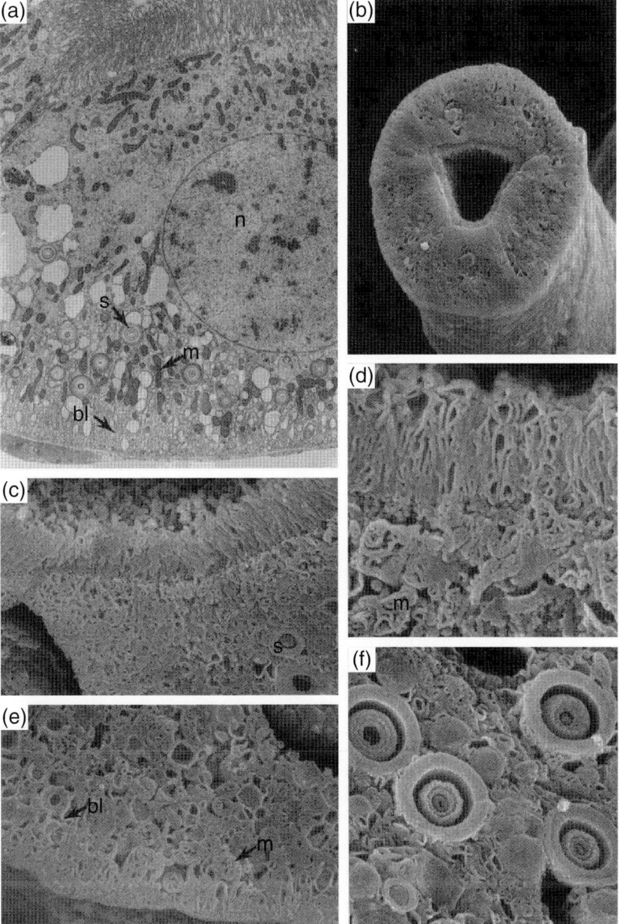

Fig. 13.5 Comparison of ultrastructural features of the cells of the Malpighian tubules of *A. domesticus* as observed by Transmission Electron Microscopy (TEM) and High Resolution Scanning Electron Microscopy (HRSEM). (a) TEM micrograph of a cross-section of a cell, showing basolateral folds, mitochondria and spherites. B-F are results obtained by using the HRSEM protocol. (b) Cross section of the midtubule revealing the monolayer epithelium with an extensive brush border. (c) Apical surface of a cell showing the abundance of microvilli extending from the apical surface, nucleus, complexity of endomembrane with mitochondria and spherites dispersed within membrane folds. (d) Higher magnification of the brush border. (e) Basolateral infolds with associated mitochondria and spherites. (f) Calcium phosphate spherites with concentric rings that appear to be separated due to the extraction of the protein matrix at the border of the lamella. Abbreviations: bl-basolateral infoldings, m-mitochondria, n-nucleus, s-spherites. (Source: Townsend *et al.* 2000. Reproduced with permission of John Wiley & Sons.)

2. Wash in alcohol.
3. Transfer to sodium hypochlorite for 2 min (0.02 qt. chlorine/100 pg H_2O) to dissolve chorion.
4. Wash in alcohol, then water.
5. Transfer through ascending alcohol series.
6. Transfer to absolute alcohol for 1 hr.
7. Transfer to benzole for 1 hr.
8. Embed in paraffin (m.p. 54°C) and section.

13.1.47 *Leopold and Palmquist technic [38]*

Application:
1. Recommended for obtaining slide-mount preparations of chromosomal aberrations.
2. Recommended as particularly useful in chromosomal studies using insect eggs.

Formula:

Solution A: Bradley-Carnoy's

Absolute Alcohol	30 pt
Chloroform	40 pt
Glacial Acetic Acid	10 pt

Solution B:

Distilled Water	100 pt
Basic Fuchsin	1 pt
Sodium (or potassium) bisulfite (or Metabisulfite)	2 pt
HCl (1 N)	20 pt
Activated Charcoal	0.3 pt

Preparation:
1. Bring water to boil and remove at once.
2. Dissolve Fuchsin and when cool to 60°C, filter.
3. Add bisulfite and HCl to filtrate.
4. Stopper and store in the dark, at room temperature, for 18 to 24 hr.
5. Add charcoal, shake well for a minute, and filter.
6. Store at 0–5°C, (solution should be yellow), keeps for several weeks.
7. Discard when pink.

Procedure:
1. Place eggs in a dilute sodium hypochlorite for 2 to 5 min.
2. Rinse in water.
3. Transfer to solution A for ½ to 2 hr.
4. Transfer through five changes of alcohol (95%) for 5 min each.
5. Transfer to a 1:1 chloroform-methanol solution overnight.
6. Hydrate through ethanol series (absolute 95, 70%, and water) for 2 to 5 min each.
7. Transfer to HCl (1 N) at 60°C for 8 to 10 min.
8. Rinse in water.
9. Transfer to solution B for 2 hr.
10. Rinse in three changes of water for 5 min each.
11. Dehydrate through alcohol series to absolute alcohol for 2 to 5 min each.
12. Clear in xylene.
13. Mount between coverslips with a synthetic mountant.

Note:
1. The dilute solution in step 1 may be obtained by diluting commercial bleach 2 to 3 times.
2. If eggs seem to require slower dehydration or hydration the addition of 50 to 30% alcohol is suggested.
3. The exposure time of the egg in solution B is a function of size, e.g., house fly eggs require 12 to 16 hr.

References

1. Brazil B.G., Lino-Neto J. & Brazil R.P. Preparation of sand fly (Diptera : Psychodidae : *Phlebotominae*) specimens for histological studies. *Memorias do Instituto Oswaldo Cruz* **98**, 287–290 (2003).
2. Sato M. & Tanaka-Sato H. Fertilization, syngamy, and early embryonic development in the cricket *Gryllus bimaculatus* (De Geer). *J. Morph.* **254**, 266–271 (2002).
3. Anderson D.S. Serial sectioning of refractory locust eggs. *Quart J. Microsc. Sci.* **105**, 379–380 (1964).
4. Slifer E.H. & King R.L. Grasshopper eggs and the paraffin method. *Science* **78**, 366 (1933).
5. Ewen A.B. An improved method for embedding and sectioning eggs of the migratory grasshopper, *Melanoplus sanguinipes Fab.* (Orthoptera: Acrididae). *Canada J. Zool.* **46**, 1084–1085 (1968).
6. Beckel W.E. Preparing mosquito eggs for embryological study. *Mosq. News* **13**, 235–237 (1953).
7. Ewen A.B. An improved aldehyde fuschin staining technique for neurosecretory products in insects. *Trans. Amer. Microsc. Soc.* **81**, 94–96 (1962).
8. Blest A.D. Some modifications of Holmes' silver method for insect central nervous systems. *Quart J. Microsc. Sci.* **102**, 413–417 (1961).
9. Larsen J.R. The use of Holmes' silver stain on insect nerve tissue. *Stain Technol.* **35**, 223–234 (1960).
10. Rowell C.H.F. A general method for silvering invertebrate central nervous systems. *Quart. J. Microsc. Sci.* **104**, 81–87 (1963).
11. Debauche H.R. A new method of silver impregnation for nerve cells and fibers. *Stain Technol.* **14**, 121–124 (1939).
12. Gregory G.E. Silver staining of insect central nervous systems by the Bodian protargol method. *Acta Zool.* **51**, 169–178 (1970).
13. Rogoff W.M. The Bodian technic and the mosquito nervous system. *Stain Technol.* **21**, 59–61 (1946).
14. Chen J.S. & Chen M.G.M. Modification of the Bodian technique applied to insect nerves. *Stain Technol.* **44**, 50–51 (1969).
15. Eltringham H. *Histological and Illustrative Methods for Entomologists.* Oxford University Press, London (1930).
16. Gray P. *The Microtomist's Formulary and Guide.* The Blakiston Company, Inc., New York (1954).
17. Dogra G.S. & Tandan B.K. Adaptation of certain histological techniques for *in situ* demonstration of the neuro-endocrine system of insects and other animals. *Quart J. Microsc. Soc.* **105**, 455–466 (1964).
18. Ittycheriah P.I. & Marks E.P. Performic acid-Resorcin Fuchsin: A technique for the *in situ* demonstration of neurosecretory material in insects. *Ann. Entomol. Soc. Amer.* **64**, 762–765 (1971).
19. Wilson S.D. *Applied and Experimental Microscopy.* Burgess Publish Company, Minneapolis, MN (1967).
20. Gaengel K. & Mlodzik M. Microscopic analysis of the adult *Drosophila* retina using semithin plastic sections. *Methods Mol. Biol.* **420**, 277–287 (2008).
21. Hess M.W. *et al.* Microscopy of the *Drosophila* facet eye: vademecum for standardized fixation, embedding, and sectioning. *Microsc. Res Tech.* **69**, 93–98 (2006).
22. Stern D.L. & Sucena E. Preparation of cuticles from unhatched first-instar *Drosophila* larvae. *Cold Spring Harbor Protoc.* 9, (2011).
23. Stern D.L. & Sucena E. Preparation of cuticles from feeding *Drosophila* larvae. *Cold Spring Harbor Protoc.* **11**, 1394–1398 (2011).
24. Millares A. Metodo rapido de simultanea figacion y coloracion selectiva de los organos internos de los culcidos. *Rev. San. Hig. Pub.* **20**, 251–257 (1946).

25. Schneider I. & Rudinsky J.A. Anatomical and histological changes in internal organs of adults *Trypodendron lineatum*, *Grathotrichus retusus* and *G. Sulcatus* (Coleoptera: Scolytidae). *Ann. Entomol. Soc. Amer*. **62**, 995–1003 (1969).
26. Cameron E. Some new methods for demonstrating the cockroach. *Proc. Roy. Entomol. Sco.* **28**, 41–44 (1953).
27. Hazelton S.R., Townsend V.R., Felgenhauer B.E. & Spring J.H. Membrane dynamics in the Malpighian tubules of the house cricket, *Acheta domesticus*. *J. Membrane Biol.* **185**, 43–56 (2002).
28. Hingst H.E. A note on making permanent preparation of anopheline mid-guts. *Amer. J. Hyg.* **22**, 278–282 (1935).
29. Saxena P.N. Staining salivary sheaths of feeding aphids; A simple Prussian blue reaction for paraffin sections. *Stain Technol.* **45**, 241–242 (1970).
30. Guthrie F.E., Campell W.V. & Bacon R.L. Feeding sites of the green peach aphid with respect to its adaptation to tobacco. *Ann. Entomol. Soc. Amer.* **55**, 42–46 (1962).
31. Kislow C., Barbosa P. & Edwards L.J. An embedding procedure for the study of aphid feeding and insect tissues. *Ann. Entomol. Soc. Amer.* **64**, 296–297 (1971).
32. Morgan P.B. A procedure for staining whole mounts of gonads of house flies. *Ann. Entomol. Soc. Amer.* **64**, 535 (1971).
33. Kramer S. A staining procedure for the study of insect musculature. *Science* **108**, 141–142 (1948).
34. Hillyer J.F. & Christensen B.M. Characterization of hemocytes from the yellow fever mosquito, *Aedes aegypti*. *Histochem. Cell Biol.* **117**, 431–440 (2002).
35. Townsend V.R., Hazelton S.R., Felgenhauer B.E. & Spring J.H. New technique for examining invertebrate membranes using high resolution scanning electron microscopy. *Microsc. Res Tech.* **49**, 209–211 (2000).
36. Trancik J.E., Czernuszka J.T., Merriman C. & Viney C. A simple method for orienting silk and other flexible fibres in transmission electron microscopy specimens. *J. Microscopy-Oxford* **203**, 235–238 (2001).
37. Kilgour D. Method of preparing microscopic sections of the eggs of *Calliphora erythrocephala*. Appendix in Scott C.M. Action of X-rays on the eggs *Calliphora*. *Proc. Roy. Soc. (B)* **115**, 100–121 (1934).
38. Leopold R.A. & Palmquist J. A method of studying early cleavage in eggs of house flies, *Musca domestica*. *Ann Entomol. Soc. Amer.* **61**, 1624–1626 (1968).

Appendix Dissecting fluids and saline solutions

A.1 Introduction

The following are saline solutions which have been recommended and used in the preliminary preparations of insect tissues. These solutions are used as an initial step or for preliminary dissections in various histological techniques.

Saline solutions used in dissection of tissues for histological preparation should be as close to isotonic with the tissues as possible. Hypo or hypertonic solutions can lead to swelling or shrinkage of the dissected tissues, leading to histological distortion. Where possible, solutions determined to be close to or in fact isotonic to the organism studied are reported below.

A.1.1 *Morgan and Labrecque's modified Ringer's [1]*

Application:
Recommended for preliminary dissection of insect tissues leading to the preparation of chromosome slide mounts.

Formula:

KCl	0.1 pt
$CaCl_2$	0.0135 pt
$NaHCO_3$	0.012 pt
NaCl	0.75 pt
Distilled water	add up to 100 pt

A.1.2 *Scott's saline [2]*

Application:
Recommended as a solution for general preliminary dissections.

Formula:

NaCl	1 pt
Distilled Water	100 pt

A.1.3 *French, Baker, and Kitzmiller saline [3]*

Application:
Recommended for tissue dissections leading to chromosomal slide preparations.

Formula:

Sodium Citrate	1 pt
Double distilled Water	100 pt

Insect Histology: Practical Laboratory Techniques, First Edition.
Pedro Barbosa, Deborah L. Berry and Christina S. Kary.

Companion Website: www.wiley.com/go/barbosa/insecthistology

A.1.4 *Belar's saline [4]*

Application:
Recommended for use in preparations of slides of mosquito chromosomes.

Formula:

Sodium Chloride	0.2 pt
Potassium Chloride	0.2 pt
Calcium Chloride	0.2 pt
Sodium Bicarbonate	0.2 pt
Distilled Water	Add to make 1 liter

A.1.5 *Modified Robb's medium [5]*

Application:
Recommended for dissection of *Drosophila* egg chambers.

Formula:

1M Calcium Chloride	1 pt
1M Glucose	10 pt
1M Hepes pH 7.2	100 pt
1M $MgCl_2$	1.2 pt
1M Potassium Acetate	55 pt
1M Sodium Acetate	40 pt
1M Sucrose	100 pt
dH_2O	To 1000 pt

A.1.6 *French, Baker, and Kitzmiller dissecting fluid [3]*

Formula:

Carnoy's	1 pt
Distilled Water	19 pt

A.1.7 *Ephrussi and Beadle Ringer's [6]*

Formula:

NaCl	7.5 pt
KCl	0.35 pt
$CaCl_2$	0.21 pt
H_2O	1000 pt

A.1.8 *Ringer's solution [7]*

Application:
Recommended as a normal salt solution for insect tissues.

Formula:

Sodium Chloride	0.80 pt
Calcium Chloride	0.02 pt
Potassium Chloride	0.02 pt
Sodium Bicarbonate	0.02 pt
Distilled Water	100 pt

A.1.9 *Glycerine [7]*

Application:
Recommended for dissections of insect tracheal systems.

Characteristics:
1. Fluid penetrates tracheae slowly so that tracheae retain their silvery appearance.
2. Tracheae are more clearly defined.

A.1.10 *Eide and Reinecke's saline [8]*

Application:
Recommended as a saline for sperm of house fly and black blow fly in which sperm remain motile.

Formula:

NaCl	0.453 pt
$MgCl_2$ $6H_2O$	0.30 pt
$NaHCO_3$	0.035 pt
Dextrose	1.155 pt
KCl	0.107 pt
$Na_2H_2PO_4$ $2H_2O$	0.040 pt
CH_3COONa	0.025 pt

Note:
1. Components of saline are in gm/100 ml.
2. The best view of the sperm is attained with hanging drops with dark field illumination (100X).

A.1.11 *Lane's fluid [9]*

Application:
1. Recommended for the dissection of adult sand flies and other small adult insects for rapid examination.
2. Recommended as solution which also acts as a clearing agent.

Formula:

Chloral Hydrate Crystals	2 pt
Phenol Crystals	1 pt
Lactic Acid	1 pt

A.1.12 *O'Shea and Adams isotonic physiological saline [10]*

Formula:

1M NaCl	140 pt
1 M KCl	5 pt
1M $CaCl_2$	5 pt
1M $NaHCO_3$	4 pt
1M $MgCl_2$	1 pt
1M Trehalose	5 pt
0.05M Trimethylaminoethane	100 pt
1M Sucrose	100 pt
Distilled Water	Up to 1000 pt

A.1.13 *Sorensen's phosphate buffer [11]*

Formula:

0.1M KH_2PO_4	3 pt
0.1 M $NaHPO_4$	3 pt
Distilled Water	94 pt

A.1.14 *C&G's balanced saline [12]*

Formula:

1M NaCl	55 pt
1M KCl	40 pt
1M $MgSO_4$	15 pt
1M $CaCl_2$	5 pt

		1M Tricine	10 pt
		Distilled Water	Up to 1000 pt
A.1.15 *Modified Weever's saline [13]*	**Formula:**		
		1M KCl	21 pt
		1M NaCl	12 pt
		1M $CaCl_2$	3 pt
		1M $MgCl_2$	18 pt
		1M Glucose	170 pt
		1M PIPES	5 pt
		1M KOH	9 pt
		Distilled Water	Up to 1000 pt

A.2 Physiological solutions

A.2.1 Calliphora vicina *saline [14]*	**Formula:**		
		1M NaCl	172 pt
		1M KCl	2.5 pt
		0.5 M $CaCl_2$	1 pt
		1M $MgCl_2$	8 pt
		0.1M $NaHCO_3$	6 pt
		0.1M NaH_2PO_4	3 pt
		Sucrose	42 pt
		Distilled Water	Up to 1000 pt
A.2.2 *Cockroach saline [15]*	**Formula:**		
		NaCl	9.27 pt
		KCl	1.314 pt
		$NaHCO_3$	0.324 pt
		NaH_2PO_4 H_2O	0.189 pt
		Na_2HPO_4	1.206 pt
		Glucose	2.7 pt
		Distilled Water	Up to 1000 pt
A.2.3 *Haemolymph-like physiological solution (HL3) [16]*	**Formula:**		
		1M NaCl	70 pt
		1M KCl	5 pt
		1M $CaCl_2$	1.5 pt
		1M $MgCl_2$	20 pt
		1M $NaHCO_3$	10 pt
		1M Sucrose	115 pt
		1M Trehalose	5 pt
		1M Hepes	5 pt
		Distilled Water	Up to 1000 pt
A.2.4 *Larval* Bombyx *hemolymph saline [17]*	**Formula:**		
		1M NaCl	130 pt
		1M KCl	5 pt
		1M $CaCl_2$	1 pt

A.2.5 *Larval Honeybee saline [18]*

Formula:

1M KCl	2 pt
0.1M $MgCl_2$	4 pt
0.1 M $MgSO_4$	6 pt
0.1 M $NaHCO_3$	5 pt
0.1M $CaCl_2$	5.5 pt
1MNaCl	0.041 pt
Distilled Water	Up to 100 pt

A.2.6 *Larval* Manduca *saline [19]*

Formula:

Solution A

1M NaCl	40 pt
1M KCl	400 pt
1M $MgCl_2$	180 pt
1 M $CaCl_2$	30 pt
Distilled Water	Up to 1000 pt

Solution B

1M PIPES	150 pt
Distilled Water	up to 1000 pt
5N KOH	Add slowly until pH is 6.5

Procedure:

1. Dissolve 83 gm sucrose and 1 gm polyvinyl pyrrolidone (PVP) in 500 ml distilled water by stirring.
2. Add 100 ml Solution A and 10 ml solution B.
3. Mix and add distilled water up to 1 liter.

A.2.7 *Locust saline [20]*

Formula:

1M NaCl	147 pt
1M KCl	10 pt
1M $CaCl_2$	4 pt
1M NaOH	3 pt
1M HEPES	10 pt
Distilled Water	826 pt

A.2.8 *Locust embryo saline [21]*

Formula:

1M NaCl	150 pt
1M KCl	3 pt
1M TES	5 pt
1M $CaCl_2$	2 pt
1M $MgSO_4$	1 pt
Distilled Water	839 pt

A.2.9 *Mosquito Hemolymph Substitute Saline (HSS) [22]*

Formula:

1M NaCl	42.5 pt
1M KCl	3 pt
0.1M $MgSO_4$	6 pt

	1M $CaCl_2$	5 pt
	1M $NaHCO_3$	5 pt
	Succinic Acid	5 pt
	Malic Acid	5 pt
	L-proline	5 pt
	L-glutamine	9.1 pt
	L-histidine	8.7 pt
	L-arginine	3.3 pt
	Dextrose	10 pt
	Hepes	25 pt

A.2.10 *Pupal honeybee physiological solution [23]*

Formula:

1M NaCl	130 pt
1M KCl	6 pt
1M $MgCl_2$	2 pt
1M $CaCl_2$	7 pt
1M HEPES	10 pt
1M Glucose	25 pt
1M Sucrose	160 pt
dH_2O	To 1000 pt
pH to 6.7	

A.2.11 Rhodinus *saline [24]*

Formula:

1M NaCl	150 pt
1M KCl	28.6 pt
1M $CaCl_2$	2 pt
1M $MgCl_2$	8.5 pt
1M $NaHCO_3$	4 pt
1M HEPES	5 pt
1M Glucose	34 pt

A.2.12 *Spider Ringer's [25]*

Formula:

1M NaCl	150 pt
1M NaCitrate	15 pt

References

1. Morgan P.B. & LaBrecque G.C. Preparation of house fly chromosomes. *Ann. Entomol. Soc. Amer.* 57, 794–795 (1964).
2. Scott T.L. Bee anatomy. Internal. Section I, Part 2. *Microscope and Entomol. Mon.* **4**, 273–276 (1941).
3. French W.L., Baker R.H. & Kitzmiller J.B. Preparation of mosquito chromosomes. *Mosq. News* **22**, 377–383 (1962).
4. Breland O.P. Studies on the chromosomes of mosquitos. *Ann. Entomol. Soc. Amer.* **54**, 360–375 (1961).
5. Swedlow J. Fixation of *Drosophila* tissues for immunofluorescence. *Cold Spring Harbor Protoc.* **8**, 931–934 (2011).
6. Ephrussi B. & Beadle G.W. A technique of transplantation for *Drosophila*. *Amer. Nat.* **70**, 218–225 (1936).

7. Kennedy C.H. *Methods for the Study of the Internal Anatomy of Insects*. Department of Entomology, Ohio State University (Mineographed) (1932).
8. Eide P.E. & Reinecke J.P. A physiological saline solution for sperm of the house fly and the black blow fly. *J. Econ. Entomol.* **63**, 1006 (1970).
9. Lane J. *The Preservation and Mounting of Insects of Medical Importance*. World Health Organization Vector Control, (1965).
10. O'Shea M. & Adams M.E. Pentapeptide (proctolin) associated with an identified neuron. *Science* **213**, 567–569 (1981).
11. Silverman-Gavrila R.V. & Forer A. Evidence that actin and myosin are involved in the poleward flux of tubulin in metaphase kinetochore microtubules of crane-fly spermatocytes. *J. Cell Sci.* **113**, 597–609 (2000).
12. Akiyama-Oda Y. & Oda H. Early patterning of the spider embryo: a cluster of mesenchymal cells at the cumulus produces Dpp signals received by germ disc epithelial cells. *Dev.* **130**, 1735–1747 (2003).
13. Carrow G.M., Calabrese R.L. & Williams C.M. Spontaneous and evoked release of prothoracicotropin from multiple neurohemal organs of the tobacco hornworm. *PNAS* **78**, 5866–5870 (1981).
14. Magazanik L.G. & Fedorova I.M. Modulatory role of adenosine receptors in insect motor nerve terminals. *Neurochem. Res.* **28**, 617–624 (2003).
15. Chiang A.S., Wen C.J., Lin C.Y. & Yeh C.H. Nadph-diaphorase activity in corpus allatum cells of the cockroach, *Diploptera punctata*. *Ins. Biochem. and Mol. Biol.* **30**, 747–753 (2000).
16. Stewart B.A., Atwood H.L., Renger J.J., Wang J. & Wu C.-F. Improved stability of *Drosophila* larval neuromuscular preparations in haemolymph-like physiological solutions. *J Comp Physiol A* **175**, 179–191 (1994).
17. Ling E., Shirai K., Kanekatsu R. & Kiguchi K. Classification of larval circulating hemocytes of the silkworm, *Bombyx mori*, by acridine orange and propidium iodide staining. *Histochem. Cell Biol.* **120**, 505–511 (2003).
18. Rachinsky A. & Hartfelder K. *In vitro* biosynthesis of juvenile hormone in larval honey bees: Comparison of six media. *In Vitro Cellular & Dev. Bio. -Animal* **34**, 646–648 (1998).
19. Riddiford L.M., Curtis A.T. & Kiguchi K. Culture of the epidermis of the tobacco Hornworm *Manduca sexta*. *TCA Manual* **5**, 975–978 (1979).
20. Zilberstein Y. & Ayali A. The role of the frontal ganglion in locust feeding and moulting related behaviours. *J. of Exp. Biol.* **205**, 2833–2841 (2002).
21. Dawes R., Dawson I., Falciani F., Tear G. & Akam M. Dax, A locust Hox gene-related to Fushi-Tarazu but showing no pair-rule expression. *Dev.* **120**, 1561–1572 (1994).
22. Clark T.M., Hutchinson M.J., Huegel K.L., Moffett S.B. & Moffett D.F. Additional morphological and physiological heterogeneity within the midgut of larval *Aedes aegypti* (Diptera: Culicidae) revealed by histology, electrophysiology, and effects of *Bacillus thuringiensis* endotoxin. *Tissue Cell* **37**, 457–468 (2005).
23. Schroter U. & Malun D. Formation of antennal lobe and mushroom body neuropils during metamorphosis in the Honeybee, *Apis mellifera* . *J Comp Neurol* **422**, 229–245 (2000).
24. Sarkar N.R., Tobe S.S. & Orchard I. The distribution and effects of Dippu-allatostatin-like peptides in the blood-feeding bug, *Rhodnius prolixus*. *Peptides* **24**, 1553–1562 (2003).
25. Casem M.L., Tran L.P. & Moore A.M. Ultrastructure of the major ampullate gland of the black widow spider, *Latrodectus hesperus*. *Tiss. Cell* **34**, 427–436 (2002).

Index

Page numbers with suffix t refers to tables and those with *f* refers to figures

Insect Histology: Practical Laboratory Techniques, First Edition.
Pedro Barbosa, Deborah L. Berry and Christina S. Kary.

Companion Website: www.wiley.com/go/barbosa/insecthistology